AF084525

CYMATICS

A Study of Wave Phenomena and Vibration

A complete compilation (2001) of the original two volumes by Hans Jenny.

Volume 1, 1967
The structure and dynamics of waves and vibrations.

Volume 2, 1974
Wave phenomena, vibrational effects, and harmonic oscillations
with their structure, kinetics and dynamics.

~~~~~~~ ❅ ~~~~~~~

This expanded edition includes a collection of "Commentaries on Cymatics" by researchers, explorers, and practitioners whose life work has added to the legacy of cymatics left by Hans Jenny.

The complete two volumes of Hans Jenny's **Cymatics** are introduced by biographical notes and a few additional essays intended to provide context for this work, as well as suggestions to help the reader who may be new to cymatics, to navigate the sometimes abstruse and challenging text. Additionally, several Commentaries on the significance of Jenny's work to other disciplines, as well as selected references and resources, appear at the end of this book.

Enlightening the Questing Eye
Awakening the Discerning Ear

In Praise of Jenny's *Cymatics*

The most influential thinkers are not necessarily those who become household names, but rather those whose ideas become part of the public dialogue and way of thinking, regardless of personal recognition and aggrandizement. We believe that Hans Jenny was such a thinker.

John Beaulieu, N.D., Ph.D.
David Perez-Martinez, M.D.

It is in the work of Hans Jenny that we can begin to see the relationship of form and sound in the physical world. Jenny's experiments have shown that sound frequencies have the propensity to call into arrangement random, suspended particles, or to organize emulsions in hydro-dynamic dispersion into orderly, formal, periodic patterns. In other words, sound is an instrument through which temporal frequency patterns can become formal spatial and geometric patterns.

Robert Lawlor
Author and researcher:
Symbology, Sacred Geometry

I met Hans Jenny in February of 1972, just four months before his death. This meeting had a profound effect on my career in mathematics....

Ralph Abraham
Read Ralph Abraham's story in the "Commentaries on Hans Jenny's Cymatics" on pg. 322

The first time I saw Hans Jenny's cymatics videos, my mind was completely blown. I saw how the sound current gave rise to form and motion, and realized the absolute truth of the statement, "*Nada Brahma*, All is Sound." I've been a student of cymatics ever since, and have even gotten my own Cymascope to bring my insights and understanding of the behavior and expression of sound waves to a whole new level.

Eileen Day McKusick
Author and Sound researcher
ElectricHealth.com

Hans Jenny's work in cymatics and the previous editions published by Jeff Volk, were instrumental in my development of a theory of music therapy based on the principles of complexity science. This fifth edition, with its extensive revisions, is an important contribution to our deepening understanding of the healing power of music.

Barbara J. Crowe, MMT, MT-BC
Professor Emeritus of Music Therapy,
Arizona State University
Author, *Music and Soulmaking* and
A Transpersonal Model of Music Therapy: Deepening Practice

Dr. Jenny's cymatic images are truly awe-inspiring, not only for their visual beauty in portraying the inherent responsiveness of matter to sound (resonance), but because they inspire a deep recognition that we are part of this complex and intricate vibrational matrix, suggesting that the subtle vibrations of our very thoughts and feelings may have great effect on what will manifest in the physical world.

Jeff Volk
Publisher

Cymatics is the science of wave phenomena that makes visible the invisible matrix underlying all physical creation. It establishes an essential bridge between the materialistic science of the last century and the vibrational science of the new millennium. Modern quantum physics has demonstrated that our world is based not only on particles, but also—and more fundamentally—on waves. What materialistic science previously believed to be solid matter has been shown to be 99.99% empty space, filled with waves providing the vibrational scaffolding around which matter forms (known in physics as the "collapse of the wave to the particle").

Hans Jenny's brilliant work has become the essential touchstone which demonstrates to the naked eye the reality and power of those oscillations behind the manifestation of physical matter. As we move from materialistic science to a new holistic "energy science," and from chemical medicine to vibrational medicine, it is cymatics that has opened the door to this transition, offering tangible proof for the existence of these invisible energy waves which are the foundation for the new paradigms of science and medicine that we so desperately need.

Robert J. Gilbert Ph.D.
Founder, Vesica Institute for Holistic Studies

The science of Cymatics opens a window of awareness into the exquisite expression of beauty which is the infinite unfolding of nature. I encourage you to sit with this book, attend to it deliberately, and explore how you might connect with the messages held within. Ponder the relevance of any paragraph that draws your attention. Feel how it might relate to your own life right now... and there's a good chance that you will glean deeper insights into the mysterious workings of creation.

Mandara Cromwell
CEO of Cyma Technologies, Inc.
Researcher on the effects of therapeutic sound since 2001
Inventor of the AMI (Acoustic Meridian Intelligence) devices
Author of *Soundflower: The Journey To Marry Science and Spirit*

During a Soundscape workshop, students point out that when they're on a blindfolded "listening-walk," their senses of hearing and smell become more acute. A similar phenomenon happens for the guide as well. It appears that when one's attention is expanded it heightens all of the senses, catalyzing a synesthetic experience. Suddenly we're not only hearing, but we're sensing sound with our entire organism.

Through *Cymatics*, Hans Jenny helps us to rediscover this experience objectively. He reminds us that we are surrounded by vibrations of exquisite intricacy and beauty which we usually cannot see. And because we don't see them, we don't pay them the same kind of attention. The more we become in touch with a global sensation of our bodies, our associative thought machine quiets, and the more possible it becomes to attune to finer vibrations, within and about us. Yet even in my typically unconscious state, I cannot help being touched by Hans Jenny's photographs. Cymatics is hard evidence of genuine magic.

Jim Metzner
Fulbright Specialist, Media & Communication
Author, *Adventures of a Lifelong Listener*.
Executive Producer of the *Pulse of the Planet* podcast, and the American Soundscape Project. His recording archive has recently been acquired by the Library of Congress in Washington, DC, and his upcoming book will explore the mystery of listening.

Putting a Human Face on Cymatics

Once one's eye is opened to the cymatic principle, you see it everywhere.

~ Hans Jenny, from his diaries

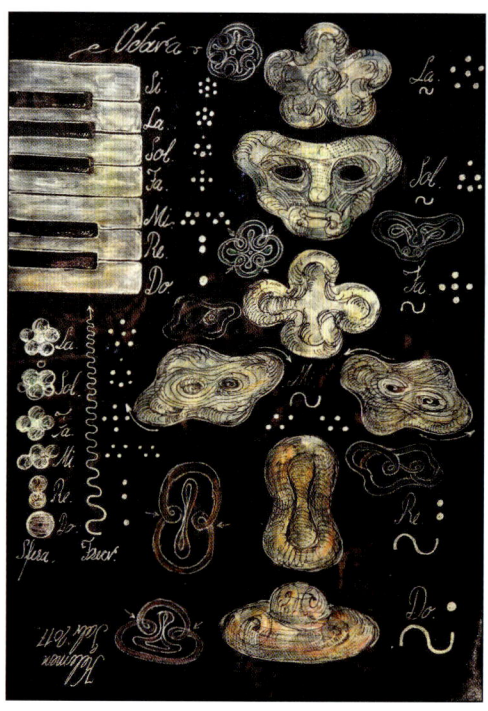

Illustration by Gabriel Kellemen

As I prepare to tie a bow around this fifth edition of our newly expanded compilation of Hans Jenny's two original volumes, (original in several senses of the word, in that Jenny both defined and articulated the science of Cymatics), it occurs to me that four decades have passed since I first published a video of one of Jenny's 16 mm films, documenting his experiments. A lot of water has rippled across the lens since then, and I wanted to share a few highlights of how this new edition came to be. But first, I need to shift focus even further into the past.

I would be remiss if I didn't acknowledge my friend and long-time collaborator, Jonathan Goldman, who introduced me to Peter Guy Manners in 1983 when he brought Peter over from his clinic in the Lakes District of England to present a series of lectures on cymatics at the Theosophical Society in Boston. While I had previously seen clips of Jenny's experiments, this meeting provided me with the opportunity to produce my first documentary, *Cymatics: The Healing Nature of Sound*. This was a three-part composite video featuring Jonathan's interview with Dr. Manners; a demonstration of Manners' therapeutic device, the Cymatic Applicator; and one of Jenny's original 16 mm films. This also proved to be a pivotal moment in my life, one that would set me on my course as a producer, a publisher, and more importantly, as a seeker after the manifold mysteries of sound— mysteries of which I had little conscious awareness at the time, but which had nonetheless, been imperceptibly directing my path.

Upon first seeing Jenny's dynamic "sound figures," it was clear to me that there was much more going on than met the ear. I was immediately awe-struck that what just a moment ago, had been an undifferentiated mass of inert material, was now transforming itself into an intricate, geometric, life-like, flowing form— all the while evolving out of this seemingly chaotic cacophony. It became apparent that there were specific unique patterns in the ways these magical manifestations of sonically-animated substances took shape. This invisible dance of creation, sustenance, and disintegration, had long ago been personified as the Hindu deities: Brahma, Vishnu, and Shiva, but I don't mean to infer that I had an epiphany— a profound revelation in which all the mysteries of this Gordian knot of manifest creation were unraveled in one fell swoop!

I was indeed moved though, by a deep, intuitive understanding— one that I have subsequently seen reflected back to me hundreds of times over the decades, on the astonished faces of awe-struck audiences, as each one witnessed these phenomena for the first time. Rather my understanding has deepened over many years of contemplating what I refer to as "universal principles of creation, transformation, and evolution," another iteration of what Jenny refers to in the final chapter of Vol. I, as The Basic Triadic Phenomenon. It continues to deepen to this day, as I see this trinity in operation everywhere around me and within me.

This little bit of copy that I wrote for the back cover of that VHS video jacket in the early 1980s, sums it up nicely:

"Dr. Jenny's cymatic images are truly awe-inspiring, not only for their visual beauty in portraying the inherent responsiveness of matter to sound (resonance) but because they inspire a deep recognition that we are part of this complex and intricate vibrational matrix, suggesting that the subtle vibrations of our very thoughts and feelings may have great effect on what will manifest in the physical world."

After nearly 40 years of intermittent consideration, I feel that it could be stated even more simply and succinctly as: **"The quality of vibration determines what shows up!"** Or perhaps, more poetically this way…

NO MATTER

The fact of the matter is...
there is no matter!
In fact, an effect
of our on-again, off-again world
makes it appear as if I were here,
speaking to you as if we were two.

When indeed it appears, that between our four ears,
there is really only one I,
that we each are a part of, yet linked at the heart of
the Ineffable... and who's to say, why?

But through this complex and hypnotic illusion
we've steeped ourselves deeply in convincing confusion
so the Truth remains hidden though plainly in sight
as if we'd forbidden ourselves to feel right
about being uniquely both one and the same
we each are the other just with a new name!

No Matter.

© 2024 Jeff Volk
CymaticSource.com

~ ~ ~ ~ ~ ~ ~ ❊ ~ ~ ~ ~ ~ ~ ~

We are slaves to what we do not know, and of what we do know, we are masters.
~ Sri Gary Olsen

MACROmedia Publishing © 2024

All rights reserved. No part of this book may be reproduced in any form by any means, mechanical or electronic, including recording, photocopying, or any storage and retrieval, without permission in writing from the publisher, except for brief passages in connection with a review.

MACROmedia Publishing
22 Sturgeon Creek Drive
Eliot, ME 03903 USA
Website: http://www.CymaticSource.com

Original Edition Credits

Volume I (originally published in 1967)
  Photos: Hans Peter Widmer and Christiaan Stuten
  English translation: D. Q. Stephenson, Basel

Volume II (originally published in 1974)
  Assistants: Christiaan Stuten and E. Schild
  Photos: Christiaan Stuten
  All photographs with Contarex-Camera and with Zeiss-Objectives
  English translation: D. Q. Stephenson, Basel

If you are unable to find this book in your local bookstore, you may order it from the publisher.
Quantity discounts for organizations are available.

Thank you to Christopher Graefe and Jessica Manley for your creativity, commitment, and enthusiasm... and for showing up at just the right moment. Chris, this book would never have happened without your technical wizardry and personal dedication, the fruits of which are evidenced in innumerable enhancements throughout this extensively revised edition. It was an inspiration, and a joy, to work with you. Lastly, I wish to extend my heartfelt gratitude to Tesa Silvestre for all she has contributed to this project. I especially want to acknowledge her thoughtful edits of Christiaan Stuten's "Biography of Hans Jenny," and John Beaulieu's "Primer," as well as her dedication to revising the translation of four crucial chapters of Jenny's original German text. My very lively conversations with her in the second half of 2022 significantly shaped the vision of what this book could become.

2001 layout and design of revised MACROmedia edition by Jennifer Dumm

First Printing (Revised Compilation), August, 2001
  Second Printing, February, 2004
  Third Printing, Spring, 2007
  Fourth Printing, Spring, 2012
Fifth Printing (Newly Revised Edition), February, 2024

Library of Congress Cataloging-in-Publication Data

Jenny, Hans, 1904 - 1972
    Cymatics: A Study of Wave Phenomena and Vibration
        p. 352
Includes: index
1. Cymatics - 2. Physics - 3. Acoustics - 4. Philosophy of Science

ISBN: 978-1-888138-10-8

Hans Jenny (1904-1972)

Hans Jenny's dedication from his original edition:

Dedicated to the memory and research of Rudolf Steiner

~~~~~~~ ❋ ~~~~~~~

Publisher's Dedication: Revised Edition, 2001

With gratitude to Maria and Ea Jenny, A. Koster, and especially Max Savin.

To the *Sound* which unifies all phenomena, generating the vast world of form in its "triadic" nature.

Table of Contents
Revised Expanded Edition

i	Developing Cymatic Perception
2	In Praise of Jenny's *Cymatics*
4	Putting a Human Face on Cymatics and *No Matter* - by Jeff Volk
10	How this "Extensively-Revised" Edition Came to Be - by Jeff Volk
12	A biographical portrait of Hans Jenny (1904 - 1972) (adapted from 2001 edition) - by Christiaan Stuten
15	Understanding *Cymatics*: A Primer for Non-Scientists - by John Beaulieu, N.D., Ph.D.
21	Foreword - by Ted Gioia
25	Publisher's Suggestions for Getting the Most from this Book - by Jeff Volk

Volume I
Original 1967 Edition

Pg.	Ch.	
28	1.	Problems of Cymatics
32	2.	Experimental Method
46	3.	Examples of Cymatic Phenomenology
74	4.	The Tonoscope
79	5.	The Action of Vibration on Lycopodium Powder
95	6.	Periodic Phenomena Without an Actual Vibrational Field
103	7.	Sound Effects in Space; Spatial Sonorous Patterns
111	8.	The Spectrum of Cymatics
131	9.	The Basic Triadic Phenomenon
137		Intermezzo: An Interlude - by Jeff Volk

Volume II
Original 1974 Edition

Pg.	Ch.	
139	1.	Cymatics
145	2.	The Wave Lattice as a Configuring Field
147	3.	Circulation in the Wave Train
151	4.	Changes of Phase in Matter at the Same Frequency and the Same Amplitude
155	5.	The Influence of Vibration on Flowable and Solid Substances
165	6.	The Oscillating Water Jet
169	7.	Vibrational Processes in a Capillary Space
173	8.	Vibrational Figures Revealed by Schlieren Photography
175	9.	Rotational and "To-and-Fro" Effects
179	10.	Ferromagnetic Masses in the Magnetic Field Under Vibration
189	11.	The Morphology of Lichtenberg Figures; A Carrier System
197	12.	Polygonally Pulsating Drops / Three-Dimensional Structures Under Vibration
207	13.	Deflection of the Electron Beam Due to the Interaction of Oscillations at Two Different Frequencies / Pathways of the Mechanical Pendulum
217	14.	Harmonic Vibrations in a Concrete Medium
249	15.	Cymatic Effects in a Wider Context
279	16.	The Biological Aspect
289	17.	Historical Review / Methodological Preview

300   Closing Comments from the Publisher - by Jeff Volk

302   Index
305   Index Notes
306   References

COMMENTARIES ON HANS JENNY'S *CYMATICS*

308   Jenny's Legacy in the Annals of Science - by John Stuart Reid
316   Shedding New Light on Sound: Validation of early glimpses into the hidden world of vibration - by Mandara Cromwell
322   Meeting Hans Jenny - by Ralph Abraham
325   Cymatic Images as Vibrational Icons for Contemplation - by David Perez-Martinez, M.D.
327   My Love Affair with Cymatics - by Jodina Meehan
330   Hard Evidence of Genuine Magic - by Jim Metzner
331   My Journey of Curiosity - by Jacob Lee Adlington
333   Cymatic Artistry - by Rachael Linton
336   Cymatics: A Felt Sense Bigger than Ourselves - by Joshua Leeds
339   Cymatics and Resonance - by Alexander Lauterwasser
342   The Universal Vibrational Field as the "World-creating Power" - by John Beaulieu, N.D., Ph.D.
                                                    and David Perez-Martinez, M.D.

CYMATICS: MAKING WAVES AROUND THE WORLD

346   Gabriel Kelemen
347   Water into Chaos: Phases of chaos and reintegration in an oscillating water sample
347   The First Museum Installation of the Cymascope
348   Christiaan Stuten and the Tonoscope
348   Cymatics in Lord of the Rings
349   Cymatics: Science vs. Music
349   Shaped by Water

How this "Extensively-Revised" Edition Came to Be
by Jeff Volk

It was coming up on Thanksgiving, 2020, the peak of the holiday gift buying season and prime time for selling expensive, coffee table books— even arcane science books "for the person who has everything." And here I was about to pull my most popular title off the market, or at least, out of what had been my primary market for the past twenty years— right in the middle of a stay-at-home mandate.

Our President had recently declared war on China, trade war that is, and China was vigorously rattling its sabers, threatening to renege on its treaty with Hong Kong, where the previous four editions of this book had been printed. As much as I didn't want to bail on my long-term partners, I had become uncomfortable printing in China, so I was on the fence as to what to do. Maybe it was time to let this old chestnut go the way of rare book collectors?

I ended up pulling the title, knowing that this would set off a feeding frenzy amongst third-party resellers on Amazon. So I prepared to cache my remaining 50 copies for those researchers and "must have" readers, figuring that this would hold me over until I made my decision. While I was certainly right about the first point, the speed with which this occurred (eight copies within the first couple of days!), totally caught me off guard as I quickly learned about pandemic publishing in the age of social media.

As things turned out, a stroke of grace and synchronicity turned my plans on their ear. Someone I had never heard of, using a platform to which I'd never given the slightest attention, incited a run on what had been a rather niche title. I soon discovered that the mysterious instigator was a jazz critic and pianist, and a multiple NY Times best-selling author, whose focus was the societal effects of music, trans-culturally and throughout history. What's more, Ted Gioia is a fastidious researcher who has uncovered and articulated profound insights about history and human nature in ways that are both enlightening and entertaining… and he is as passionate as he is eloquent, about his subject matter. He was the perfect person to appreciate, and lionize, a consummate observer and sound researcher, such as Hans Jenny.

With all the buzz that Gioia's posts stirred up*, I was reassured that there would still be a market for this book, even at a higher price due to increased printing and shipping costs. Little did I know what was to occur over the next couple of years as the full impact of the pandemic spread around the globe like a tsunami! But back in 2020, I was encouraged, and astounded, to see the premiums that *used copies* of this now rare book were fetching on the commodities exchange known as Amazon— selling quickly and often at several times the retail price.

---

*See Ted's revised commentary as the Foreword to this edition.

So after deciding to take a gamble and reprint the fifth edition in the States, I called John Beaulieu and asked him if he would write a new "Commentary on Cymatics" for this revised edition. He immediately agreed, so I asked another long-term colleague, Barbara Crowe, for a similar favor, to which she promptly replied: "*After many years as Chair of the Graduate Dept. of Music Therapy at ASU, and authoring two books on the subject, my writing days are over… but I will write a blurb for your new edition.*"

In her pithy little statement of the impact that Jenny's work had upon her own research, two rather innocuous words stood out to me as if prophetic.

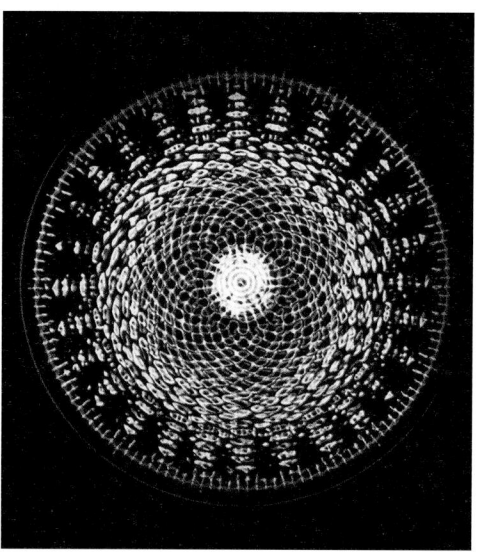

*Hans Jenny's work in cymatics and the previous editions published by Jeff Volk, were instrumental in my development of a theory of music therapy based on principles of complexity science. This fifth edition, with its **extensive revisions** (emphasis added), is an important contribution to our deepening understanding of the healing power of music.*

Barbara J. Crowe, MMT, MT-BC
Professor Emeritus of Music Therapy, Arizona State University
Author: *Music and Soulmaking*, and *A Transpersonal Model of Music Therapy: Deepening Practice*

So I set out to compile an assortment of compelling testimonials from a diverse array of authors and acousticians, artists and musicians, researchers and technicians, philosophers and physicians— all of whom had been profoundly influenced by Jenny's substantial and wide-ranging body of work— until what finally emerged was this expanded edition, with a select collection of *Commentaries on Cymatics* added at the end of the book after Volume II.

A biographical portrait of Hans Jenny (1904 - 1972)

by *Cymatics* photographer and assistant, Christiaan Stuten, adapted from 2001 edition

Hans Jenny was a Renaissance man whose diverse callings as a musician, painter, doctor, and scientist were seamlessly woven into a dynamic personality marked by great intensity and a deep sense of competency in all he undertook.

As a youngster, his affinity for music and his aptitude for the piano made a musical career seem a likely path for him. His father held various positions in the Evangelical church, and throughout grammar school, Hans would play the organ for church services. His musical tastes grew as varied as his talents, and he became equally at ease improvising jazz or performing a piano sonata. He expressed his great love of the arts in many different ways, including his admiration for the likes of Mozart and Wagner, his continuous study of Leonardo da Vinci and Rafael, and his passion for drawing and painting. Besides being artistically gifted, he was proficient in both Greek and Latin, developing a life-long interest and formidable command of classical and modern history, and was a dedicated and rigorous student of the philosophical works of Plato, Aristotle, Heraclitus, and Nietzsche. Goethe's approach to science also had a powerful influence on Jenny's mode of observation, and it strongly shaped the methodology which later guided his cymatic experiments.

Jenny was a tireless observer of the natural world. His keen eyes became his primary tools as a researcher, and as an artist who, early on, loved to paint the natural world around him, especially animals. His acute perceptivity was already evident as a schoolboy when he would go on ornithological excursions in the Swiss countryside outside of Basel, where he quickly learned to recognize and identify every bird in the vicinity by both its call and appearance.

It was on one of these excursions, bicycling in the hills overlooking the Birs Valley near Dornach, that his gaze first fell upon the original Goetheanum, a unique and impressive wooden structure built under the inspiration and direction of Rudolf Steiner. The fourteen-year-old lad was fascinated by the twin-lobed structure. And as fate would have it, he later had the opportunity to accompany his parents and a small group of educators and public officials on a guided tour of what was to become the locus of Anthroposophical teaching—a tour led by none other than Rudolf Steiner himself.

The second Goetheanum, rebuilt 1923-28

Although he had acquired a solid knowledge of the bible in his youth, Hans Jenny was largely indifferent toward church religion. He did, however, find a particular affinity with the spiritual science of Anthroposophy. And thus began his studies. He greatly admired Steiner, and read an enormous amount of his extensive writings, though he did not always resonate with "Anthroposophists." It was Steiner, with his emphasis on the super-sensible, spiritual dimension of life, who was to influence Jenny most profoundly, providing both a direction for his insatiable curiosity and a framework for his inner strivings. Not surprisingly, he proved to be a good student and started giving lectures on Anthroposophy before he even finished medical school. After completing his doctorate, he taught science at the Rudolf Steiner School in Zurich for four years.

Hans Jenny was a most entertaining man whose sharp wit was tempered with heartfelt compassion. This was to serve him greatly in what would become his primary profession as a family physician in the village of Dornach. His practice spanned more than 30 years and ranged from making house calls to local farmers (occasionally even treating their animals), to seeing prominent socialites at his clinic in the nearby city of Basel. With his breadth of interest, sense of humor, and magnetic personality, he would quickly build rapport with just about anyone. His vibrancy was so contagious that no sooner would he enter the room than his patients would feel better.

Even with a bustling practice, Hans Jenny still found time for his many other passions: painting, traveling and lecturing, keeping up with the latest advances in the natural sciences, and experimenting in the field of scientific endeavor which is now known as cymatics. Jenny's predilection for painting animals and their environs grew from a hobby in his younger years to an absolute inner necessity as he grew older. He painted with oils on canvas as well as on composition board for larger works. He never copied nature. Instead, he absorbed impressions whose motifs then gestated within him, sometimes for years, until they emerged in a controlled explosion of creativity, color, and form. Even these larger works seldom took more than an hour or two for him to complete. Executed in a vital, expressive manner, unique in the realm of "animal portraiture," his paintings captured the soul of the animal while at the same time reflecting powerful archetypes within the psyche of man. The resulting body of over two thousand paintings firmly established Jenny as a fine artist, with numerous exhibits throughout Europe and as far afield as Argentina.[1]

One of his most prized possessions was a pair of binoculars that his father had given him as a boy. Although they were nothing fancy, he always carried them with him on his many travels around the world. He had an insatiable desire to deeply know each of the places he visited while lecturing and traveling. Wherever he was, he never missed an opportunity to visit zoological gardens, whether in his hometown of Basel in Switzerland or in some distant city. Two of his favorites were the town of St. Maurice in the Engadine region, in the Swiss Alps, and, quite appropriately, the town of Florence in Italy, which, in its day, was the center of the Renaissance.

Hans Jenny was nearly 60 years old when he published, *Das Gesetz der Wiederholung* (The Laws of Repetition) in 1962. The cymatic explorations introduced in that book were only rudimentary precursors of the more sophisticated experiments he later documented in Volumes I and II of *Cymatics* published in 1967 and 1974, respectively. And yet, he was already beginning to reveal a vision full of daring and inspiration. He would frequently begin his lectures by stating that he hoped his research into cymatics would open the eyes of others to the underlying periodic phenomena

---

[1] An exquisite compilation of 275 full-color reproductions of Jenny's original oils, entitled *Tierlandschaften* (Animal Landscapes), is available through MACROmedia Publishing. It includes commentaries by art historians, biographical sketches, and excerpts from Jenny's diaries, giving unique insights into the significance of his artistic work (See page 352).

in nature that he so clearly perceived. The outer form of things, their skin or surface appearance, held no boundary to his penetrating gaze and insightful mind. Jenny liked to quote an aphorism of Heraclitus: "All is flow. All is in flux." More than a favorite maxim, it was a very apt and pointed characterization of how he perceived the world, and what he most deeply wished to make perceptible to all.

It was a tremendous privilege to work side by side with Dr. Jenny for those many years, and I only hope that this brief synopsis will convey some semblance of the character of this most remarkable man.

Christiaan Stuten was a talented photographer who spent fourteen years assisting Jenny in documenting the fascinating experiments featured in this book. The biographical portrait of Hans Jenny he offered above was originally published as the foreword to the 2001 edition of *Cymatics* and revised for this 5th edition.

 View a short video of Christiaan Stuten demonstrating how the human voice can create harmonic patterns using Jenny's tonoscope. • https://vimeo.com/843711920/380f21f468

Understanding *Cymatics*: A Primer for Non-Scientists
by John Beaulieu, N.D., Ph.D.

Dr. Jenny's work first came to my attention in 1974, while I was working as Supervisor of Music Therapy at Bellevue Psychiatric Hospital in New York City. One day, a colleague placed a copy of Volume I of *Cymatics* on my desk. I still remember the joy of looking at the cymatic sound patterns and the wonderful sense of "Yes!" that rippled through my being. Each picture was worth more than a thousand words and I felt as though I were reading volumes in minutes. Since my graduate studies and work at the hospital were based on applied systems theory, it did not take me long to recognize that Dr. Jenny was a dedicated and rigorous systems researcher. As I dove more deeply into his book, I found his writings to be just as exciting as the images!

45 years of "wow!" and then "how?"

Having shared and taught cymatics for nearly five decades, I know that those drawn to Dr. Jenny's work are often "right brain" by nature. Many artistic, intuitive, and creative types are immediately taken by the beauty of Hans Jenny's cymatic images, and can readily sense their meaning, and intuit their broader implications. And yet, they can just as easily feel intimidated and daunted by Dr. Jenny's writings, and for good reason. Beyond offering a collection of mesmerizing images, *Cymatics* is also a thorough phenomenological study of vibration, befitting an accomplished physicist and systems researcher dedicated to rigorous left brain science. His writings about cymatics are embedded in scientific rigor and clarity and supported by exacting methodology and procedure.

I still remember the first time I introduced cymatics to a group of composers and musicians in New York City in the mid-1970s. I began by explaining to them how Dr. Jenny's experiments and images were produced. And I did so in about two minutes because I assumed it would be their right brain's limit for theoretical input before growing restless. Next, we viewed slides of cymatic photographs, and the room filled with expressions of wonder and astonishment: "wow!", "far out!", "yeah!", "hmmm!" After the presentation, I invited questions, but everyone just sat there in what I took to be an appreciative silence. Since no one had any questions, I ended the presentation and everyone went home.

This was my first "Music and Sound in the Healing Arts" class, and I was not sure what to make of what unfolded, but I decided to trust it. To my surprise, I came home to an answering machine full of questions from participants, and more questions continued coming in throughout the rest of the week. To this day, I still get the occasional question about Dr. Jenny's work from a student who attended that very first class.

Sooner or later, the left side of our brain says "What about me?" and wants to understand. So all these years of hearing "wows" followed by "hows," have inspired me to write this primer.

My intention here, is to communicate the essence of Dr. Jenny's work based on years of "right brain" students' questions. I hope to inspire more people to engage with the sophisticated scientific observations and insights that Hans Jenny developed in response to his experiments. To be clear, I am not seeking to offer scientific precision. If you are scientific by nature and want science, I suggest that you dive right into Dr. Jenny's writings. My goal is to help artists and creative people who are not versed in science and systems theory, to find meaningful doorways into this material, and to develop a greater and more balanced appreciation of what Dr. Jenny has given us.

I'll begin this introduction in the same way that I opened my first class on cymatics: by explaining, as simply and concisely as possible, the process by which cymatic experiments are performed and the images created.

How Dr. Jenny's experiments and images were produced
Dr. Jenny performed many of his experiments by putting substances such as sand and other powders and fluids, on a metal plate. The plate was attached to a crystal oscillator that was controlled by a frequency generator capable of producing a broad range of vibrations. By turning a dial on the frequency generator, Dr. Jenny could cause the plate to vibrate at many different frequencies.

Let me explain further. Oscillators are devices that produce vibrations. They are often called "vibrators" in the popular marketplace. Imagine an electric massage device or go to any department store and ask to see a massager (which is a simple oscillator). Turn it on and it will vibrate or oscillate. Next, place the massager on a bone. Feel how the vibrations are amplified throughout your body. Applying the massager to your bone is the equivalent of Dr. Jenny attaching his oscillator to a metal plate. The density of the steel plate amplifies the vibrations of the oscillator in a similar way as do your bones. A massager creates, at most, a handful of vibrations that can be perceived audibly as a hum. The hum you hear with your ears and the vibration you feel from the massager are the same. In contrast to a massager, Dr. Jenny's oscillator was driven by a frequency generator that was capable of producing thousands of different tones. He could turn a dial and instantly change the vibrations moving through the plate and by so doing, he could observe the effect of different vibrations on different substances. While he watched the sand or other substances organize into different patterns on the plate, he could also hear the sounds produced not only by the oscillator, but also, as you may notice in the photo above, through a small speaker that was hooked up to the oscillator. And if he lightly touched the plate, he could even feel the vibration in his fingertips.

The pervasiveness of periodicity and the basic triadic phenomena
Much of Dr. Jenny's work aimed to show the pervasiveness of periodicity in the natural world. Periodicity refers to the rhythmic pendulation between opposite poles which most animate and inanimate objects can be seen to exhibit upon close observation. Examples of such periodicity include inhalation and exhalation, the cardiac rhythms of the heart, waves of the ocean, etc. Again and again, he would experimentally demonstrate that periodicity was an essential

element of the design and organization of "Nature, animate or inanimate," and the behavior of all natural, biological, and psychosocial systems.

More specifically, Jenny's books (as well as his films) aimed to introduce his readers and viewers to a manifestation of periodicity which he called "Triadisches Urphänomen," translated into English as the Basic (or Primal) Triadic Phenomenon. Over and over again, his experiments showed that, under the stimulus of vibration, the substances he used as indicators (powder, iron filings, etc.) would continually shift from (1) a "fixed" or seemingly static mode where they appeared as specific patterns and shapes easily visible to the eye, to (2) a phase of movement where everything was in flux (dissolution, chaos, re-organization). He referred to these two alternating poles as the figurate and the kinetic.

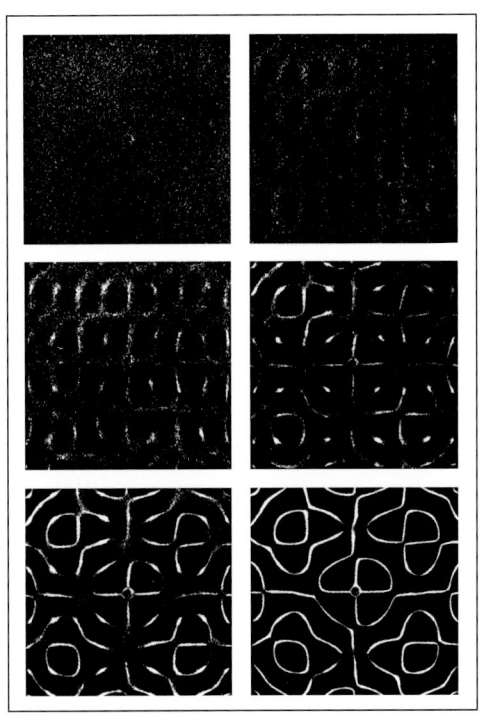

What was the third part of the triadic phenomenon then? For Jenny, it was "the periodic" or "the whole," which was nothing less than the vibrational field itself, sometimes expressed in seemingly static patterns (the figurate) and sometimes expressed in flowing movements (the kinetic).

In addition, it is essential to understand that, for Jenny:

> *"These three aspects— the periodic, as the fundamental field, with the two poles of the figurate and dynamic— always appear as one. They are inconceivable without each other. It is altogether impossible to take one or the other away, without the whole thing ceasing to exist."*

Direct observation as a portal into perceiving and immersing into a vibrational universe

In Chapters 1 and 2 of Volume 1, Dr. Jenny spells out the essentials of his methodological approach and stresses the importance of "getting inside" phenomena with empirical and systematic research that could be readily and repeatedly observed.

Although vibrational phenomena can be perceived in three very different ways (acoustically with our ears, visually with our eyes, and kinesthetically with our physical touch or sensory perception of vibration), Jenny favored the visual sense and offered this explanation at the beginning of Chapter 2 in Volume 1 (The Experimental Method):

> *"In attempting to observe the phenomena of vibration, one repeatedly feels a spontaneous urge to make the processes visible and to provide ocular evidence of their nature. For it is obvious that, by virtue of the abundance, clarity, and conscious nature of the information communicated by the eyes, our mode of observation must be visual. However great the power of the ear to stir the emotions, however wide-ranging the information it receives, particularly through language, the sense of hearing cannot attain that clarity of consciousness which is native to that of sight."*

Dr. Jenny's greatest desire was to influence others to become cymatic phenomenologists themselves, and to create their own lines of research and artistic expression. He wished that we could all become keen observers of the ubiquity of vibratory phenomena and let nature "speak" for itself. As he makes very clear in Chapter 1 of Volume 2, one of his primary aspirations is to encourage the development of one's perception of periodic phenomena:

> *"The most important thing, then, is to develop an actual organ of observation and perception of rhythmic and periodic systems. Training this ability is also the purpose of this second volume, and our documented excursion into the field of cymatics will hopefully serve this goal."*

And a fundamental aspect of skillful perception, according to Dr. Jenny, was the capacity to perceive the whole.

Seeing the whole: Dr. Jenny's wholistic approach to scientific exploration
Dr. Jenny always sought to observe the whole and to understand the behavior of parts in relationship to the whole. At one point, he asks: "What is the status of the parts, the details, the single pieces, the fragments?" And he answers: "in the vibrational field, it can be shown that every part is, in the true sense, implicated in the whole."

Because the experience of the whole is most important in cymatic research, it is the whole phenomenon upon which we concentrate as we observe and follow nature unswervingly with our eyes, ears, and brain.

A basic law of systems is that the whole is greater than the sum total of its parts. And a metaphor that is commonly used to illustrate this law is that of a team of scientists studying an elephant. The problem is that the scientists initially have no idea that they are working on an elephant. One scientist is observing and measuring the behavior of the foot; another is focused on the tail, and the velocity at which it wags; and yet another is focused on observing one toenail and its chemical composition. Their views are fragmented. And while they are each publishing papers about their research in prestigious scientific journals in different disciplines, they have no idea that their work is related.

One day a scientist comes along and accidentally "sees" the whole. She calls the whole "an elephant." She clearly sees the relationship of the parts to the whole and how they move together, but everyone thinks she is crazy and the specialists start to fight over her idea of "an elephant." Over time, more and more people begin to see "the elephant" until, one day, it becomes perceptible to all, and "there is an elephant." Once this happens in a given field of exploration, many isolated areas of research can subsequently be explained in a larger context.

Dr. Jenny's writings are always inviting us to focus on the whole and not get distracted by the behavior of the parts.

Hans Jenny's Influence on Energy Medicine

Although he was a medical doctor, Dr. Jenny never intended for his work to be applied therapeutically. Rather, he wished to demonstrate the primacy of vibration and its ever-present effects throughout the entirety of nature. And yet, he made an enormous contribution to the emerging fields of "Sound Healing" and "Energy Medicine." Through his painstaking experimentation and acute observation, he was able to articulate a conceptual basis that has had significant implications for the broad reach of scientific endeavor.

Cymatic research is a "sound" example of the principles underlying vibrational medicine. If you were to be present during a cymatic experiment, you would simultaneously hear a sound and see a pattern forming out of a substance that had been placed on a vibrating rubber membrane or steel plate. What you first perceived as an inert "blob" of sand or water, without movement, form, or pulse, would instantly morph into an animated, pulsating form as soon as the plate or membrane was excited by the vibrational field created by the oscillator.

Now, let us assume that, for some reason, you could only see the static result of this process, i.e. the pattern that had formed. You had not been able to hear or see the dynamic process by which this had emerged. The fact that this form which you perceive as solid was generated by a vibrational field emitting a sound would never have occurred to you, and the very idea would seem preposterous. If you were, however, to suddenly witness someone accidentally brushing the sand on the plate, disrupting the pattern, you would be astounded to see it return to its original shape within seconds. And how would you explain that? Jenny's research repeatedly demonstrated that under similar conditions (i.e. type and size of plate, type and amount of substance used, etc.), a given vibrational frequency will always produce the same pattern.

The above example can serve to illuminate the basic differences between conventional, materialistic medicine, on the one hand, and energetic healing, on the other. Let me explain. "Energy medicine" seeks to understand people as unified energy fields or, in Dr. Jenny's words, as "wholes." From that perspective, the physical body, emotions, and thought processes are like cymatic forms which are organized by underlying vibrational fields, with the densest (i.e. the physical) being animated by the subtler vibrations (i.e. emotions and thoughts).

In energy medicine, the underlying vibrational field is called an energy field. The health practitioner seeks to perceive, evaluate, and support the energy field rather than focus on a specific symptom. The practitioner's goal is to use therapeutic modalities such as homeopathy, acupuncture, touch, color, music and sound (tuning forks, singing bowls, gongs, the voice, etc.), to effect and change the energy field. As the person shifts into resonance with a more coherent field, and a more harmonious pattern emerges, their array of symptoms may disappear.

The idea of energy fields is both new and ancient. Physicists have sought to explain the strange behavior of quantum particles via the existence of a unified field. Albert Einstein said: "We may therefore regard matter as being constituted by the regions of space in which the field is extremely intense… There is no place in this new kind of physics both for the field and matter, for the field is the only reality."

Dr. Jenny used phenomenology and systems theory as research modalities for observing the effects of whole vibrational fields. He wanted students of cymatics to understand that the phenomena they witnessed on the vibrating surface were always the product of a larger field, and that any changes in the frequency of the vibrational field would immediately alter the phenomena being observed. In the latter chapters of *Cymatics, Vol. II*, Dr. Jenny sought to illustrate a connection between cymatic vibrational fields and the behaviors of biological, weather, and social systems. He was not saying that cymatics was the cause. Rather, I believe what he meant was: "If you look at the whole, you will come to new understandings. Let these cymatic experiments inspire your imagination to deeper insights into the universal principles of nature."

While conventional medicine is largely characterized by specialization and compartmentalization, energy medicine takes a more generalized approach to healing, based on the understanding of energy fields. When the energy medicine practitioner evaluates the energy field, he/she can recommend specific "vibrational therapies" (e.g. music, sounds, movement etc.) to support a shift in the field. The result will be a new energetic field in which the old symptoms can no longer exist.

This is a "transformation" as opposed to "fixing a part." The old energy field will still be available, yet we will now have developed the ability to shift into a new field. Ultimately, we learn that we have the freedom to create and choose different fields. And ideally, we would find ourselves on a continuum of fields, always observing and entering into a greater whole, a greater experience of wholeness.

An Invitation

Dr. Jenny's dedication to perceiving the "whole picture" is a discipline from which there is much for us to learn. We can be inspired to see ourselves as "wholes," with the capability to shift into different vibrational fields at any time. Cymatics ultimately teaches us that we are limitless beings with immense creative and healing powers. Dr. Jenny exhibited many of these gifts in his own life. May you feel so moved as well to embrace them for yourself, as you explore the vast implications of his work.

I hope that I have inspired you to make the effort to go beyond the "wow" experience of admiring the mesmerizing photos, and to devote the time and attention necessary to carefully consider his writing. This will take some deliberate and focused concentration, but if you apply the necessary discipline to understand Dr. Jenny's profound scientific perspective, you will realize that it was well worth the effort. And perhaps you might even end up agreeing with me that Dr. Jenny's written words form a beautiful mosaic that is just as profound as his cymatic pictures.

John Beaulieu, N.D., Ph.D., is a naturopathic physician, composer, and musician. A world-renowned pioneer, researcher, and teacher in the field of sound healing, he's the author of several books including *Human Tuning: Sound Healing with Tuning Forks*, and *Bellevue Memoirs: My Patients, My Teachers*. For the last 50 years, he has taught very diverse audiences about the art of sound healing for personal wellness and for integration into all professional therapeutic practices. You may learn more at http://www.biosonics.com.

Foreword

by Ted Gioia, Author of *Music: A Subversive History* and *Healing Songs.*

In mid-November of 2020, I tweeted a video of Japanese artist Kenichi Kanazawa engaged in an unusual musical performance.[1] The stark and unadorned footage, recorded with a mere cellphone, only lasted a hundred seconds but it was absolutely riveting, and it went on to receive several million views over the following weeks.

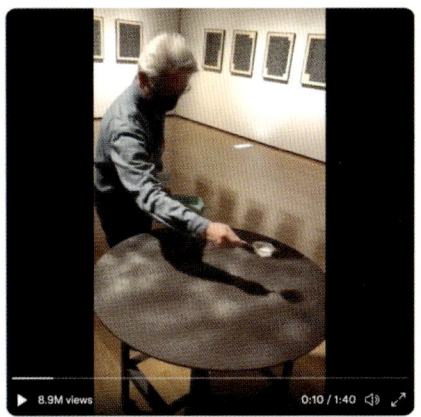

A visual demonstration of the power of sound to create order out of chaos.

The video shows Kanazawa in an art gallery, sifting a white powder over a steel tabletop in a random pattern that looks like the kind of mess a child might make toppling the sugar bowl. Then he begins rubbing the edge of the tabletop with a small rubber mallet, producing a repeated high tone at a rate of approximately 120 beats per minute. The friction created by the mallet rubbing against the steel causes the tabletop to vibrate, and the particles to start dancing across the surface. Gradually, they form into a complex star-shaped pattern, both beautiful and symmetrical. It all happens in a matter of seconds, but the pattern produced looks as if it had been carefully constructed by an artist over the course of hours.

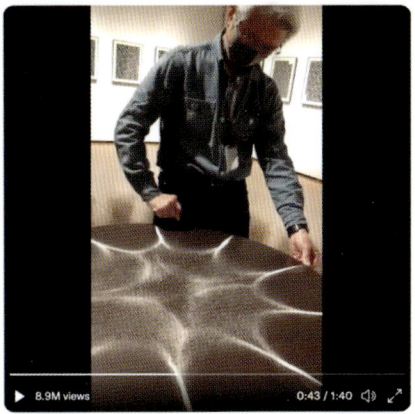

Kanazawa then repeats the process with a larger mallet which produces a lower bell-like tone at a faster tempo. And now, the pattern magically shifts into a circular shape, ornamented with ten symmetrical protruding knobs. Once again, the figure has emerged almost instantaneously in response to a specific tone, but it looks like the sort of beautiful sand sculpture a skilled artisan would have taken a long time to create.

As this video went viral, I received hundreds of comments from Twitter followers. Most were astonished, awed, and curious and some assumed it was some kind of sorcery or magic trick. How could such mysterious sounds possibly possess the power to rearrange the material universe in the ways shown in that short clip? How could something so intangible as music or sound produce such a distinctive physical effect? Surely, there had to be some gimmick involved.

---

[1] https://twitter.com/tedgioia/status/1327511170190864384

In reality, the science behind Kanazawa's performance has been known since the late 17th century, if not earlier. In 1671, Robert Hooke (1635-1703) created geometric patterns by rubbing a violin bow against a thin plate coated with a fine layer of flour. A hundred years later, physicist and musician Ernst Chladni (1756-1827) undertook further experimentation along these lines, and to this day, these 'sound-effects made visible' are still commonly called Chladni figures.

These pioneers certainly deserve credit, but today, our understanding of the science and aesthetic of the process of making sound visible primarily comes from one individual: Dr. Hans Jenny (1904-1972) whose writings and photographs you now hold in your hands.

As you may have read in Christiaan Stuten's intimate biographical portrait of Jenny on page 12, the founder of cymatics was trained as a medical doctor, but his skills encompassed a much wider array of disciplines, from art to philosophy to zoology. His many interests guided his life-long quest to understand the organizing principles of both nature and society.

Jenny devoted much of his life to exploring and documenting the process through which sound and vibration reveal the beauty and precision of the implicate order that underlies the physical universe.

Before him, this type of experimentation was mostly regarded as a parlor trick. Jenny, on the other hand, considered it a promising new field of scientific inquiry, and gave it a name: "cymatics," from the ancient Greek "*kyma*," which means "wave." And for him, this new discipline could and should involve a great deal more than just moving particles about a surface. He saw cymatic research and experimentation as having the potential to encompass and illuminate the core processes by which all things in the natural world take form and evolve. If we could unlock the mystery of waves and vibrations, Jenny believed, we would not only open our eyes to the sublimity of the universe but might set off revolutions in everything from medicine to the arts.

In service of this vision, he advocated a very specific methodology marked by a strong orientation toward the direct observation of natural phenomena; and this phenomenological approach deserves close attention on its own merits.

The primary focus of Jenny's research, and the most impressive, was to make sound visible to the human eye. Equipped with new technologies, he was able to research and document the effects of vibration with a depth and precision unattainable to his predecessors. These more sophisticated tools included the frequency generator, the crystal oscillator, and the stroboscope. In addition, he invented a piece of experimental equipment which he named the tonoscope. This unique tool allowed him to demonstrate the power of the human voice in novel ways, as I discussed in my book *Healing Songs*.

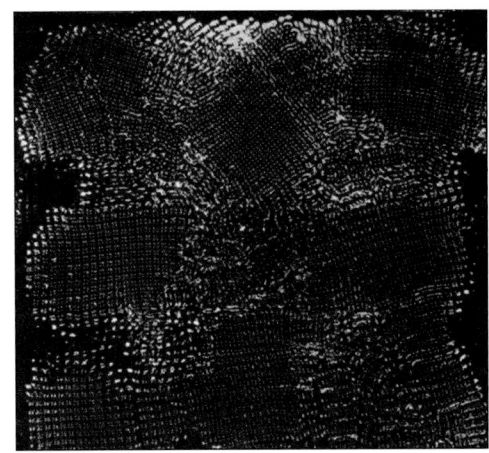

Under his guiding hand, sand, powders, and particles, as well as pastes, gasses, and fluids were shown to respond to vibrations, and most interestingly to me— to music. In one of his experiments, Mozart's Jupiter Symphony revealed its ability to reshape liquid into a delicate lace-like pattern, worthy of an artisan's workshop. And Bach's Toccata and Fugue in D minor demonstrated its power to create an extraordinary mosaic of byzantine intricacy (See figures 70-73 on pp. 77 and 78).

No one, until Jenny, had done this with music. He literally revealed a new dimension of composition before our eyes. And, at the same time, his work helped us grasp anew the oldest lore of creation myths— where matter is formed by deities and demiurges by means of their world-making songs. We encounter this type of narrative in every culture— Hindu stories of Shiva's world-making drum; Australian Aboriginal stories about the musical origins of landscapes in a part of the world where, to this day, certain pathways are still called songlines; or Biblical passages proclaiming the formative logos. It is also said that, back in Ancient Greece, in the early days of Western philosophy, Pythagoras would hold up a stone and tell his followers: "This is frozen music." Jenny did not set out to validate any of these belief systems, but his work did give them a greater empirical grounding than they had ever possessed.

Jenny extensively documented his work in both photographs and films. The images and 16 mm film footage he left behind are so awe-inspiring that, looking at them, it would be easy to believe that we have entered the realm of abstract art of the highest purity and intensity; and, in a way, we have. However tempting it might be to focus exclusively on the mesmerizing visuals of cymatics, let us not make the mistake of neglecting to read and contemplate Jenny's extensive commentaries, and his reflections on the wider implications and ramifications of his work.

Jenny bridged scientific and humanistic realms, left and right brain, and worldviews that had typically existed in isolation from one another. Not long ago, Tam Hunt and Jonathan Schooler put forth a new theory of consciousness, based on the hypothesis that rhythm might be the missing link between mind and matter. These researchers invite us to consider that, in a universe in which everything is vibrating and oscillating— both in our bodies and the external environment, our very sense of identity and agency may be a quasi-rhythmic, musical phenomenon. Hunt and Schooler describe their work as a "resonance theory of consciousness," and its initial impulse is clearly evidenced in Jenny's *Cymatics*.

How much power does sound really possess? Let me share one last story. A research team at UCLA recently announced that they successfully brought a 25-year-old man back from a coma, through the simple application of pulsating ultrasound. This impressive result emboldened them to try out the procedure with more severely afflicted patients. They achieved further success with a 56-year-old man who had been minimally conscious for more than 14 months and began to revive after just two treatments. They went on to revive a 50-year-old woman who had been in a profound coma for more than two-and-a-half years, following a cardiac arrest. After she received the ultrasound treatment, she could recognize objects and respond to spoken commands for the first time in years. All this was done with non-invasive procedures relying solely on sound. In this instance, the ultrasound was produced by a small device, the size of a coffee cup. The medical world would never call it a musical instrument, but why wouldn't we? This is the high-tech healing music of the 21st century, and we are just beginning to tap its possibilities. Hans Jenny would hardly have been surprised by these developments. In a way, he foresaw them decades ago.

I find it astonishing that, 50 years after Jenny's death, both his name and body of work remain largely unknown in the world of science, and even more so, in the world of music. Yet, more than any other figure of his time, he showed how these two spheres could be seamlessly integrated.

It is my hope that the years to come will not only grant Hans Jenny's work the visibility and appreciation it deserves, but that it will also inspire further exploration from a diverse range of scientists, musicians, and artisans, capable of further building upon his substantial legacy.

Ted Gioia is an American jazz critic and music historian who is the author of twelve books, including *Music: A Subversive History*, *The Jazz Standards: A Guide to the Repertoire*, *The History of Jazz*, and *Delta Blues*. He is also a jazz musician and one of the founders of Stanford University's jazz studies program. He has written about Hans Jenny and cymatics in his book *Healing Songs* (2006).

Publisher's Suggestions for Getting the Most from this Book
by Jeff Volk

Hans Jenny was by no means the first, nor was he even the most well-known investigator of the study of wave phenomena and vibration. A greatly abridged timeline of the evolution of this field, leading up to Jenny's explorations that culminated in his coining of the term "cymatics," appears as the opening to John Stuart Reid's commentary at the end of this edition.

Understanding the scope and magnitude of Jenny's extensive contributions to the study of wave phenomena is no easy task. Likewise, explaining his observations and conclusions in a language that would be accessible to a reader not as well-versed as he, is equally daunting. Perhaps the only thing more challenging than this is to *be* that reader, struggling to make sense of many unfamiliar concepts as well as modes of perception that may have yet to be developed. Having first re-published this material over 20 years ago, I can speak from all three of these perspectives!

One of Jenny's most important contributions was to expand our understanding of acoustics by "making the invisible visible." Similarly, our task as publishers became one of "making the incomprehensible, comprehensible!" This too was no small matter, and it can be likened to diving down a rabbit hole, into a land that is both strange, yet strangely familiar.

As you work your way through this book (and at times it will indeed feel like work!), I think you'll be well rewarded by the clarity of Jenny's observations and his articulation of what he was able to perceive. At the same time, many of the technical terms, and the concepts they describe, will likely take a bit of time to assimilate. To further support your understanding, we have included several resources in the back of the book, after the conclusion of Jenny's original text of Volume 2.

One suggestion that may provide a quantum leap in your appreciation of Jenny's work and its wide-ranging implications, is to view the video clips on our website and perhaps some of the other videos listed in the References section at the end of this edition. Two longer videos of Jenny's original 16 mm films are described in the color pages at the very back of the book.

Lastly, I would like to suggest that you approach this material in a relaxed and receptive fashion rather than trying to absorb everything in one reading, sequentially as it is presented. Jenny states emphatically that there is no way you will understand all of this upon first reading! For example, the final chapter of Volume 2 gets pretty weighty as Jenny takes a whirlwind tour of Plato's *Timaeus*, where "Timaeus the Pythagorean describes the creation of the world and tells us in detail how Nature and man came into existence."

You might find it rewarding to approach this book as a curated "armchair expedition" into the unseen world of creation. Jenny is a fabulous tour guide, but he can be hard to keep up with. So I thought I'd offer you a head start by giving you a contextual overview of the labyrinth that you're about to enter. A labyrinth is a journey of transformation that demands patience and perseverance and in exchange, offers deep insights into one's own nature through personal reflection. Light and sound are essential media for such a journey, which is indeed transformative. The reflective quality of light (knowledge) offers at best, a superficial understanding; You must strike the glass to reveal its authenticity.

So what is called for here, is a process of developing our latent capacities of perception to permit us to look behind the curtain of habitual "awareness." This is a process that has been described in many ways in numerous esoteric teachings over the millennia, all of which involve these basic tenets:

Four Facets of Conscious Awareness

1. Focusing Attention
2. Developing Concentration
3. Clarifying Perception
4. Expanding Awareness

This is indeed a process that entails discipline and commitment, perhaps even years of hard work, but it need not entail arduous vows of austerity and self-denial; a distasteful protocol imposed upon us by someone else, or that we impose upon ourselves.

Might we imagine this instead as a love affair; filled with the exciting allure of curiosity, awe, and fascination; allowing one's mind to relax its incessant need to understand; and to relinquish its demand to remain apart from whatever one is observing. In other words, you are hereby invited to use an objective scientific protocol to open yourself to a new way of perceiving, or perhaps an *old* way, the way in which, as a child, you may have once experienced the newness of everything, as if for the very first time.

So, may you grant yourself the gift of patience as you take these next faltering steps on your unique journey of self-discovery, steadfastly following the trail you're blazing. It's a well-worn path, even though few have actually followed it to its culmination.

Through Jenny's study of the extensive works of Rudolf Steiner, and his absorption into the concepts and imagery articulated both by Steiner and his great influence, the philosopher and natural scientist, Johannes Goethe, Jenny did something that very few scientists are willing to do: and that is to explore the realms of spirit as equally valid as that of the material domains, and to do so in a meticulously scientific and methodological manner. In so doing, he did not shy away from the soul nor shunt it off into the catacombs of religious dogma or metaphysical speculation. Rather he embraced it full-on, as a genuine and divine presence in all aspects of the natural world, while recognizing the elegant and elaborate process of evolution that carries each soul onward toward its ultimate realization. So as a scientist, as a spiritual scientist, Jenny chose to culminate his magnum opus on cymatics with this most enticing footnote: "Those who have a vital understanding of the idea of development, cannot agree that there are already human beings; instead they will say that men will for the first time become men."[1]

In my estimation, that is a most worthy aspiration for all men and women, upon which to set their sights.

May you enjoy the journey!

---

[1] *Cymatics* Volume 2. p. 299

CYMATICS
Volume I

The structure and
dynamics of waves
and vibrations

1

Problems of Cymatics

Wherever we look in Nature, animate or inanimate, we see widespread evidence of periodic systems. These systems show a continuously repeated change from one set of conditions to another, opposite set. This repetition of polar phases occurs alike in systematized and patterned elements and in processes and series of events. A few physiological examples may be mentioned in brief. The great systems of circulation and respiration are virtually controlled by such natural periods or rhythms. Inspiration and expiration of the lungs, systole and diastole of the heart are only these basic rhythmic processes writ large. In the nervous system the impulses occur serially and may therefore be described as frequencies. Much the same applies to the active muscle system which is actually in a state of constant vibration. The more closely one examines these functions, the more evident do these recurrent sequences become. Events then, do not take place in a continuous sequence, in a straight line, but are in a continual state of constant vibration, oscillation, undulation and pulsation. This also holds true for systematized structures. On the largest and smallest scale, we find serial elements, repetitive patterns— and the number of fiber stromata, space lattices, and reticulations is legion. If we turn our eyes to the great natural domains, periodicity expands to include the ocean itself. The whole vegetable kingdom, for instance, is a gigantic example of recurrent elements, an endless formation of tissues on a macroscopic, microscopic and electron microscopic scale. Indeed, there is something of a periodic nature in the very concept of a tissue. Again, periodic rhythms are a dominant feature of the animal kingdom. The metamerism of the various phyla is a case in point. It is the operation of this law that gives many worms, arthropods and vertebrates their special characteristics. From one specific point in the development of the germ onwards, the principle of organization is repeated on a grand scale in the segments. Every system is affected: skeletal, muscular, nervous, vascular, renal system, etc. But this principle is most clearly seen in the cellular character of organisms. Organs are not homogeneous masses, but tissues of the utmost delicacy which go on developing and repeating themselves indefinitely. Linked up with this is the sequence of generations, invariably a regular sequence of alternating polar-like phases. Even conditions inside the cell— the processes of cell division and the mechanics of the gene systems— are subject to this principle of oscillation. However natural these things may seem, they are, in fact, not. It must be realized that this periodicity represents an aspect of the world, and at first its mysteriousness always inspires a feeling of the greatest astonishment. In organisms, of course, we then find pure oscillatory phenomena rising to a higher plane in the formation of sound; and language itself appears on a still higher plane within this same field. If an inventory were to be drawn up of periodic phenomena in the realm of the organic, it would have to include the whole scope of morphology and physiology, biology and histology. But we must not forget the inorganic world. In this field we shall merely mention some typical examples, recalling known facts, with particular reference to physics. Here we encounter vibrations in a pure form, more specifically in waves. In the vast spectrum extending from gamma radiation, through the ultraviolet and visible light to infrared (heat rays), to electric waves (microwaves and radio waves), we have a field which may be termed periodic in the purest sense of the word. Then there are waves in the various states of matter, acoustic vibrations, ultrasound and hypersonics. Again, the lattice structures of matter in the crystalline state are also periodic. Periodic structure is a salient principle in say, the space lattices of mineralogy. What insights into vibration and periodicity have been gained in the vast range extending from the cosmic systems (rotations, pulsations, turbulences, circulations, plasma oscillations, periodicity of many kinds in both constituent elements and the whole) down to the world of atomic or even nuclear physics (shell model of nucleus; nucleon structure; organization of meson clouds)! Here again, the idea of periodicity is all-embracing. The few examples we have given here will serve as signposts. But to reveal the systematic, universal character of periodic phenomenology a great deal would, of course, have to be added: structural chemistry, colloidal chemistry, phenomena of mechanical tension such as appear in the isochromatic and isoclinic fringes of photoelasticity, and all the families of associated trajectories, to name only a few examples. Also of interest here is the problem of matter waves (L. de Broglie). Diffraction patterns have in fact been produced by material particles (atoms and molecules) in experiments. Thus these particles also display a wave-like behavior. This law of periodicity is particularly evident in the fault systems of geographical formations, which affect immense areas of rock.

Solar physics is another field in which oscillatory and wave processes are prominent. Our mental picture of the sun can accommodate serial structures, actual acoustic waves, plasma oscillations, turbulences, tendencies to recurrence of many kinds, periodic dynamics, etc. Moreover, many of the systems we have mentioned are polyperiodic in character. The rhythms and vibrations interpenetrate. But in every case periodicity is constitutive of their nature; without periodicity they would not exist at all.

Each of the fields we have mentioned would, of course, require a monograph of its own if it were to be properly described from the point of view of periodicity. The brief examples given here, in which reference is made to known facts, are only intended to give *some* idea of the inventory which, as suggested above, might be drawn upon the basis of periodicity. Hence it might be said without hesitation that the systems available to our experience are essentially periodic and that phenomena appear to be periodic throughout. However different the objects concerned, however different their causes and functional mechanisms, they have in common rhythmicity, oscillation and seriality. Nonetheless it must be realized that this conclusion does not take us to the real heart of these rhythms and wave processes. Indeed, it is only the discovery of the ubiquitous character of waves in the world that confronts us with the precise question: How actually do these vibrations function in a particular environment, a particular medium, a particular material? Even if we know whether we are dealing with hormonal influences, neural impulses, or mechanical or chemical factors, the actual problem still remains: What is really happening in all these periodic phenomena? What actually takes place in the periodic field? Now in view of the extreme variety of things affected, the extreme variety of systems coming into consideration, we have to seek out the rhythmic or serial where it is most characteristic, study it carefully, and observe its intrinsic character. Considering, then, that the repetitive alternation of opposed phases is common to all these phenomena, can we in this way obtain a description of periodicity which will reveal a basic phenomenon and afford a clear picture of its most fundamental nature? One example, which stands for many, will bring the problems involved into sharper focus. Let us consider the striated muscle in action. When the skeletal muscle is fully contracted, it displays what is known as tetanization. It is then seemingly in a state of continuous contraction. Closer examination and measurement, however, reveal an entirely different picture. It has been shown that in tetanization there are in the muscle, oscillations which can be demonstrated mechanically, optically and acoustically; they correspond to the frequencies of the impulses transmitted to the muscle. The "muscle sound" audible when the muscle is contracted is due, therefore, to the rhythm or frequency of the "miniature contractions at maximum tetanus" (Reichel 1960). Whatever happens, then, in the active muscle takes place in these rhythms. Let us consider exactly what this means. The numerous and vastly complex processes in the active muscle are all subsumed in this periodicity. It is in this vibratory field that all the bioelectric, chemical, mechanical, energetic, thermal, structural, kinetic and dynamic processes take their course. What are the effects of this oscillatory process in all these sectors? What are the kinetic effects of vibration on liquid systems? How do chemical reactions take place when they are enacted in media whose processes are without exception periodic in character? These are questions which follow directly upon actual observation. As we said above, the example of muscles in action must stand for many others. However, this organic system involves structures of the greatest intricacy; their very complexity forbids simple discussion. And yet they are a clear invitation to explore their peculiar nature, their dynamics and kinetics, their structure and texture as revealed in their periodicity. It is these problems which are the focus of our research. What we are concerned to do then, is not to formulate hypotheses about backgrounds and final causes, but rather to press on step by step with our exploration into this field and to find methods of giving tangible expression to this phenomenology.

Observation must begin, however, with relatively simple processes; many variations must be made in experimental conditions; and the object itself must be allowed to point the way from one set of experiments to another. It must be stressed that it is not a question of demonstrating the periodic and the rhythmic as such, or eliciting it from the complexities of its world according to the criteria of wave theory. The contrary is the case. It must be detected in its own world, its own environment, so that its specific effects are discovered and its multifarious operations recognized. Only by "getting inside" the phenomena through empirical and systematic research can we gradually elicit systems in such a way that mental constructs can be created which will throw a light on the ultimate realities. For it must be stated quite categorically that we have to proceed on strictly empirical and phenomenological lines and that all interpretative or analogical thinking will be out of place. If a name is required for this field of research it might be called cymatics (*to kyma*, the wave; *ta kymatika*, matters pertaining to waves, wave matters). This underlines that we are not dealing with vibratory phenomena in the narrow sense, but rather with the effects of vibrations. Our documentation is primarily concerned with the experimental demonstration of phenomena in the acoustic and lower ultrasonic range. Examples will also be interposed showing periodic phenomena occurring without an actual vibratory field in order to afford a view of the general field of periodicity or, in other words, of cymatics in the broader sense.

2

Experimental Method

In attempting to observe the phenomena of vibration, one repeatedly feels a spontaneous urge to make the processes visible and to provide ocular evidence of their nature. For it is obvious that, by virtue of the abundance, clarity, and conscious nature of the information communicated by the eyes, our mode of observation must be visual. However great the power of the ear to stir the emotions, however wide-ranging the information it receives, particularly through language, the sense of hearing cannot attain that clarity of consciousness which is native to that of sight. Who can reproduce a symphony after only one hearing, or even recall all its themes? But how many are there who, after looking at a picture, can in principle describe its main elements. It is not surprising then, that workers in experimental acoustics should have striven to make its phenomena visible during important periods of the development of the science. Special mention might be made of E. F. P. Chladni (1756-1827) who discovered the sonorous figures named after him while he was investigating Lichtenberg figures. With a violin bow he stroked metal plates sprinkled with powder and was thus able to make the vibration processes visible. The vibratory movement caused the powder to move from the antinodes to the nodal lines, and Chladni was thus enabled to lay down the experimental principles of acoustics (e.g. die Akustik, 1802). Work on this basis was not easy and, more particularly, the conditions of the experiment did not allow a sufficient range of observation since they could not be freely varied while the experiment was in progress. Thus the first necessity was to elaborate methods enabling the conditions of the experiment to be accurately fixed while still allowing free variation within these limits. One such method, which utilizes the piezoelectric effect, deserves special mention. Many crystals are distorted by electric impulses, and conversely they produce electric potentials when they are distorted. If a series of electric impulses is applied to the crystal lattice, the resulting distortions have the character of real vibrations. We will not go further here, into the complexities of vibrating crystal space lattices. Suffice it to say that these vibrating crystals afford a whole range of experimental possibilities. First of all, the number of impulses can be precisely determined with the generator exciting them. Thus we can always know the frequency (number of vibrations per second) and also the strength of the impulse (excursion or amplitude of the vibrating body). Moreover, the precise site of stimulation can be known in every case. Most important of all, however, is the fact that the experiment is not limited in time. The frequency and the amplitude can *both* be altered during the experiment. Hence it is possible not only to produce vibration patterns and investigate the laws to which they continuously conform, but also, and more especially, to make a close study of the transitions as one figure gives way to another. The experiment can be discontinued at any stage and each phase observed. Figs. 1-6 show a sonorous figure forming under the piezoelectric effect of a crystal oscillator.

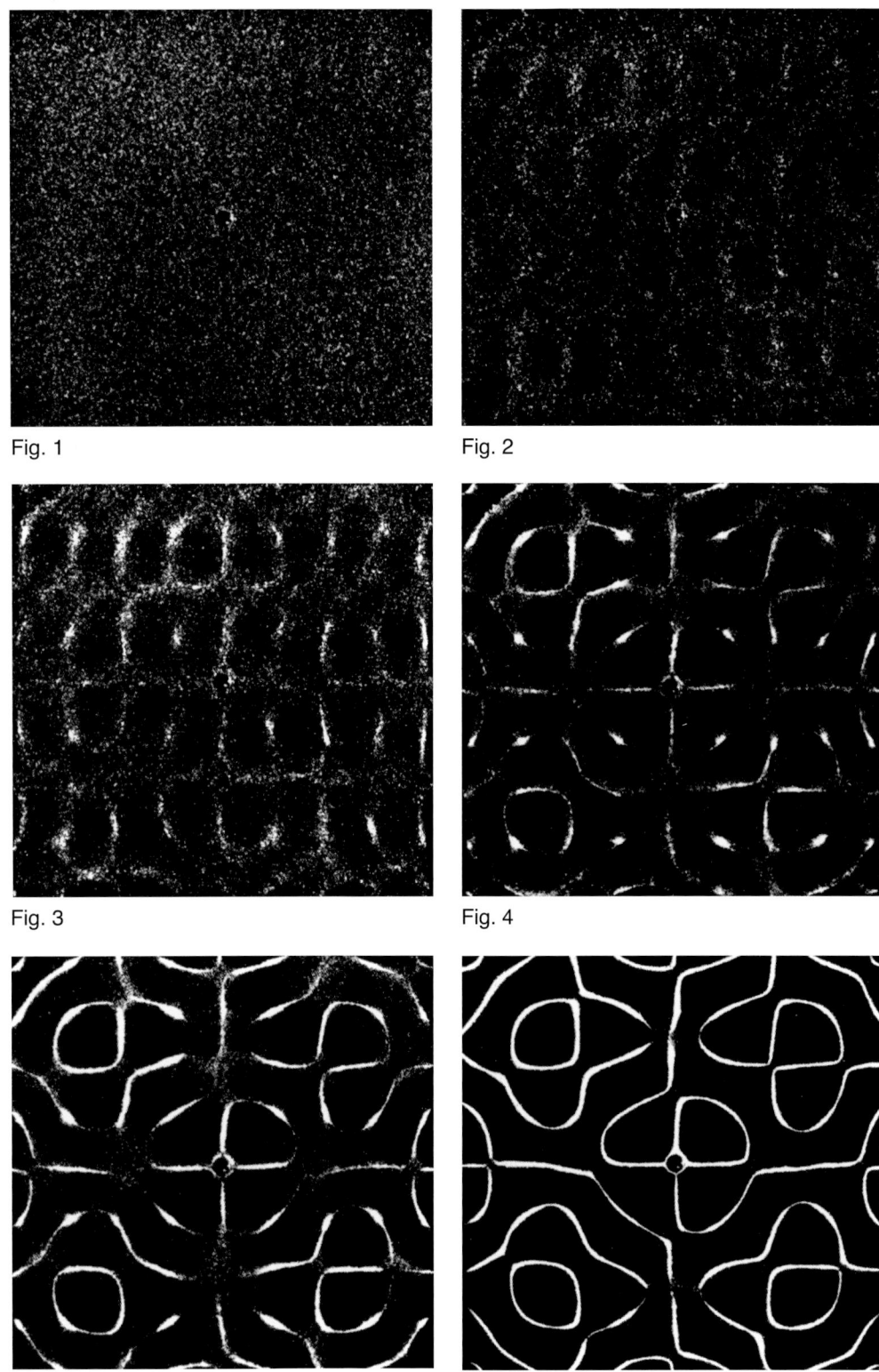

Fig. 1

Fig. 2

Fig. 3

Fig. 4

Fig. 5

Fig. 6

Figs. 1-6
The illustrations show a simple sonorous figure taking shape under the action of crystal oscillators (piezoelectric effect). Steel plate 31x3l cm. Thickness 0.5 mm. Frequency 7560 cps. The material strewn on the plate is sand which has been calcined to purify it.

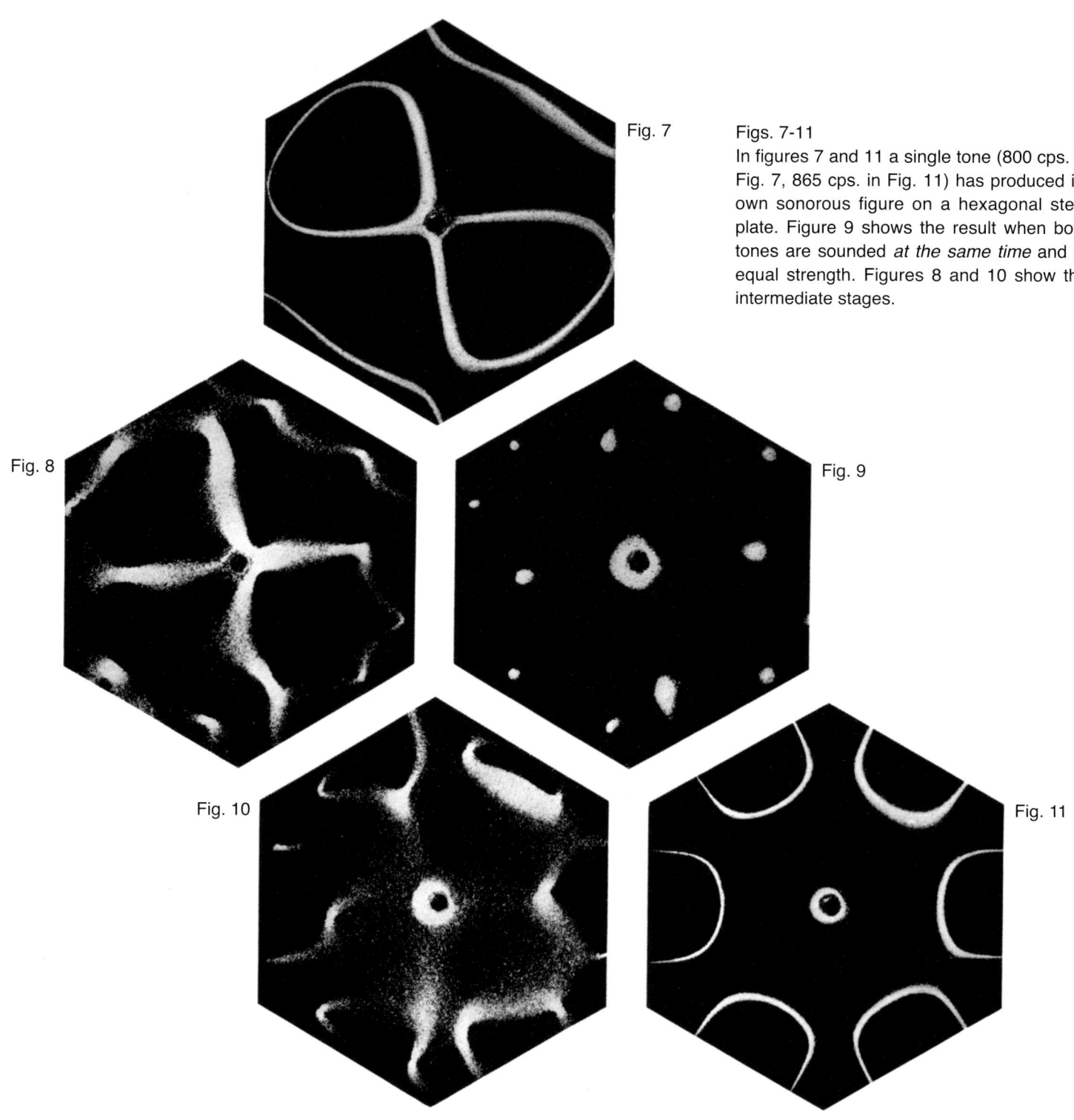

Figs. 7-11
In figures 7 and 11 a single tone (800 cps. in Fig. 7, 865 cps. in Fig. 11) has produced its own sonorous figure on a hexagonal steel plate. Figure 9 shows the result when both tones are sounded *at the same time* and at equal strength. Figures 8 and 10 show the intermediate stages.

It is a steel plate (size 31x3l cm, thickness 0.5 mm), upon which calcined quartz sand has been sprinkled. The oscillator is fixed to the underside. Fig. 1 shows the situation before the exciting impulse. In Fig. 2 the impulse has started and gradually the vibrational patterns of the plate are rendered visible. In Fig. 6 the sonorous figure has formed. It must be realized that in Figs. 2-6 the whole process is also audible. The exciting note can be heard continuously throughout the various stages (frequency 7560 cps). It would be really true to say that one can hear what one sees and see what one hears. Some experiments will now be described in order to show what this method is capable of achieving.

It is of course possible to attach more than one crystal oscillator to the same body and observe the effect. Fig. 7 shows a sonorous figure (frequency 800 cps) and Fig. 11 a second figure (frequency 865 cps). Each is individually excited. In Fig. 9, however, they are simultaneously excited, i.e. both notes can be heard. The figure shows the resultants of the two vibrations. From here it is only a step to making beats (interferences) visible. They have already been demonstrated by using a plate with irregularities in its material (Zenneck). If the characteristic tones of this material were proportional to the beat, the loops could be seen to change regularly in diameter. By using crystal oscillators a great variety of interference conditions can be selected. If two notes are produced with frequencies giving rise to beats, the whole Chladni figure pulsates or moves to-and-fro. Again, the phases can be changed in the course of the experiment.

The stages arising between the actual sonorous figures are of particular interest. Currents appear. The sand is moved around as if it were fluid. Nevertheless the organization of the vibrational fields persists in as much as these currents of sand move in the same or opposite directions. Fig. 12 shows such a process; the arrows indicate the direction in which the current is moving. Fig. 24 also shows this experiment and it is seen in greater detail in Fig. 25. Naturally the sand simply serves as an indicator. The actual events in vibrating plates and diaphragms are of extraordinary complexity. In the fields, for instance, areas appear which the indicator reveals to be in rotation. The powder congregates in small circular areas which continue to rotate regularly as long as the note is sounded. This rotational effect is not merely adventitious, but appears systematically throughout the vibrational field. Fig. 13 shows a plate with a number of rotating areas in which the direction of rotation of each area is contrary to that of its neighbors. The arrows show the direction of rotation. The circulation continues steadily and its course can be followed by marking with colored grains. Phases can also be demonstrated in which both currents and rotation are present. The following phenomenon is then observed: a small round heap which must be imagined to rotate has sand flowing towards it from two sides; the sand joins the heap, however, at

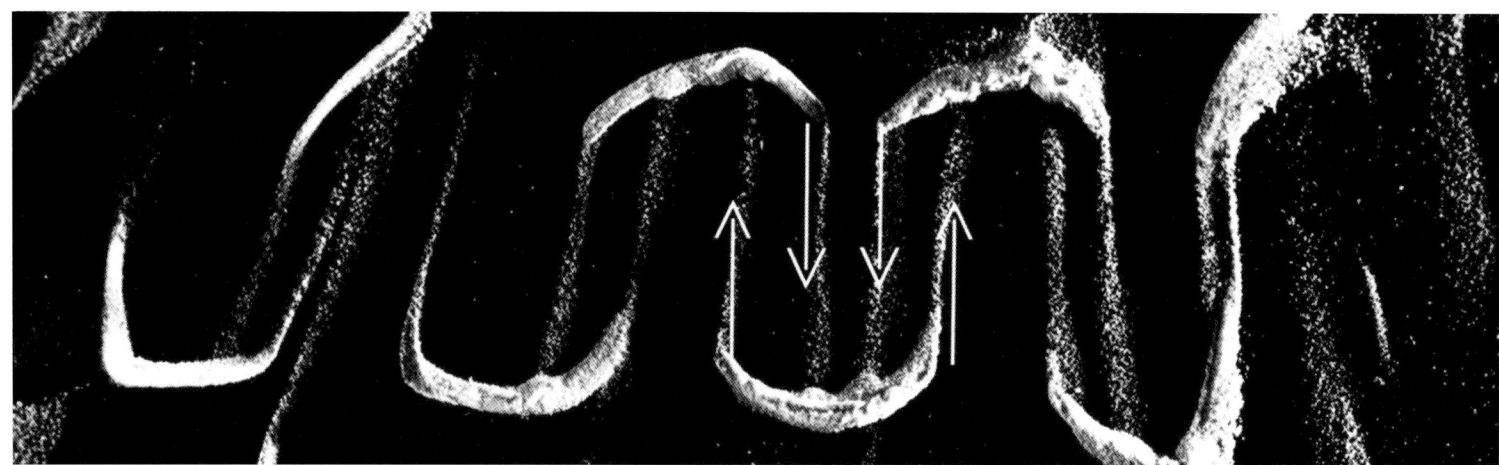

Fig. 12
The arrows indicate the direction of flow of the streams of particles. The material is lycopodium.
Plate 25x33 cm. Thickness 0.5 mm. Frequency 8500 cps. This picture shows a detail.

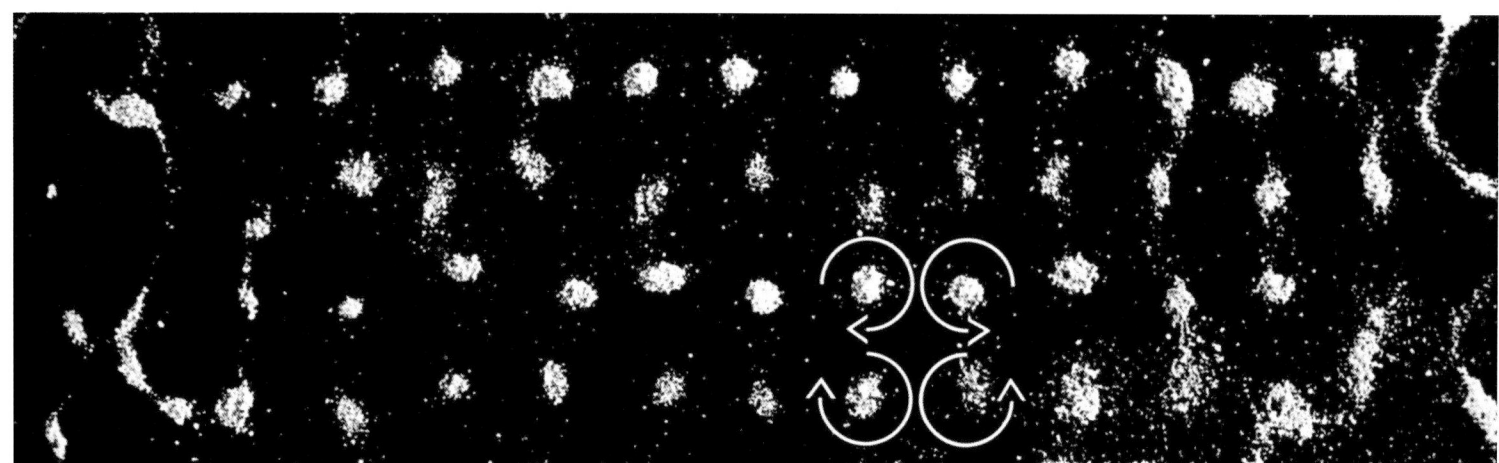

Fig. 13
Numerous rotational effects. The small round areas are in constant rotation. The arrows indicate
the direction of rotation. Steel plate 31x3l cm. Frequency 12,460 cps. The material is quartz sand.

opposite points on its circumference and is absorbed into it. Fig. 26 shows this happening. We thus have circular heaps which are joined together by bridges of flowing sand, or put another way, each circular heap has two flowing arms which move towards it, turn, and flow into it. These rotary effects can also be seen in "quite ordinary" Chladni figures. They point the way to still further investigations into the real vibratory processes in these bodies. At the same time they are a reminder that we should watch for such processes in vibrational fields. Indeed these currents, centers of rotation, revolving heaps with influent streams and connecting flows must actually be expected, and the material occupying the field will indicate the vibrational pattern prevailing there. There can be no doubt about the occurrence of these formations.

It is obvious that all these processes (interference, flows, rotation) could be more appropriately documented by cinematography; photographs can only stimulate the mind into grasping these processes imaginatively.*

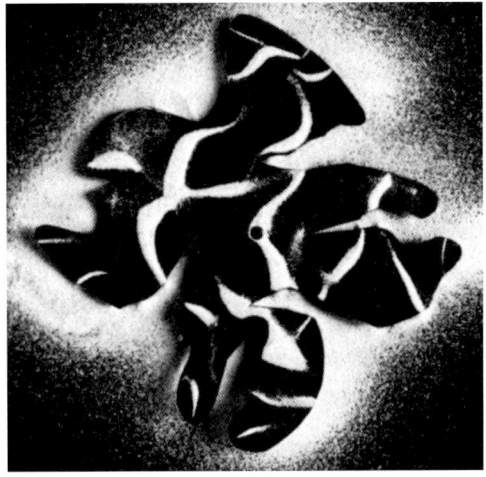

Fig. 14
A steel plate deliberately cut in an irregular shape. Maximum diameter 23 cm. Thickness 0.5 mm. Frequency 4100 cps. The material is quartz sand.

Fig. 15
A symmetrical plate with complicated subdivisions. Distance across at widest part 30 cm. Length 18 cm. Frequency 21,400 cps. Excitation is from the top downwards.

*Editor's note: Dr. Jenny did make several films of his experiments, highlights of which are now available on video. See promo pages at the end of this edition.

Fig. 16
Guitar excited by a tone of 520 cps. The sand shows the nodal lines of the mode of vibration.

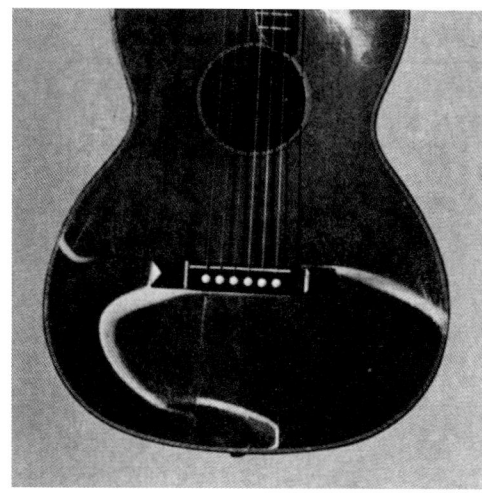

Fig. 17
The vibrational pattern of a brass plate. Oscillatory processes in complicated solids can be rendered visible by means of crystal oscillators. Frequency 10,400 cps. The material is lycopodium powder. Diameter of the plate 20 cm, depth 3 cm, thickness 0.3 mm.

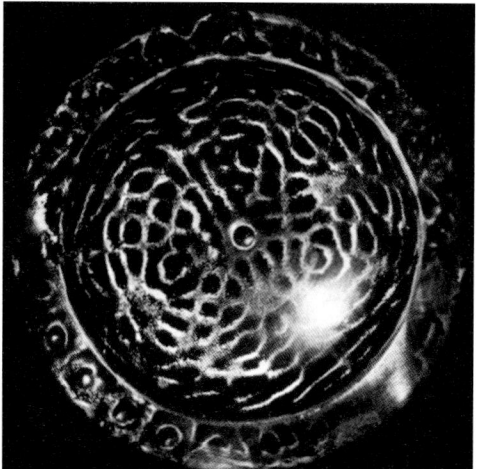

With these selected methods the vibration of complicated bodies can also be rendered visible. The vibrations of structures whose mode of vibration cannot be calculated at all, or only approximately, can in this way be made accessible to experience. Fig. 14 represents a sonorous figure on a steel plate of arbitrary shape. The conditions revealed on the variously formed lobes can be studied. Fig. 15 also reproduces a complicated figure. It is precisely in figures of this kind that symmetrical relationships between the various parts can be found. It might be mentioned in passing that in such structures as these both planar and axial symmetry enter as factors. In Fig. 16 we see a guitar energized by sound (520 cps). The indicator substance reveals the real wave events in the wooden body. If the material contains irregularities, say cracks in the wood, there are corresponding changes in the pattern of oscillation. Vibrational behavior in forms whose modes of vibration cannot be calculated can also be rendered visible in this way. In Fig. 17, a brass plate has been made to vibrate by sound and its pattern of vibration can be clearly discerned.

An interesting observation may be made at this point. The masses affected by a tone are, of course, naturally forced into the form corresponding to the vibrational effect. While the tone impulse persists, liquids and viscous masses will remain in their place if the diaphragm is tilted, or even held vertically. If the vibration is discontinued, the masses slip down under the force of gravity. If the resumption of the tone is not delayed too long, the masses return to their position, i.e. they climb up again. In a sense it would be legitimate to speak of an antigravitation effect. In Fig. 18 the vibrating diaphragm is arranged obliquely; the mass will not slip down so long as the tone is present.

Even the ringing of a bell can be rendered visible. Fig. 19 shows the vibrational pattern that appears.

In order to take investigations into the vibration of complicated bodies a step further, a photoelastic technique was developed. The stroboscope, which is an instrument for rendering visible the phases of rapid periodic motion, is used as the source of light. The stroboscopic light is polarized and penetrates the transparent model which is made to vibrate. The analyzer enables the vibrational process to be observed as a photoelastic phenomenon. This technique makes feasible the study of even such complicated forms as musical instruments (e.g. the violin) in a state of vibration, at least within the limitations imposed by the use of a model.

If a liquid is used as an indicator instead of sand, an entirely different picture is obtained. The nodal lines disappear and the antinodes appear as wave fields. Fig. 20 is a sonorous figure. In Fig. 21 we have the same plate, excited at the same frequency, but covered by a sheet of liquid. The same pattern can be recognized in both pictures. In Fig. 20 the antinodes

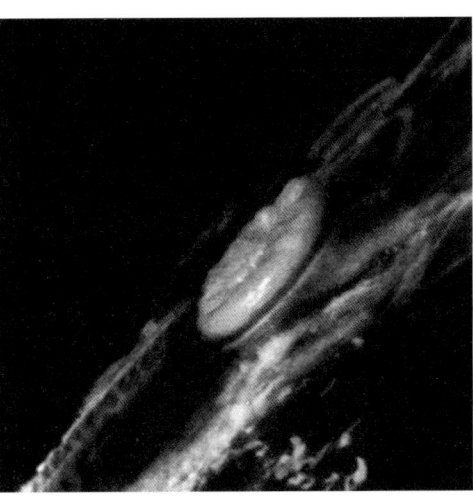

Fig. 18
The circular shape has been created out of a paste-like mass by a tone. The vibration holds the paste in its field even when the diaphragm is held obliquely or vertically. If the tone were discontinued in figure 18, the paste would flow downwards.

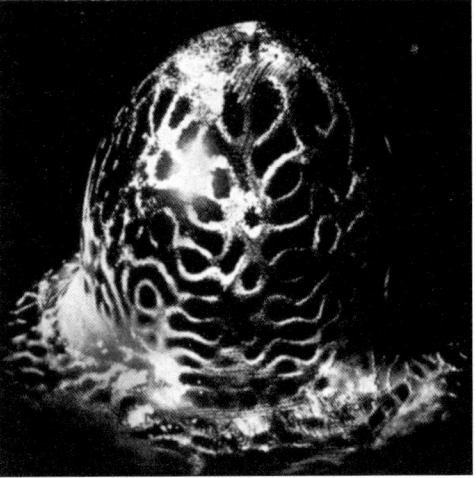

Fig. 19
The mode of vibration of even such a complicated (non-calculable) form as a bell can be rendered visible. Frequency 13,000 cps. Diameter of the mouth of bell 15 cm, height 11 cm, thickness 0.3 mm. Lycopodium powder was used.

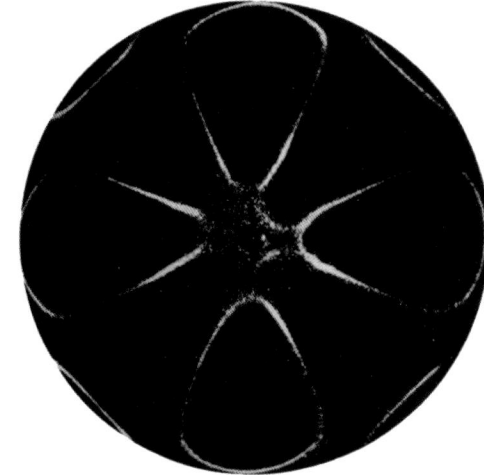

Fig. 20
Sonorous figure on a circular plate. Diameter 16 cm, frequency 1060 cps. The nodal lines are made visible by sand.

Fig. 21
The same plate excited by the same frequency as in figure 20 but covered with a liquid. Now the areas of movement (antinodes) are visualized directly as wave fields.

are empty, but the nodal lines are shown in sand. In Fig. 21 the fields of movement are rendered visible by the wave lattices; there is nothing to be seen where the nodal lines are. The most varied forms of wave trains are to be seen in lattice areas. The movements at the margins of the fields are striking. If lycopodium powder (spores of the club moss) are strewn on the surface, violent movements can sometimes be seen. A turbulent zone consisting of unstable wave phenomena often forms there. Thus in the one case (Fig. 21) the moving elements are shown; in the other (Fig. 20) those parts which exhibit no movement are shown. Thus the *actuating* is, as it were, opposed to the *actuated*, the *creating* to the *created*. This draws attention to the fact that the patterns taking shape must be understood in terms of their environment, that patterns in general are, as it were, an expression of the movement and energizing process. One might speak of a *creans/creatum* relationship. Thus there are many conditions under which the mind might be said to be directed to the environment, to the circumambient space, to the field from which space lattices, networks, etc. take their rise in the first place. In other words: observation of organized patterns and the milieu creating them raises questions as to the processes incidental and precedent to the formation of such patterns. This nexus of problems is one that merits further investigation, e.g. in mineralogy, colloid chemistry, in periodic or rhythmic precipitations, and in the field of chemical reactions in general. What happens before fibers, fibrils and crystals are separated out? Since such systems are shown to be periodically textured or patterned, periodic processes of a corresponding kind must be present in the preceding stages. But they always have to be verified in the concrete case.

Vibrating materials naturally react quite unequivocally to heat conditions. Not only the crystal oscillators, but also the plates and diaphragms change their vibrational characteristics to a remarkable degree. They might truthfully be described as sensitive. The following experiment is representative of many. Fig. 22 is produced by a tone with a frequency of 1580 cps. If the outermost edge of the steel plate is heated for a few seconds by a flame, the whole vibrational pattern changes at once, as is shown in Fig. 23. If the tone is continued during cooling, the original sonorous figure returns after a few moments (Fig. 22). Not only the effects of the vibration, but also the vibrating media themselves have highly specific characteristics.

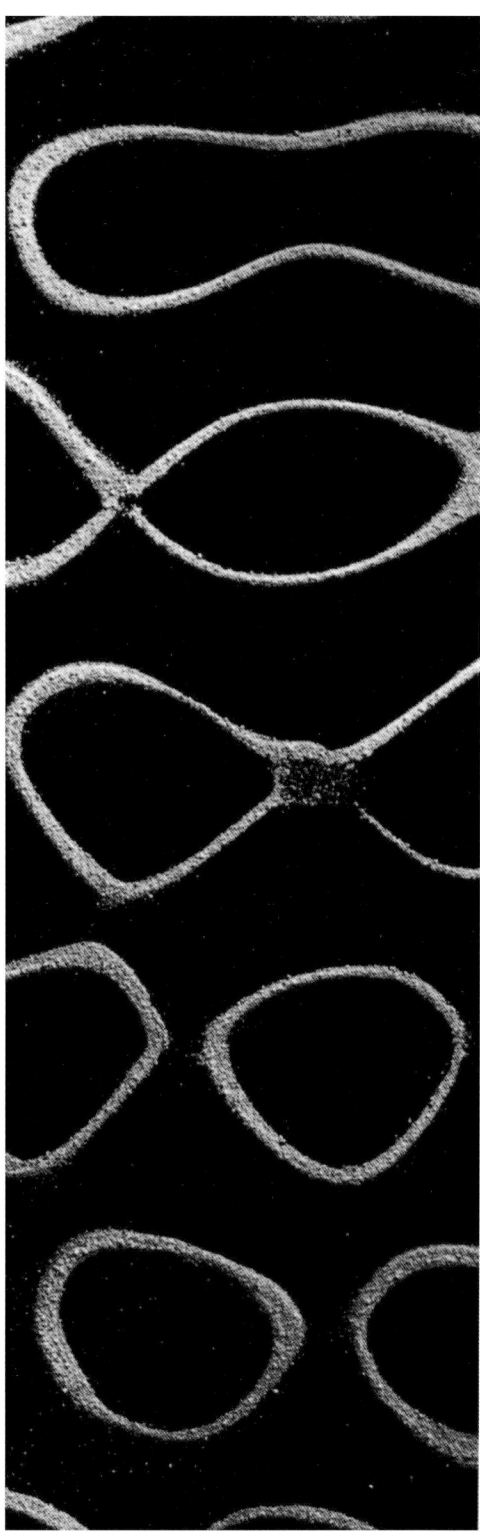

Fig. 22

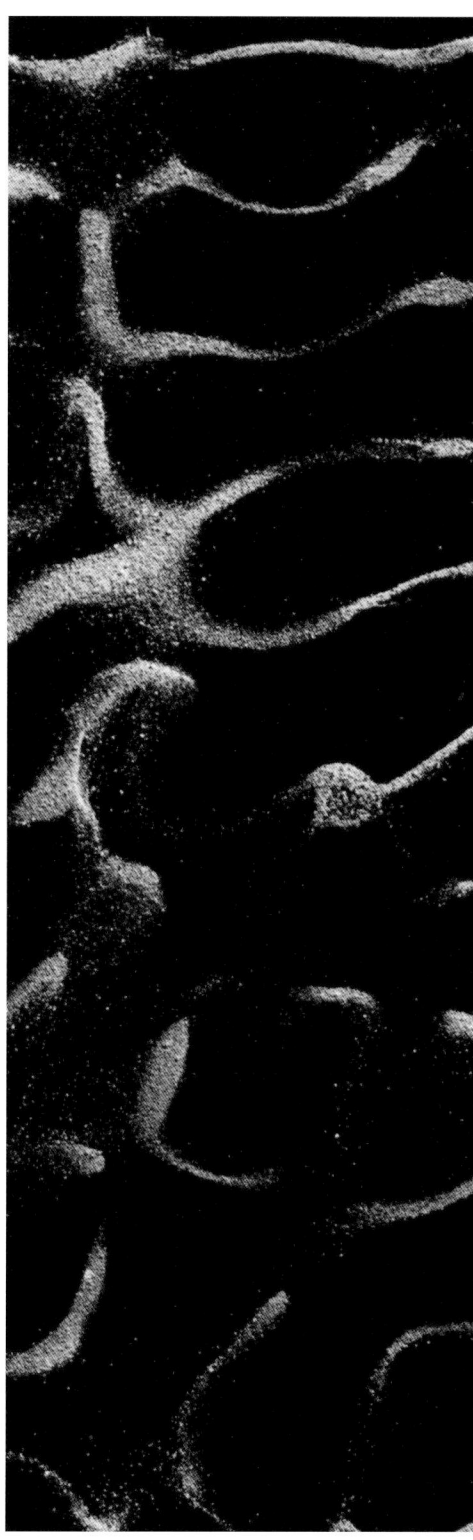

Fig. 23

Fig. 22
Sonorous figure. Plate 24.5 x 32.5 cm, thickness 0.5 mm, frequency 1580 cps. *before* the effect of heat.

Fig. 23
The same experimental conditions as in Fig. 22 except that an extreme corner of the plate was touched by a flame for a few seconds. Immediately the whole shape is distorted. Fig. 22 reappears once the plate has cooled.

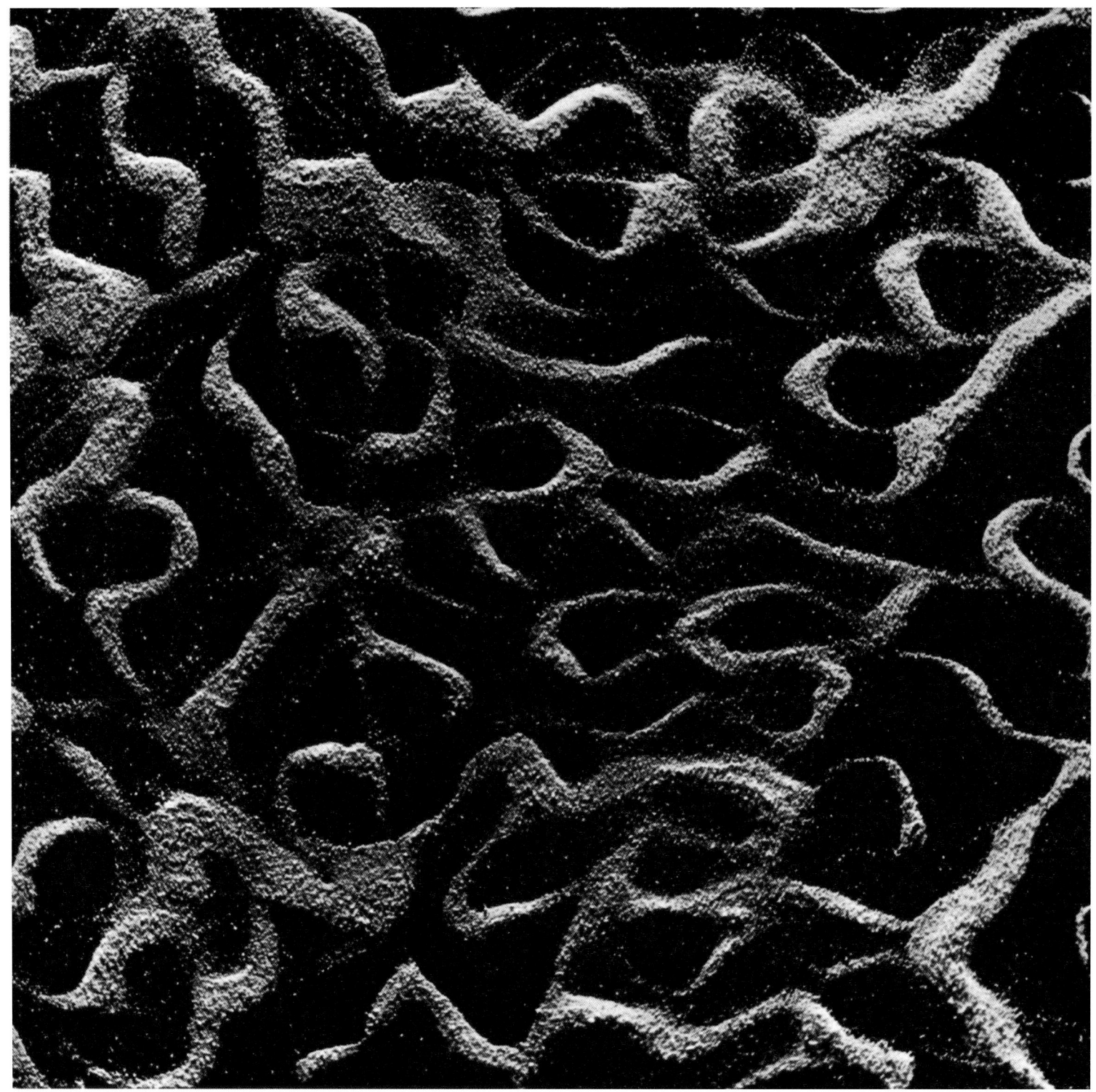

Fig. 24
The sand figure is not a sonorous figure in the usual sense; the particles of sand are in a state of flow. Excitation is by crystal oscillators. Steel plate 25x33 cm, thickness 0.5 mm, frequency 10,700 cps. (Cf. Fig. 12.)

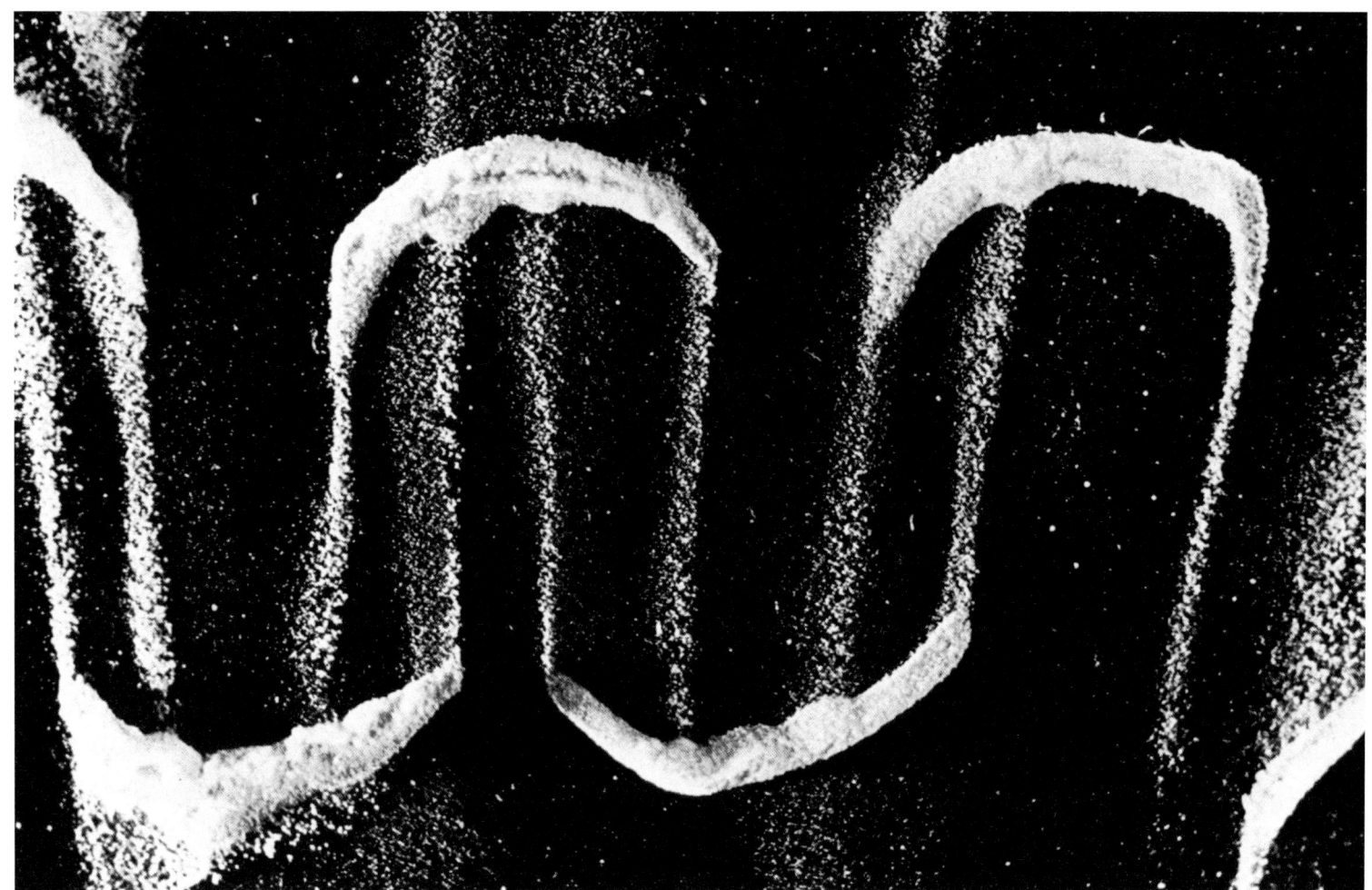

Fig. 25
This again is a photograph of a "flowing figure." Lycopodium (spores of the club moss) is used to indicate the currents. Frequency 8500 cps. (Cf. also Fig. 12, detail)

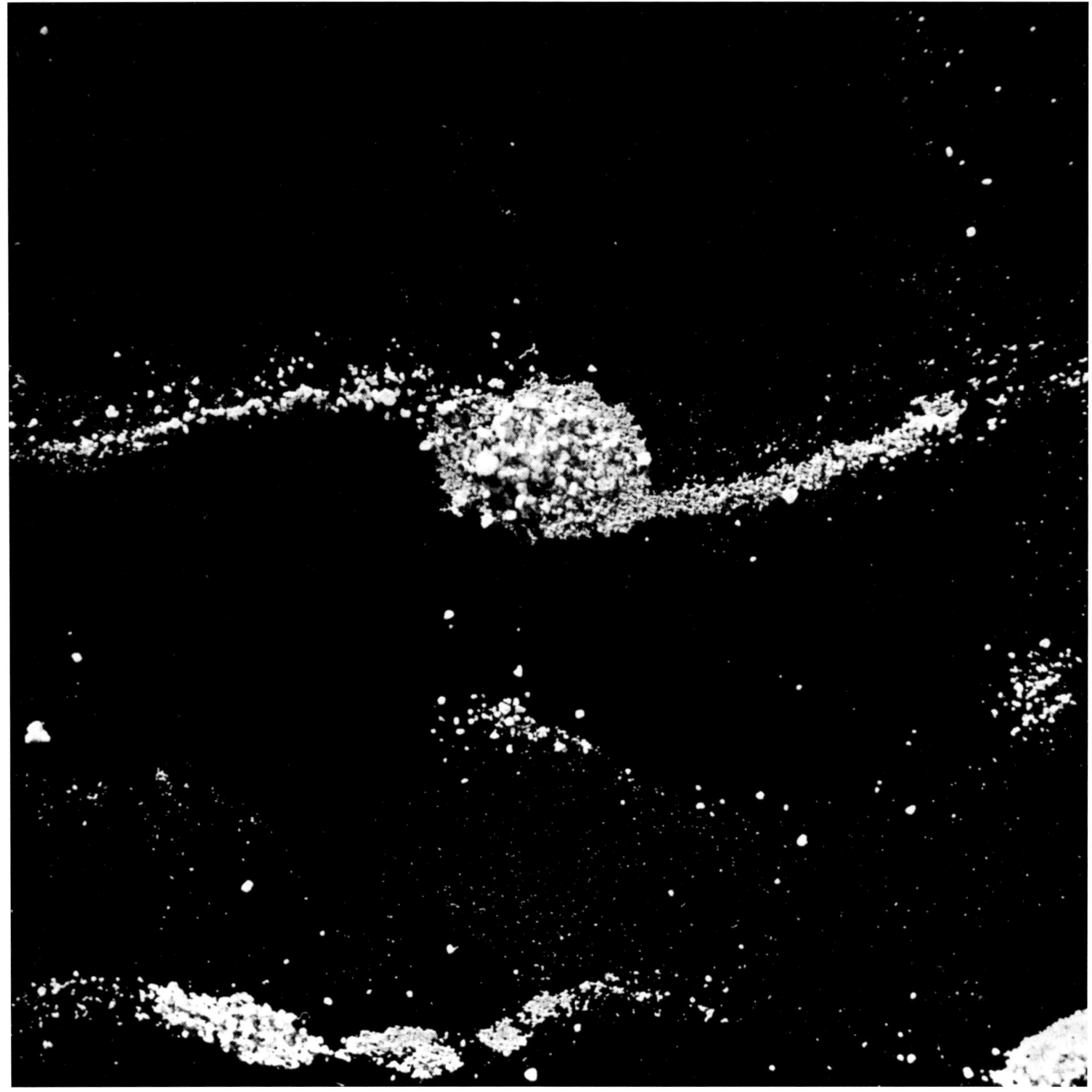

Fig. 26
Rotational effect. The round heaps of sand must be imagined in rotation. The sand is flowing in the two longitudinal areas; it is flowing towards the round shape and joining it at opposite ends. This exceptionally interesting phenomenon is, of course, reproducible. It may be termed a rotational system with two bridge arms. Frequency 12,470 cps.

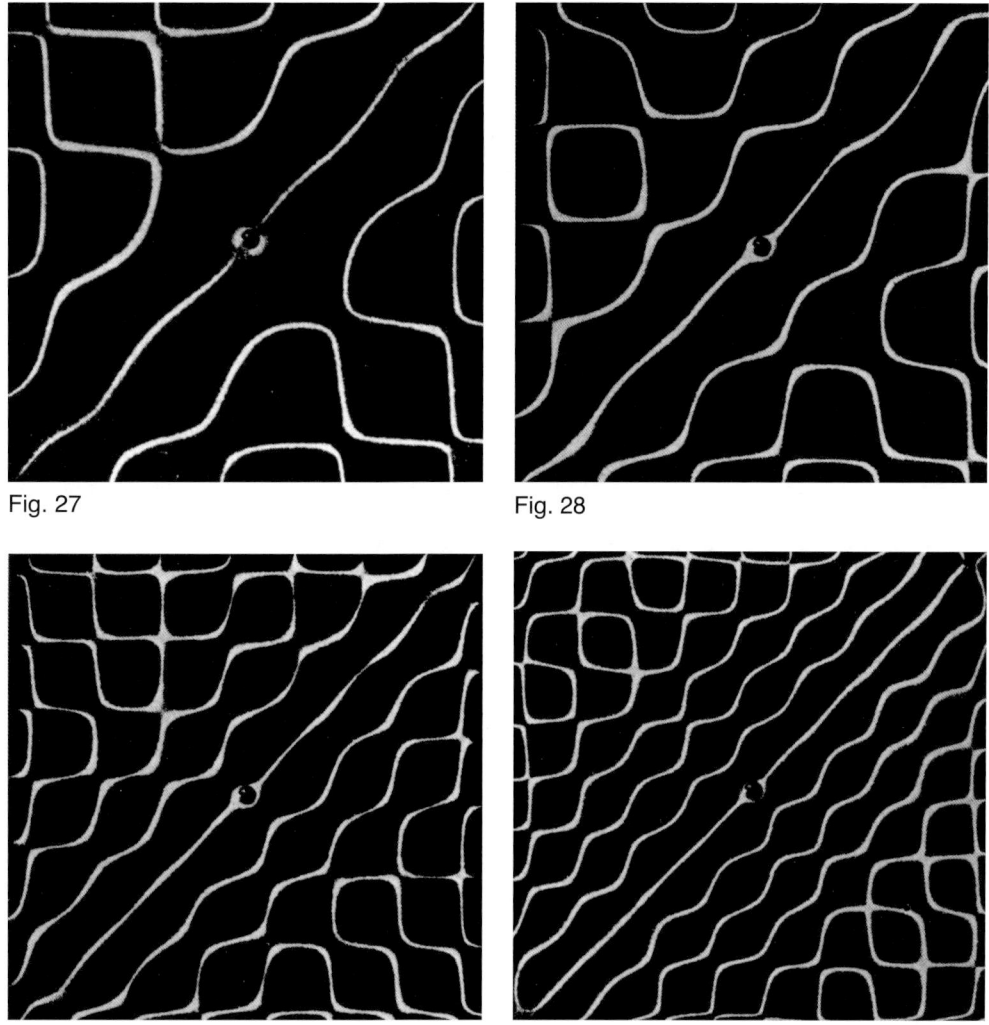

Fig. 27

Fig. 28

Fig. 29

Fig. 30

Fig. 27-30
The four figures reveal approximately the same pattern but the number of elements grows as the pitch increases. In Figs. 27 to 30 the frequencies are 1690, 2500, 4820, 7800 cps., respectively. Steel plate 23x23 cm. Thickness 1mm.

3

Examples of Cymatic Phenomenology

Some examples will now be given of vibrations rendered visible. What we actually witness here is the effect of vibration. The effects produced by vibration in this or that material, in this or that medium — that is what is to be demonstrated. We are in fact present at the very site where the oscillating process takes its effect. First we shall simply pass under review a series of such phenomena. The classes of phenomena appearing, and the relationships existing between them will be revealed little by little by the things themselves.

First we see some sonorous figures (Figs. 36-42). All these experiments were performed with crystal oscillators. This type of experiment enables the vibrational patterns to be produced in series and compared. It is notable in these serial experiments that the same formal pattern recurs at increasing frequencies, but that the number of constituent elements also increases at the higher frequencies. It is apparent from the series Figs. 36-42 that the figures at the higher frequencies display many more elements.

In Figs. 27-30 we see a formal pattern of which similar versions are repeated as the frequency increases. In Figs. 27-30 the frequencies are 1690, 2500, 4820, and 7800 cps., respectively. Here again use was made of a steel plate (23x23 cm. and 1 mm thick). The size of the plates can, of course, be varied at will. We have used plates ranging from the size of the ear drum (approx. 7x9 mm) to 70x70 cm. Also the material of the plate can be selected as desired (glass, copper, wood, steel, cardboard, earthenware, etc.). Apart from the rule that the number of elements increases with the frequency, a variety of other observations may be made of the sonorous figures (Figs. 36-42). Since the process can be stopped at any stage, the formation of nodal lines can be followed step by step. The observer can see how the particles are transported and how two lines draw together. In the marginal zones turbulent areas can be seen in which the particles form vortices. One point is particularly noteworthy. In the series of vibrations one metamorphosis is quite literally followed by another. The variety of resultants and products of co-acting forces is inexhaustible. Continuous variations are to be found on the basic concentric/radial pattern in round plates and the parallel/diagonal pattern in rectangular plates. But there is an additional factor to consider. One formal element is repeated in one and the same sonorous figure. We are therefore actually dealing with the same configuration. Now, no plate is completely uniform throughout, and as a result the energy imparted to it by the frequency is not evenly distributed. The consequence of this is that the pattern is imperfectly formed in some places and fails to appear completely; but more important still is the fact that the elements of form vary in the sense that they become, as it were, variants of a form (see e.g. Fig. 38). At one moment a pattern is closed and separate; at another the same element is open and linked up with its environment. Again, patterns become united which, elsewhere on the same plate during the same experiment, are disjoined. Simply by reference to these principles of resultant systems, metamorphosis and variability, an atlas of sonorous figures could be compiled. But amidst all these figurate patterns the kinetic factor keeps on appearing, particularly in the "continuous waves" we have already mentioned and the rotational

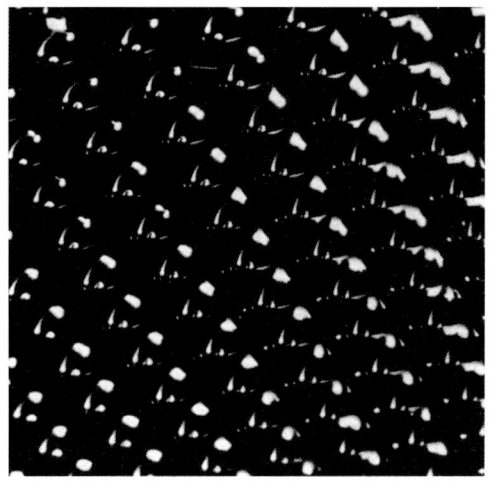

Fig. 31

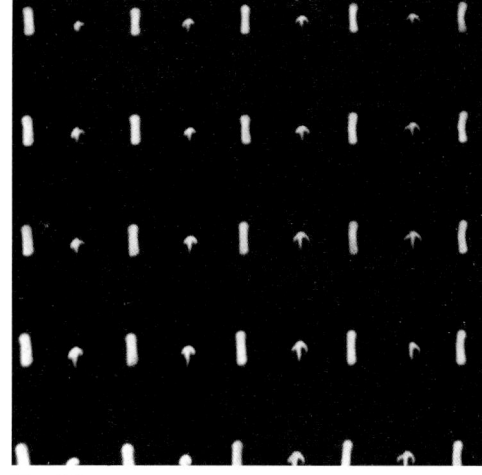

Fig. 32

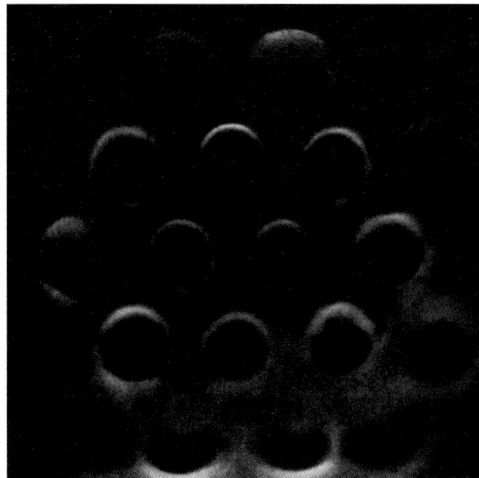

Fig. 33

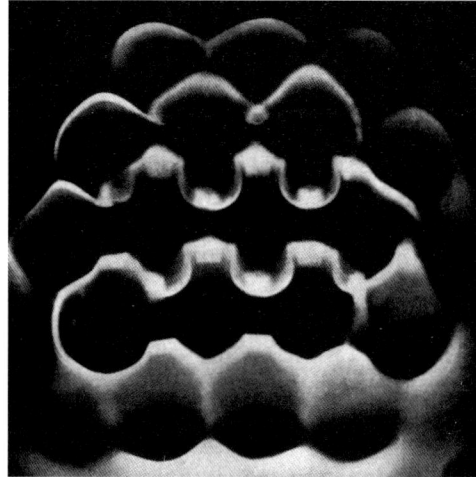

Fig. 34

Figs. 31-34
A liquid is made to vibrate on a diaphragm with a diameter of 28 cm; the modes of vibration give rise to a multiplicity of patterns and textures. Sometimes standing, sometimes traveling, they change as the frequency is modulated. Certain of the formations are reminiscent of Lissajous figures.

Fig. 35
A train of waves generated in glycerin on a vibrating diaphragm. There are no transverse waves.

effects. In such cases the whole figure is in a state of flow or comprises a number of rotational centers with adjacent centers invariably revolving in opposite directions. Figs. 13 and 24 must therefore be conceived not as figures, but in kinetic terms. Fig. 25 is a detail from a flow phase.

Thus when figure, configuration, form, pattern and texture, etc. are referred to in this connection, we are using empirical descriptions. The woof and weft of these phenomena is so continually shifting that they cannot be pinned down to any rigid definitions. It will be apparent in the course of our disquisition that these various categories of phenomena interlink and interpenetrate to such an extent that definitions would inevitably dismember something which represents a closely knit unity throughout. This applies in particular to the effects of vibration in liquids. Under the action of oscillations at frequencies as high as the lower ultra-sonic range, it will be noted that the surface of liquids crinkles to form a wave field. This effect can be traced in detail if the reflected image of a cross or lattice of threads on the surface of the liquid is observed. In the reflected light it appears to be distorted. The extent of the distortion corresponds to the exciting frequency. Figurate patterns, however, also appear in liquids. Fig.44 shows such an oscillation pattern in glycerin; likewise Fig. 45. We must realize that in these and the following photographs a great diversity of things must be observed at the same time. Configurations (standing waves) can be recorded, but at the same time there are currents which impart flow to the patterns. Eddying regions point to turbulences. Now and again an interference oscillation flits across the field, etc. When various areas of oscillation move towards each other, there is a tendency for Lissajous figures to form with more or less clarity.

Figs. 31-34 and 46-49 show a transition from figures to organized arrays. Hexagonal, rectangular and imbricate patterns strike the eye as honeycombs, networks and lattices. Occasionally a picture forms but then the textures shift and change into every imaginable display. Figs. 46-49 and 31-34 may be taken as examples. Once again there are reminiscences of Lissajous figures (e.g. Fig. 49).

It is hardly necessary to add that interference phenomena can once again be demonstrated in these experiments with liquids. A pulsing movement may be imparted to the whole sheet of liquid while the wave lattices, depending on the phase conditions, may, for example, appear and vanish with lightning rapidity.

Mention should be made at this point of an unusual experiment. The curvature of the surface already referred to, causes the mass of the liquid to shift. This brings about a change in the oscillatory character of the whole system. The procedure is as follows. A film of water some 3 cm. in diameter is energized by a frequency of approximately 2400 cps. A delicate wavelike pattern is

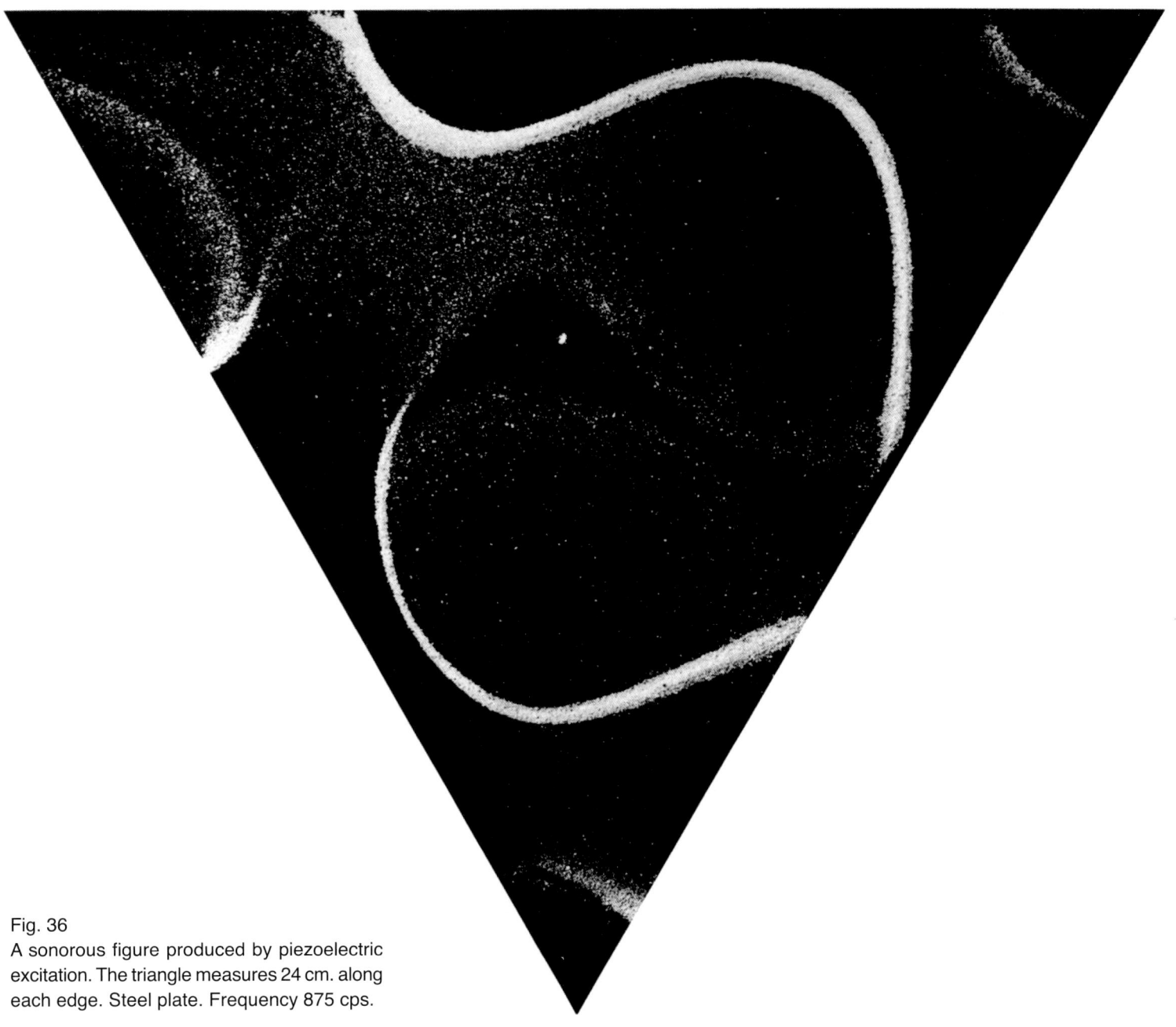

Fig. 36
A sonorous figure produced by piezoelectric excitation. The triangle measures 24 cm. along each edge. Steel plate. Frequency 875 cps.

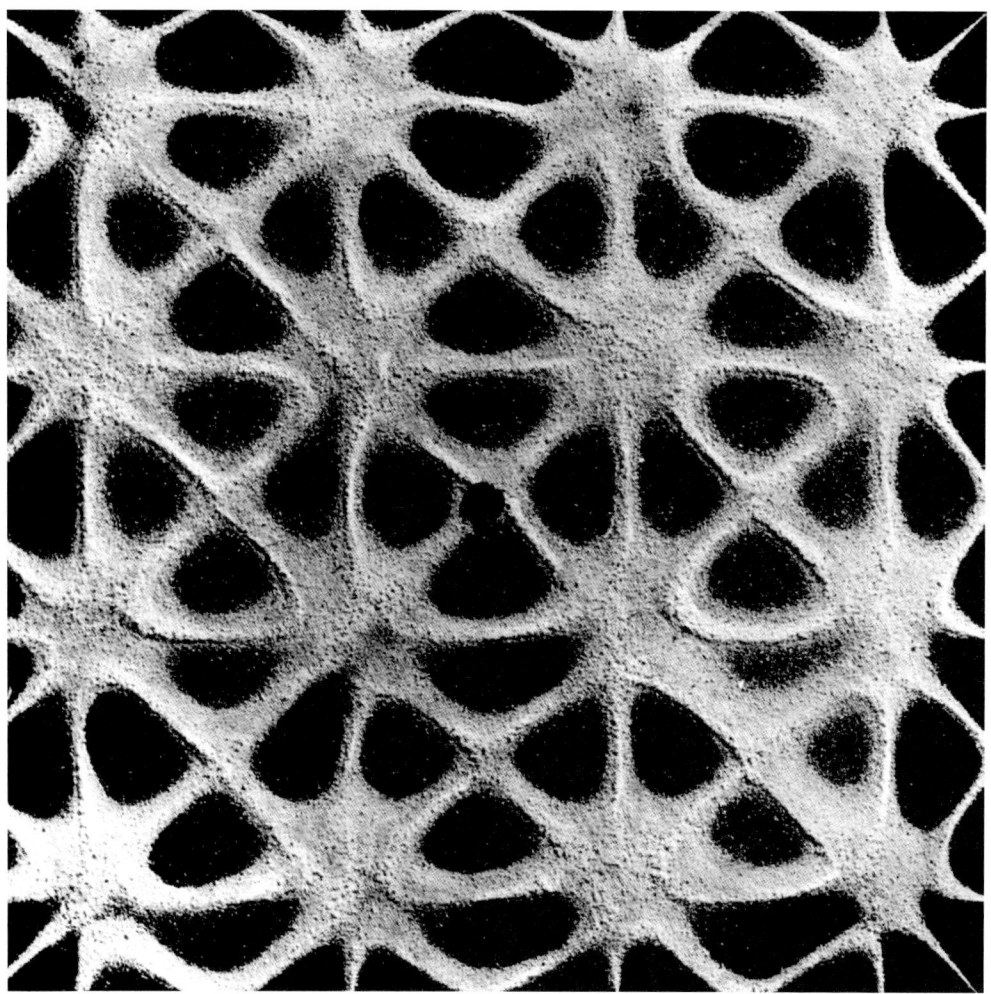

Fig. 37
A sonorous figure, also produced by piezoelectric excitation. Square steel plate 23x23 cm, thickness 1 mm. Frequency 6700 cps. In such experiments as these, the formation of nodal lines of sand can be accurately observed. They often pile up like dunes.

formed. But at the same time a pulsing movement is set up which affects the whole. The curvature of the surfaces and the ensuing changes in mass and position cause the liquid to slip away from the frequency impinging upon it. The factor causing the curvature is thus removed. The mass now returns to the initial position and once again comes under the influences of the energizing tone. The process is repeated and this pulsation becomes regular. In the experiment described, the rate of pulsation was 80 per minute. This phenomenon exemplifies a pulsating system.

As we proceed with our series of experiments with liquids, we obtain organized patterns of ever increasing delicacy. The impression created is one of textures. Figs. 50-53 show what happens. The pictures look like closely interwoven fabrics. Appropriate changes in the strength of the vibration (amplitude modulation) can be used to show how these wave "fabrics" are woven out of wave trains. By adjusting the movement of a vibrating diaphragm it is possible to produce only one train of waves. The wave may be of the standing or traveling variety (Fig. 35). Further undulations can be generated by changing the amplitude, whereupon a grid-like pattern appears. In Fig. 54 this process can be seen in the nascent state. Some of the large waves are beginning to undulate in themselves. Signs of incipient transverse elements can be seen more particularly in the central parts of the photograph.

Whereas these figures, whether standing or traveling, have a certain constancy of appearance as organized patterns and textures, there are also phenomena in a constant process of change. Figs. 55-58 are turbulences generated by vibration, showing this type of dynamic formation. These formations, which are constantly coming and going, appear in a marginal zone, where a vibrational process meets a zone of inactivity. They are thus entirely different from the turbulent processes appearing in flow phenomena. Nor should they be confused with vortices (to which we shall be proceeding in a moment) which are hydrodynamic phenomena produced by vibration. These "wave curls" must be conceived as being in motion, but in an unstable state. They disappear but appear again in a similar configuration; thus they are irregularly regular or regularly irregular. They are also in evidence along the borders of the wave fields noted in the sonorous figures produced in liquids. They show up particularly well in fluorescent liquids. In ultraviolet light, for instance, they figure as "light sources" in contrast to the non-fluorescent areas. This curious phenomenon with its vortices and waves which tend to vanish and then return, is thus rendered wholly visible. Pictures such as Figs. 57-58 may look so similar that they might be thought to be photographs of the same situation but with some variations. It may happen that the waves disappear between the photographs and then reappear.

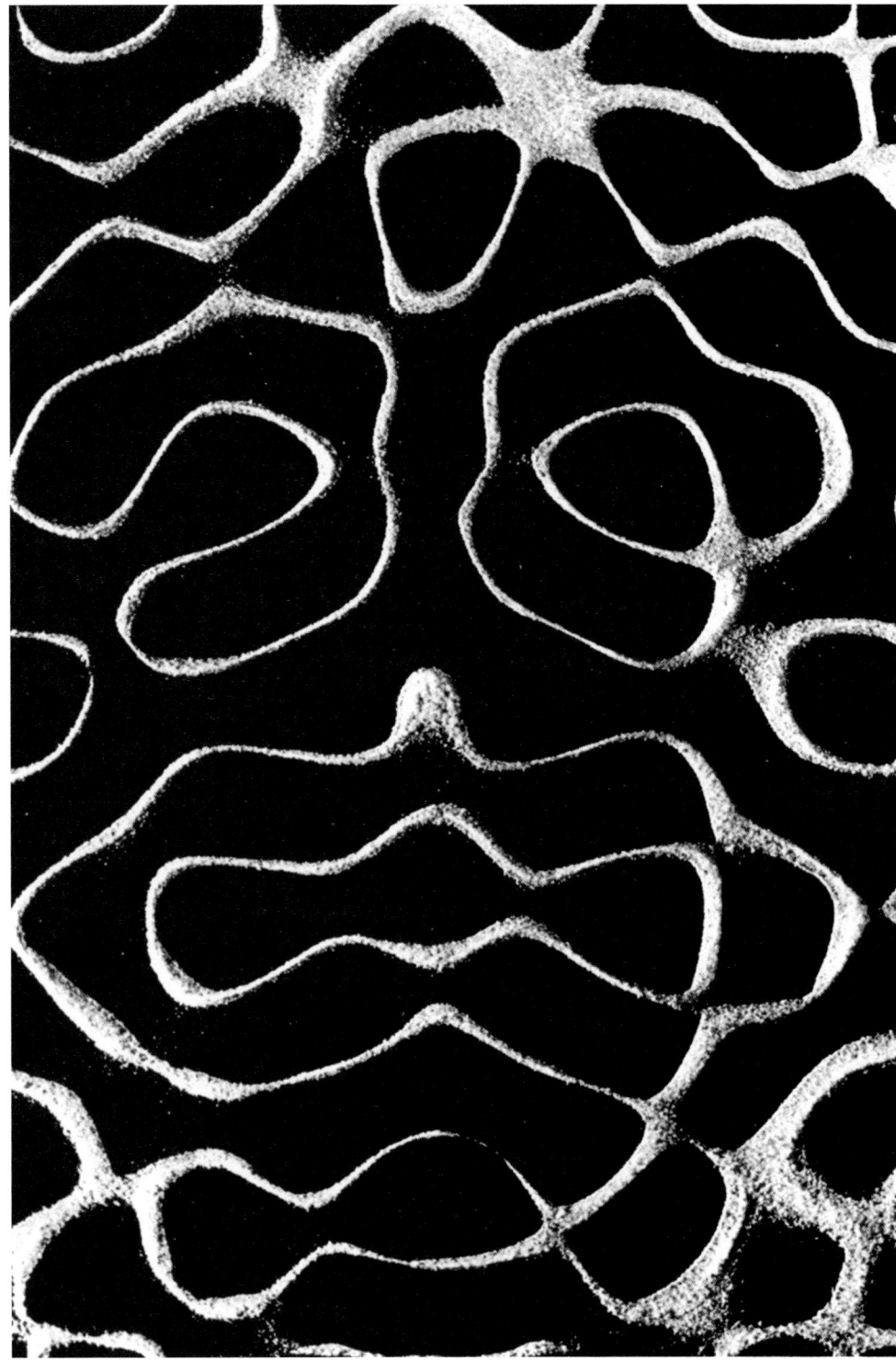

Fig. 38
Sonorous figure. Plate 25 x 33 cm. Thickness 0.5 mm. 5330 cps. Configurations like these enable us to follow the variety of forms which one and the same pattern can assume. The basic pattern undergoes metamorphosis.

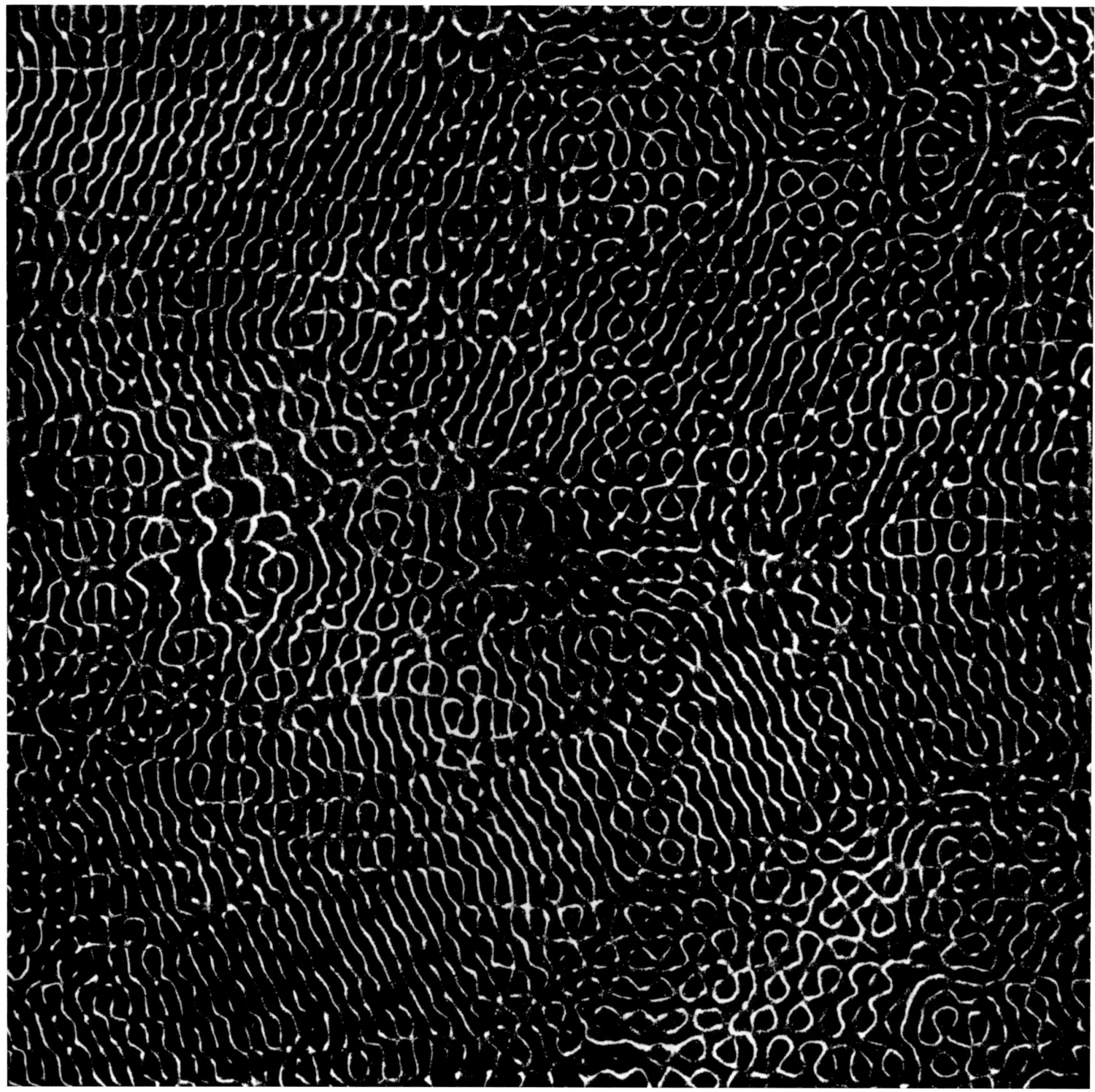

Fig. 39
A sonorous figure excited by crystal oscillators. Steel plate 70x70 cm.
Frequency 17,600 cps. The high frequency results in a large number of fields.

In 1928 G. von Békésy discovered in experiments with models of the cochlea of the ear that vortices are formed by vibration. Two vortices were formed at the same time at a specific site depending on the frequency. Von Békésy was then able to observe these eddies in the ear itself and also their transmission in the spiral of the cochlea. He found that as the frequency increased, the two eddies moved towards the stapes in precise conformity with the laws of hydrodynamics. The direction in which the eddy rotates is constant; the speed of rotation increases with the strength of the vibration; the location, as explained, is determined by the frequency. These therefore are hydrodynamic phenomena caused by vibration.

In the experiments described here, vortices are also produced by tones although in an entirely different way. The simplest case is shown in Fig. 59. This is a film of liquid a few centimeters in diameter. To make the pattern of currents visible, a black dye was used as an indicator. It will be seen that a pair of eddies is formed symmetrically. Here again the speed of rotation is proportional to the amplitude. Fig. 60 shows the process by which the eddies are generated. This experiment, however, is performed with a sheet of liquid about 23 cm. across which is made to vibrate by a tone (frequency 100 cps). Trains of waves corresponding to this tone can be seen in various parts of the photograph. The black dye which originally covered the whole sheet

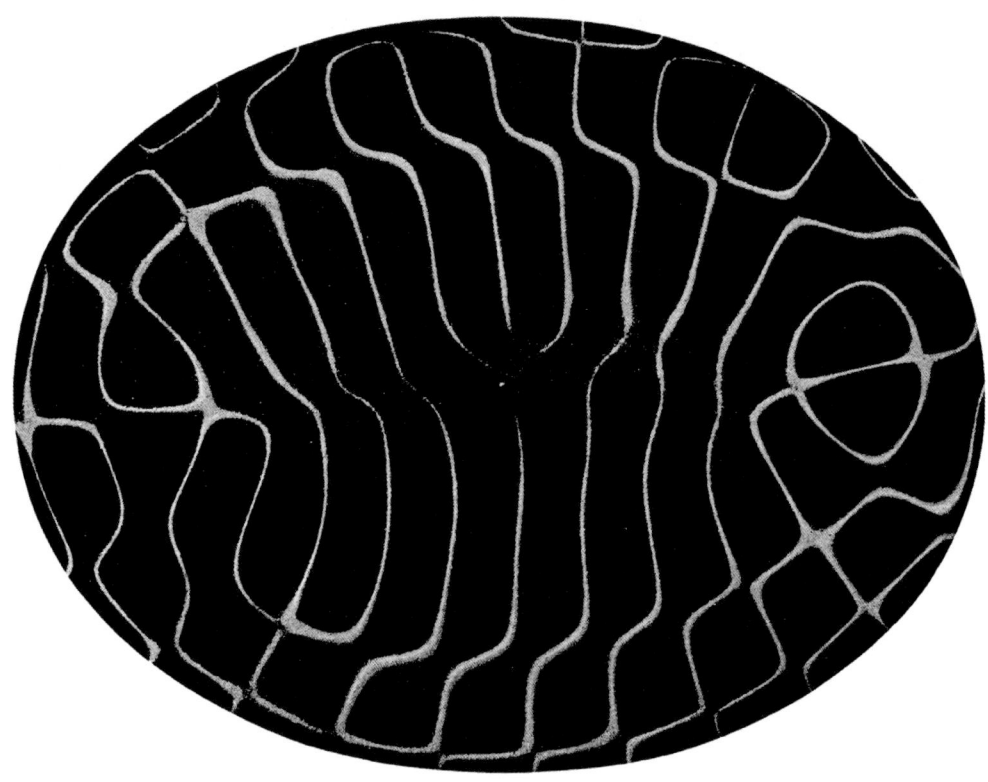

Fig. 40
A sonorous figure on an ellipsoid steel plate. Greatest diameter 31 cm, 3095 cps. Excitation by crystal oscillators. Quartz sand was strewn on the diaphragm.

begins to form eddies; in other places it is still unbroken. What we see in this photograph is really the nascent state. In Fig. 61 the process has progressed much further. Some pairs of eddies have formed. In such cases there is always a strict bilateral symmetry. On the face of it, this is astonishing. This eddy formation, however, is found to be directly connected with the vibrational pattern and its organization. Individual wave trains can also be discerned in this vibrational pattern. Additional pairs of eddies are formed at the sides, and these again are strictly symmetrical. In the marginal regions rotary currents with a tendency to form vortices can be seen everywhere. In passing, it is interesting to note the numerous turbulent areas which form in the marginal zones. Figs. 62 and 63 give details of this phenomenology of eddy formation. Fig. 62 affords a picture of the very center of this symmetrical pattern. The precision of the flow is clearly visible; the axis of symmetry runs from the top to the bottom, and the dividing line between the quadrants runs from left to right. Below right the eddy turns in a clockwise direction, and below left in a counterclockwise direction. It is particularly interesting to see in Figs. 62 and 63 how the currents move in layers. The dye indicator is distributed in laminar fashion. Fig. 64 shows once again a vortex in the form of four quadrants; once more there is symmetry and laminar flow. In all these pictures it must be imagined that the liquid is flowing quietly in a vortex while the energizing sound can be heard. If the experiment is performed with an opaque liquid, say mercury, some lycopodium can be strewn over it in order to make the currents visible.

These experiments have similarly revealed that not only are proper vortices formed, but also curious rotational currents such as may be seen in Fig. 65. The first beginnings of vortices can be discerned in the upper and lower section of the energized liquid. In the center there are indications of a figure-eight form, whose periphery is formed by the black fluid made to flow by the energizing sound. There is, of course, no division in the form of a lemniscate; here again the rotational flow is steady.

This particular field of cymatic studies calls for further extensive research. How do currents flow in three-dimensional space? Currents and countercurrents must be demonstrated, and above all, the operation of these hydrodynamic phenomena must be studied at the minutest scale, as we have already seen in the hydrodynamic theory of hearing.

If we allow these examples of cymatic phenomenology to file past our inner eye, we are struck by the extraordinary richness and diversity of the forms and processes. Yet, even at this stage, certain significant features stand out from the background pattern. Figures, organized patterns and textures appear, real wave processes arise, then the whole is set in motion and fluctuates, giving rise to eddies, flows and streams. However, still further categories of phenomena will become apparent to us; and we shall then see whether we can gradually educe the outlines of basic phenomena.

Fig. 41
A sonorous figure excited on a circular steel plate. Diameter 50 cm, 6250 cps. In pictures like these the patterns resulting from radial and concentric nodal lines can be studied.

Fig. 42
The same plate as Fig. 41 except that the frequency of vibration is 16,000 cps. By using the piezoelectric method, patterns of vibration can be produced in series. The fact that the frequency is higher than in Fig. 41 can be inferred by the larger number of fields.

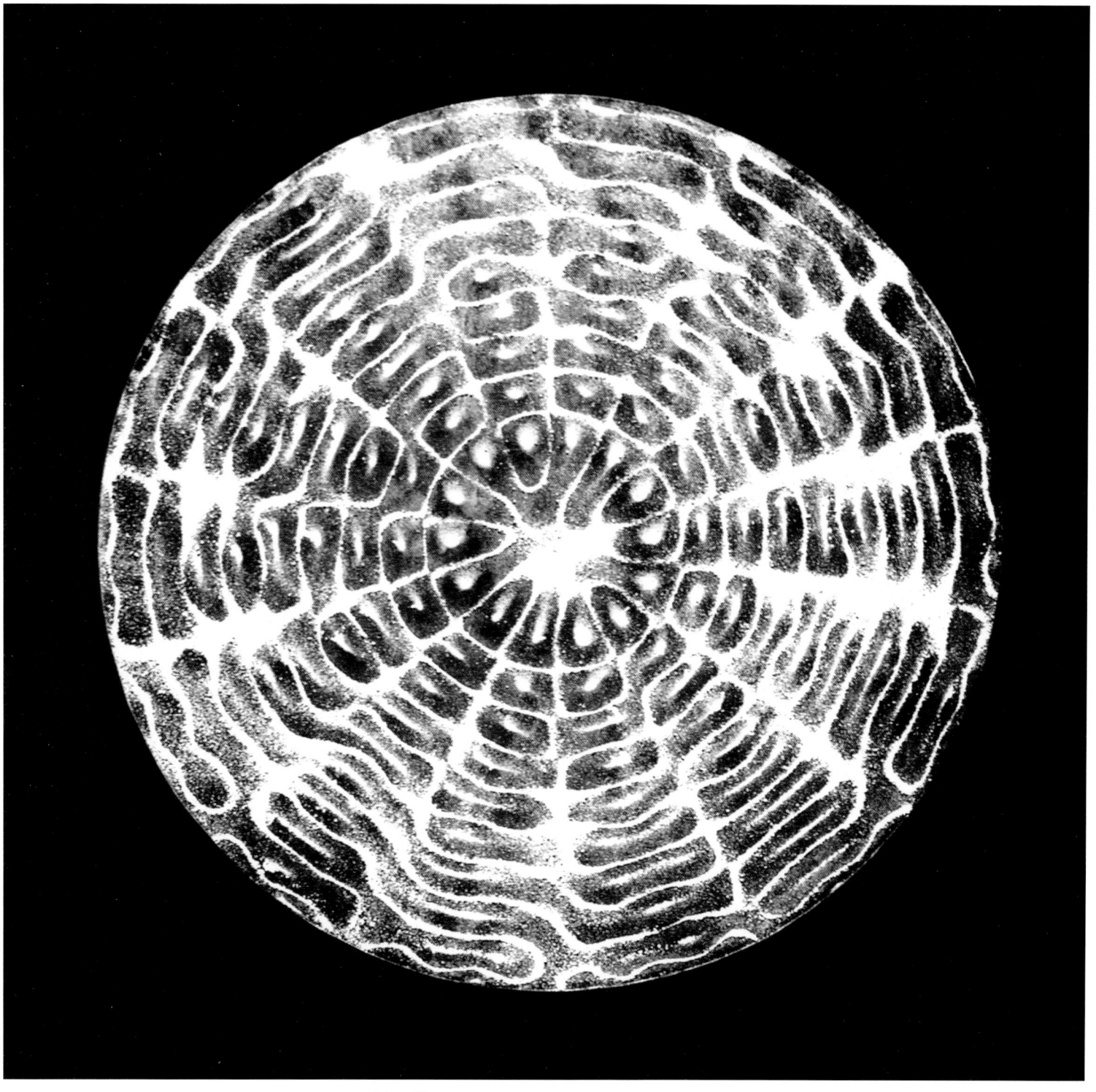

Fig. 43
Sonorous figure. Plate: diameter 32 cm, thickness 0.5 mm, the material is strewn quartz sand, frequency 8200 cps.

Fig. 44
A layer of glycerin excited by the oscillation of a membrane. Figures can also be seen in the liquid.

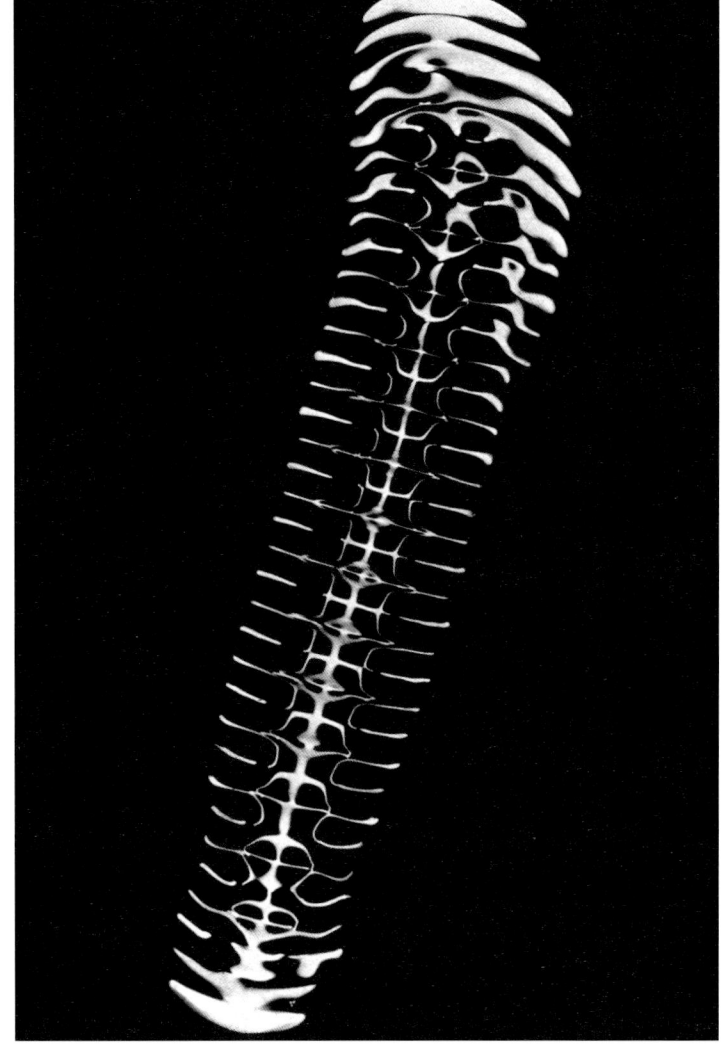

Fig. 45
As in Fig. 44, a layer of glycerin has been made to vibrate by a tone acting upon a diaphragm. The result is a continuous formal pattern.

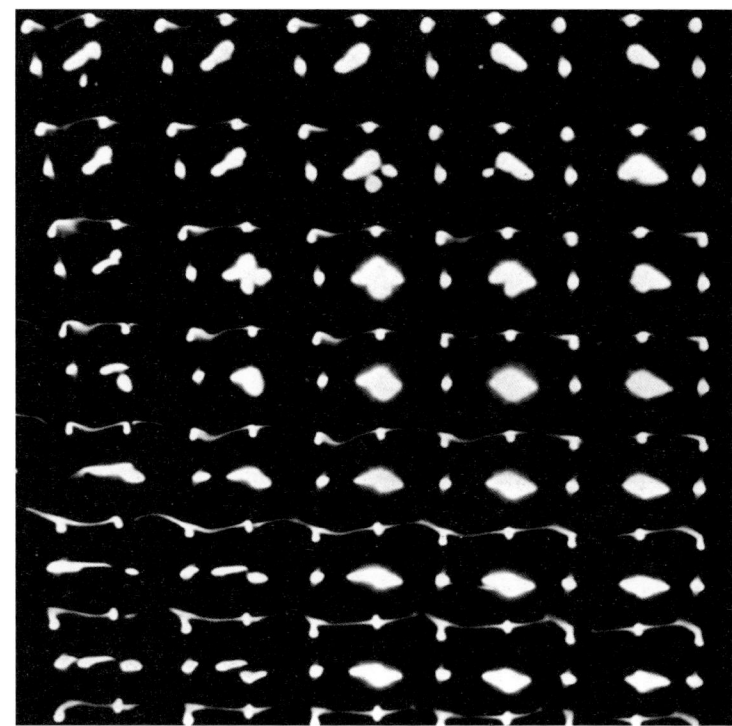

Fig. 46

Fig. 47

Figs. 46-48
Numerous varieties of "woven" patterns are found in various materials, particularly liquids. These forms may have a cellular (Fig. 46), honeycomb (Fig. 47), or imbricated (Fig. 48) appearance and are produced in the liquid by vibration.

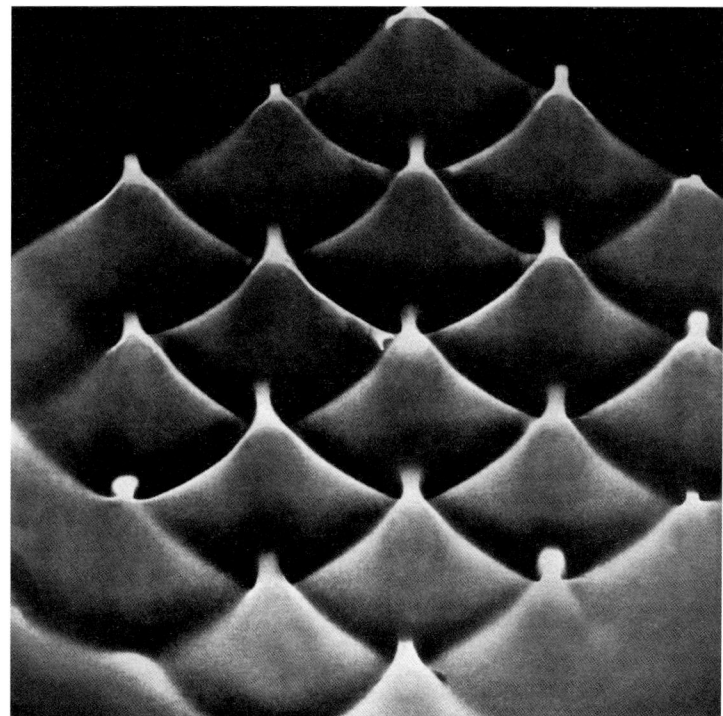

Fig. 48

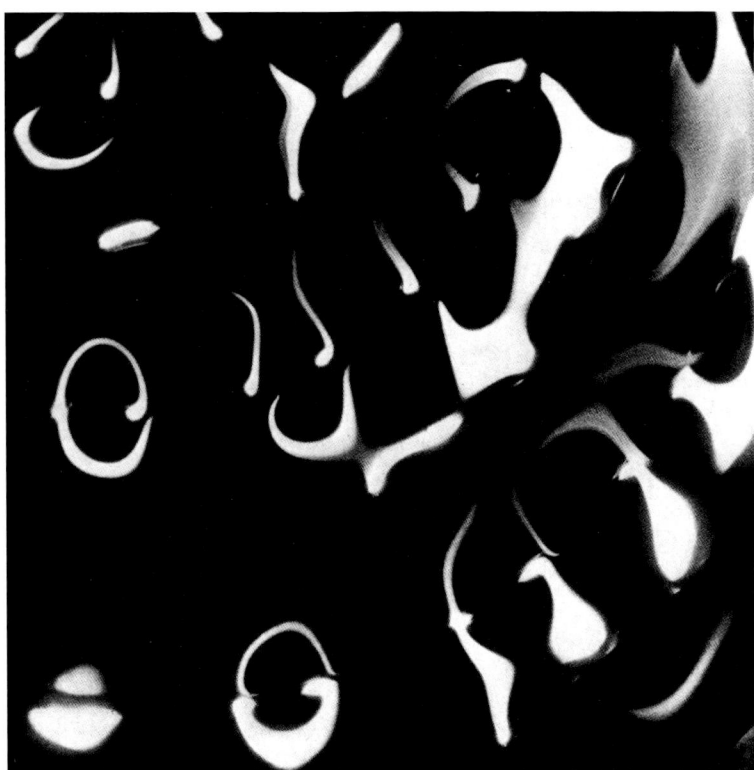

Fig. 49
Here the "forms" are seen to be flowing and pulsing. The shifting figures produced in this sheet of liquid by vibration are reminiscent of Lissajous figures.

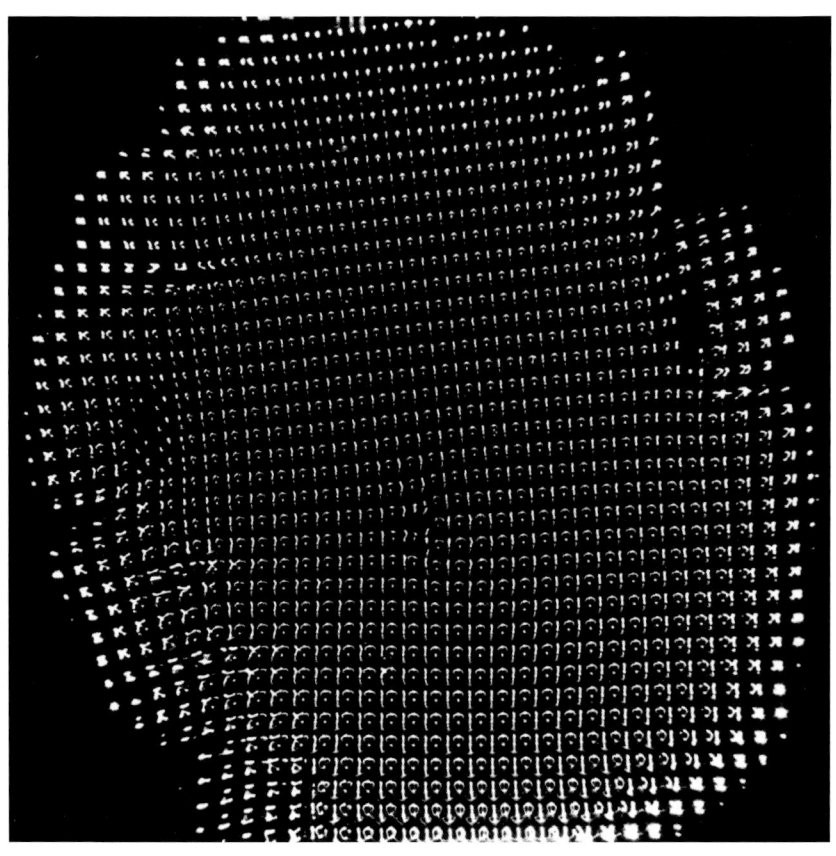

Fig. 50
"Standing" lattices showing great regularity in their features may appear in oscillating sheets of liquid. Glycerin has been poured onto the diaphragm as a liquid layer.

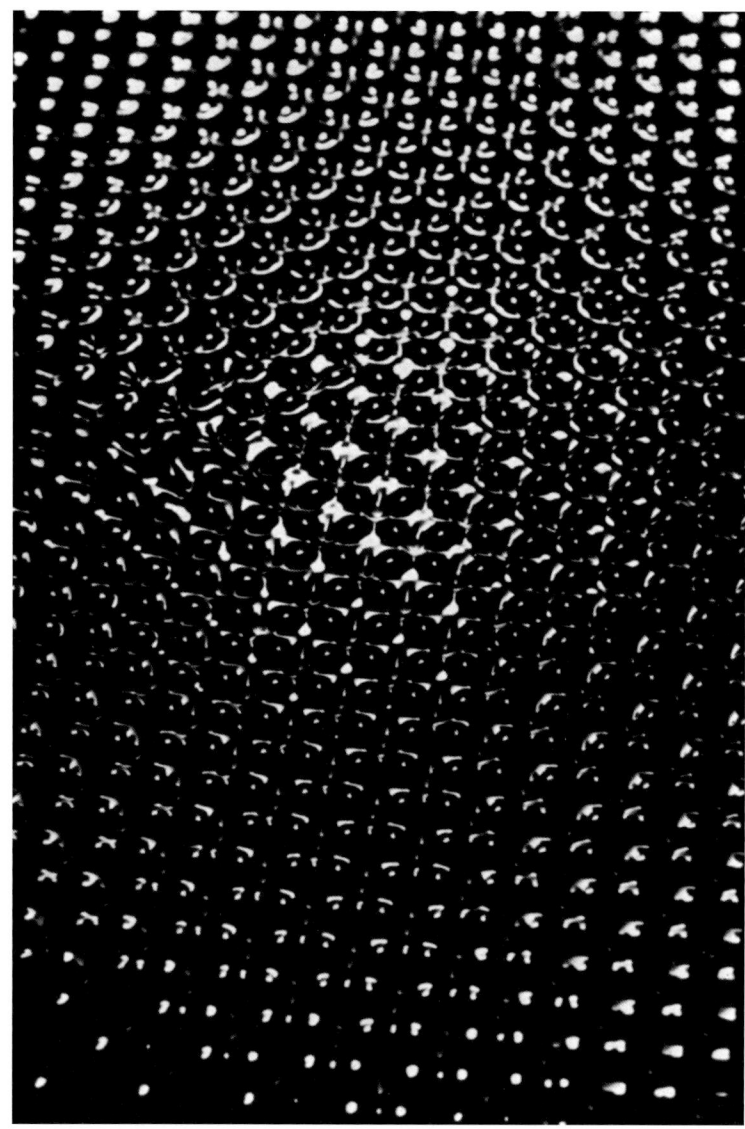

Fig. 51

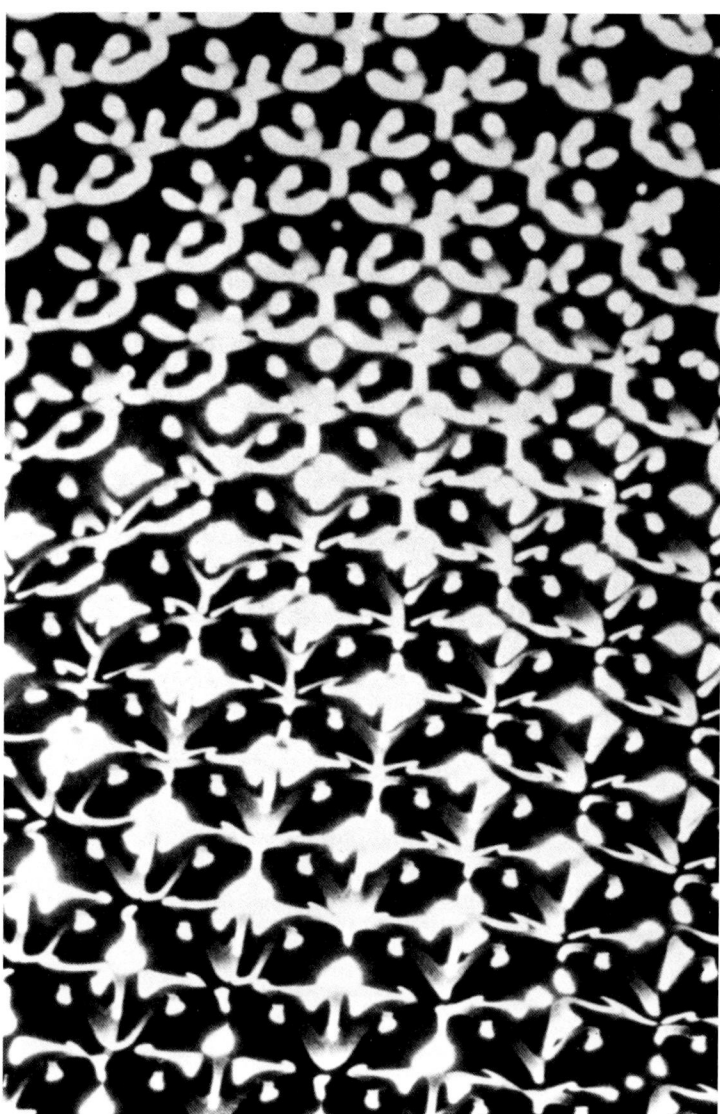

Fig. 52

Figs. 51-52
The pictures change dramatically with even slight modification of the exciting frequency or amplitude. They must be conceived as being in motion and also subject to interference phases. (Fig. 52 is a detail of the central part of Fig. 51.)

Fig. 53
When the pattern produced by vibration is as delicate as this, it may be appropriately termed a texture. These wave fields are produced in a layer of glycerin on a diaphragm excited at a frequency of 300 cps. Here again many parts must be imagined in motion. Labile waves appear, particularly in the marginal areas of the textures, and often look like turbulences.

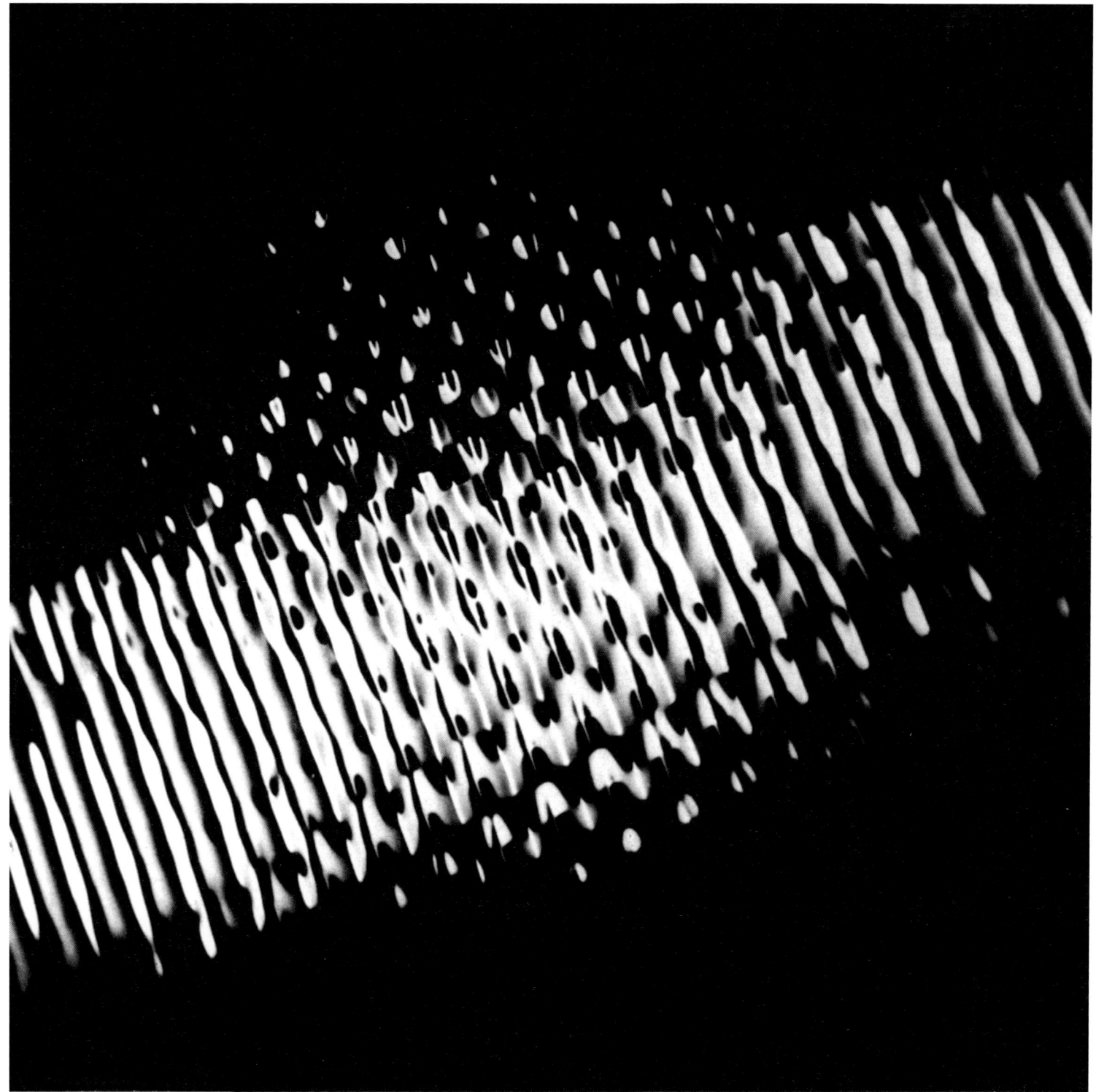

Fig. 54
This wave structure is a development from the pattern in Fig. 35. There we saw simply a train of waves. Here the amplitude has been increased (which corresponds to a crescendo). This causes the formation of a lattice which can just be made out in Fig. 54. Some waves are just beginning to reveal in themselves the effects of cross-motions: wave lattices in the nascent state.

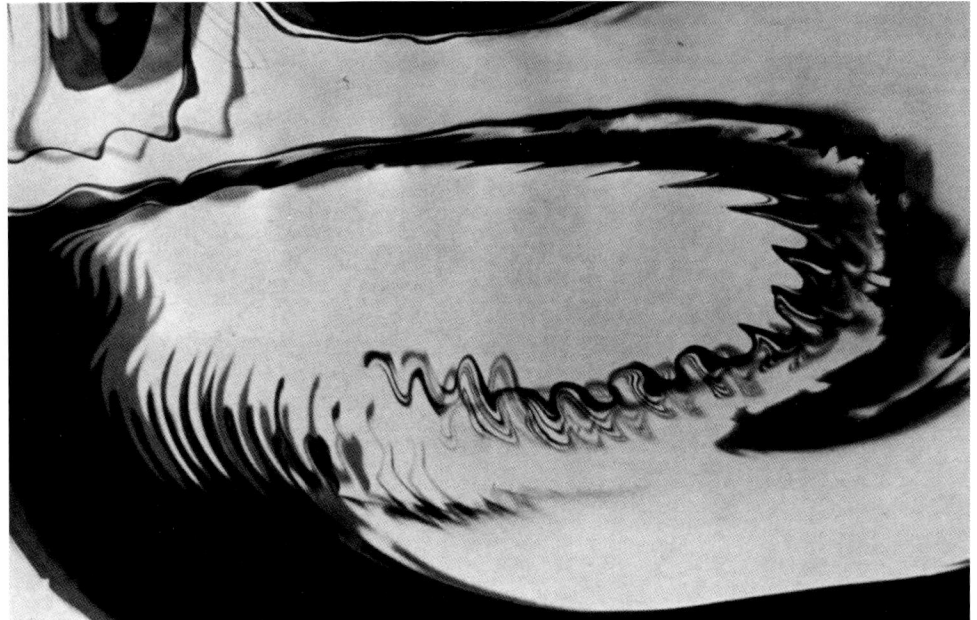

Fig. 55

Figs. 55-58
These are photographs of turbulences. They are not turbulences as understood in terms of hydrodynamics, rather they have been created by vibration. They appear, fade away and reappear. They are therefore labile waves which have a complex and vortical appearance. They appear more particularly in and near marginal areas, between two different areas of movement, or between dynamic and static areas.

Fig. 56

Fig. 57

Fig. 58

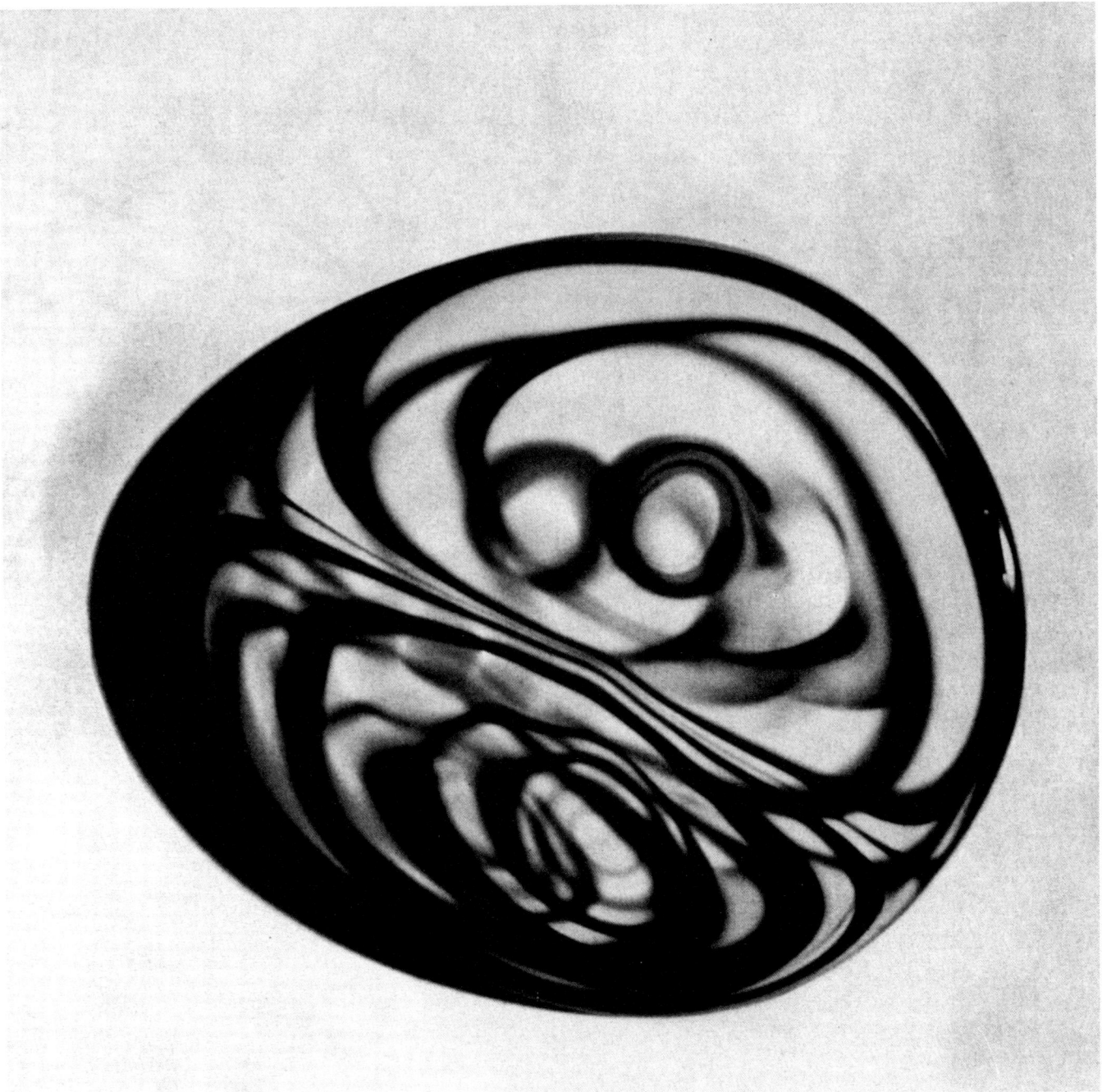

Figs. 59-66
In these photos we see hydrodynamic phenomena caused by vibration. If liquids are excited under suitable experimental conditions, vortices are formed which remain remarkably symmetrical in their behavior. Vortices are invariably formed in pairs and rotate in opposite directions.

Fig. 59
The simplest case. Diameter about 2.5 cm. One pair of vortices has formed. A black dye was used as an indicator.

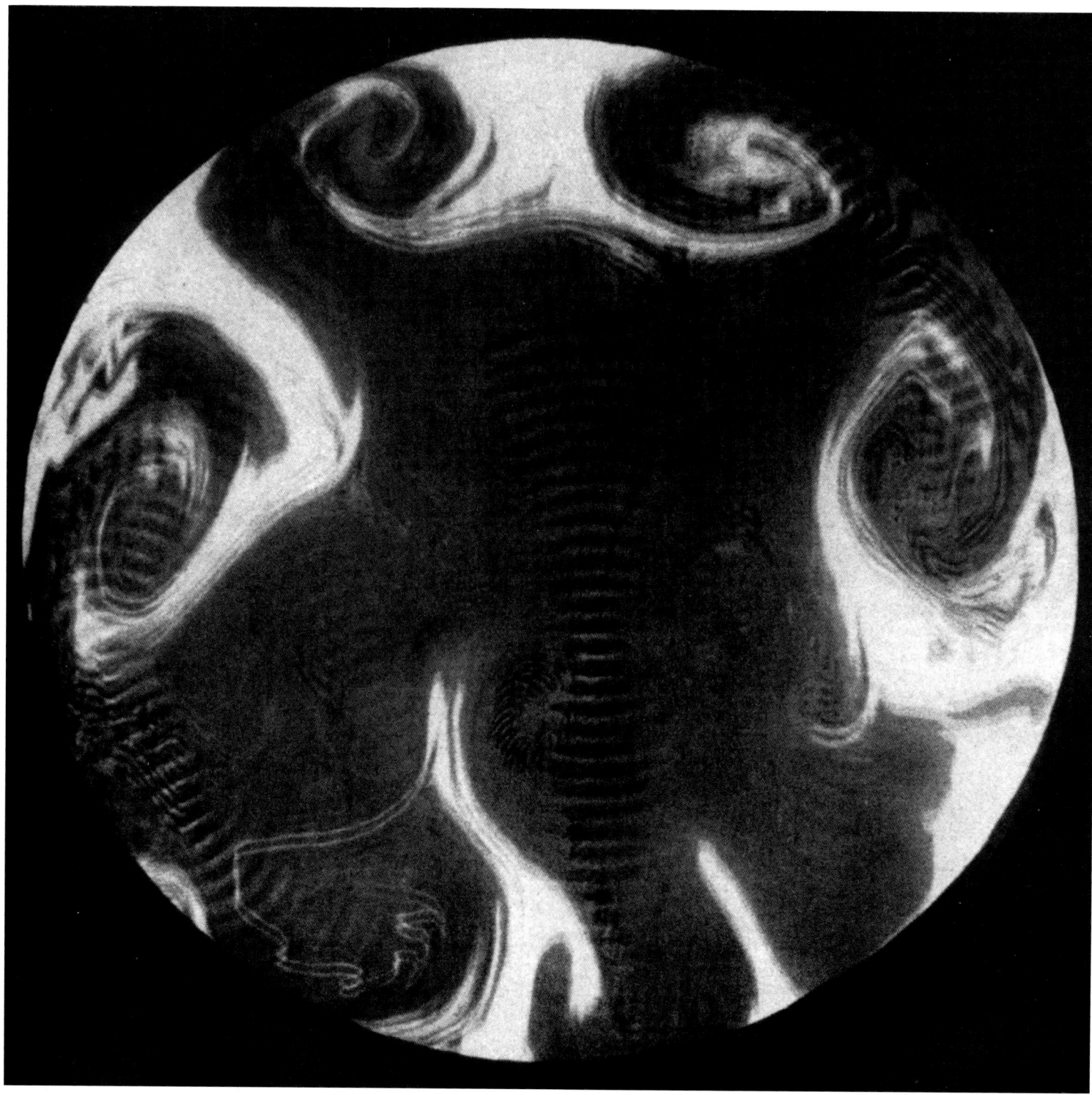

Fig. 60
The process is in the nascent state, i.e. the vortices are just forming. Some wave sequences can be observed. These are a sign that vibration underlies the whole process. (Diameter 23 cm., 100 cps.)

Fig. 61
A stage further in the process. Several pairs of vortices have formed, each strictly bilateral in its symmetry. The vibrational field measures 28 cm. across.

Fig. 62
Detail from an experiment with vortices. We can see the point at which four quadrants abut. The axis of symmetry runs from the top to the bottom. The flow— these vortices must be imagined in constant rotation— is laminar in character, i.e. it proceeds in layers.

Fig. 63
Another detail from an experiment with vortices. A pair of vortices, the one on the right turning in a clockwise direction and the one on the left in a counterclockwise direction. This experiment is also audible.

Fig. 64
A pair of vortices, each pair situated in a quadrant. Again there is bilateral symmetry. The louder the tone (i.e. the greater the amplitude), the more rapid the rotation.

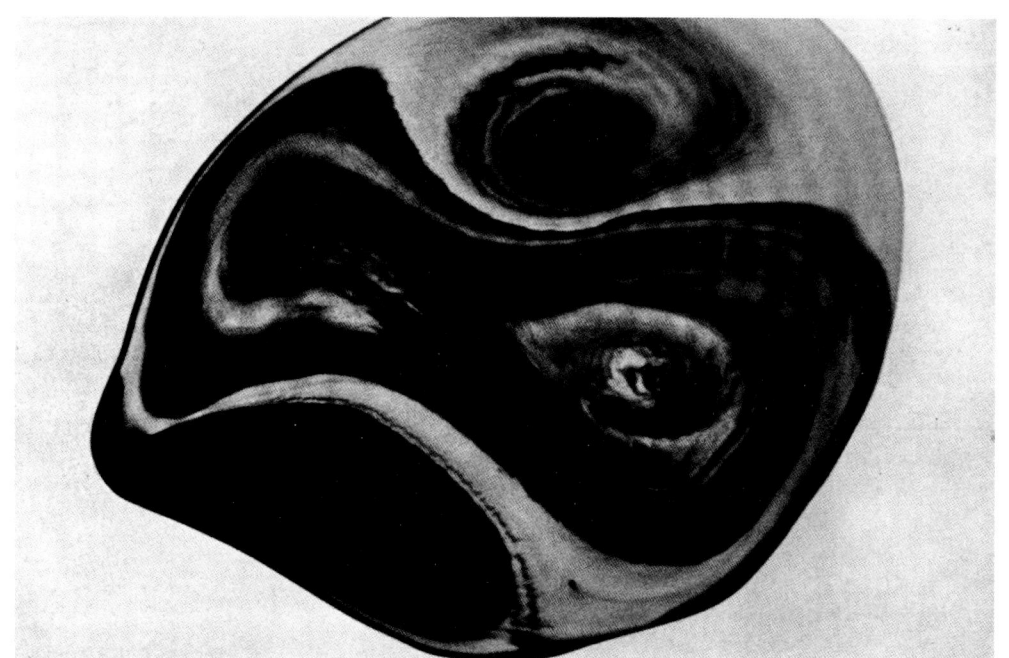

Fig. 65
Hydrodynamic phenomena originating in vibration are a rewarding field for observation. In addition to vortices other flow patterns can be found. In Fig. 65 a figure-eight form can be discerned; there is however no intersection as in a lemniscate. The effects of sound have produced circulation.

Fig. 66
Detail of a vortex. All these forms are in the strictest sense hydrodynamic phenomena originating in vibration.

4

The Tonoscope

Observing the action of the human voice on various materials in various media suggests itself as an obvious procedure. For this purpose the tonoscope was developed. This is a simple apparatus into which the experimentor can speak with out any intermediate electroacoustic unit. Thus vibrations are imparted to a diaphragm on which sand, powder, or a liquid are placed as indicators. Speaking actually produces on this diaphragm figures which correspond, as it were, to the sound spectrum of a vowel. Figs. 67, 68 and 69 are photographs of these vowel figures. The pattern is characteristic not only of the sound but also the pitch of the speech or song. The indicator material and also the nature of the diaphragm are, of course, also determinative factors. However, given the same conditions for the experiment, the figure is a specific one. One can sing a melody and not only hear it, but also see it. As the photographs show, organized patterns and configurations are quite apparent. They are there for the eye, man's most sensitive organ of sense, to see. A pattern appears to take shape before the eye and, as long as the sound is spoken, to behave like something alive. The breath alone can cause it to move; a texture of forms is created by the fluctuations of the voice. The eye can also see variations as the voice is raised or lowered. During continuous speech the patterns metamorphose continually. The purpose of these observations was to show that such figures and patterns can give rise to a visual experience which can be fully equated to the aural experience. This is an achievement which not only opens up a new world to those with normal hearing, who see for themselves that their speech actually involves the production of vibrating patterns which continually penetrate and fill space, but also, and most important of all, it enables the deaf to experience what they are actually producing with the speech they have learned and been trained to use. They do not hear the sound they create although they no doubt see the speech movements of others. But they have no sensory experience which is equivalent to hearing a tone, a vowel, a word. A truly figurate pattern, however, speaks a language of its own; a picture is a perceptual experience complete in its own terms. Experience has shown that these tonoscope pictures are highly stimulating to the mind of the deaf-and-dumb. If a deaf-and-dumb patient, for instance, speaks an 'O', the sound element contained in the speech of the deaf-and-dumb subject also appears in the vibrational pattern. Now he can *see* the 'O' which a normal person speaks. The difference between the two pictures is very marked, but he is now given an opportunity of practicing until he has achieved the form of a purer 'O' sound. Everything he does becomes visually apparent to him. As soon as he can produce the form of a purer 'O', we can hear that he is also speaking a purer 'O'. The same holds true for pitch. Here again he is enabled to practice until he can achieve nuances of speech which he is unable to experience through the sense of hearing. He is now open to visual experiences which can be equated with the stream of speech, the pulsing of the air mass, the air stream, etc. He can *see* the to-and-fro motion of his sounds, his words and sentences, and also the flow patterns made by good speech. Since deaf-and-dumb patients have very sensitive vision (and a very fine sense of touch) they can familiarize themselves with the visual speech of the tonoscope and train with its aid.

An electroacoustic variant of the tonoscope is suitable for visualizing not only speech but also music. The sonorous patterns are directly impressed on, say, a liquid, and not only the rhythm and volume become visible but also the figures which correspond to the frequency spectra exciting them. These sonorous patterns are extraordinarily complex in the case of orchestral sound. Figs. 70-73 show the visual versions of musical excerpts: 70 Bach, 71 Bach, 72 Mozart, 73 Mozart.

If we look at these passages on a silent film, we can at first make nothing of them. If, for example, we *see* the Jupiter Symphony without being able to hear the music, we should never guess from the visual effects with their flowing and constantly changing patterns, that we were seeing Mozart. The rests in particular, which are inscrutably and essentially part of the music, appear at first to the beholder as black holes, as a cessation of activity. But if the sound is turned on, everything flashes to significance for the eye. Hearing is added and restores to experience its full content. It takes training to be able to "hear by seeing." The tonoscope is a device with which one can train oneself in this art.

In this particular context it may be said that the sounds of the human voice have specific effects on various materials in various media, producing what might be called corresponding vocal figures. These again are patterns produced by vibration and as such are appropriate objects for cymatic study.

Figs. 67-69
These three forms have been produced by speaking directly into the tonoscope. This is a simple apparatus with which a spoken sound can be rendered visible on a diaphragm. Figs. 67-69 show the vibrational patterns produced by vowels, reflecting the sound spectra and pitch of each particular sound. The material of which the diaphragm is made and the indicators used (powder, sand, liquid, etc.) are also important factors; the pattern also changes according to the quality of the individual voice. But the form remains the same as long as there is no change in the conditions of the experiment.

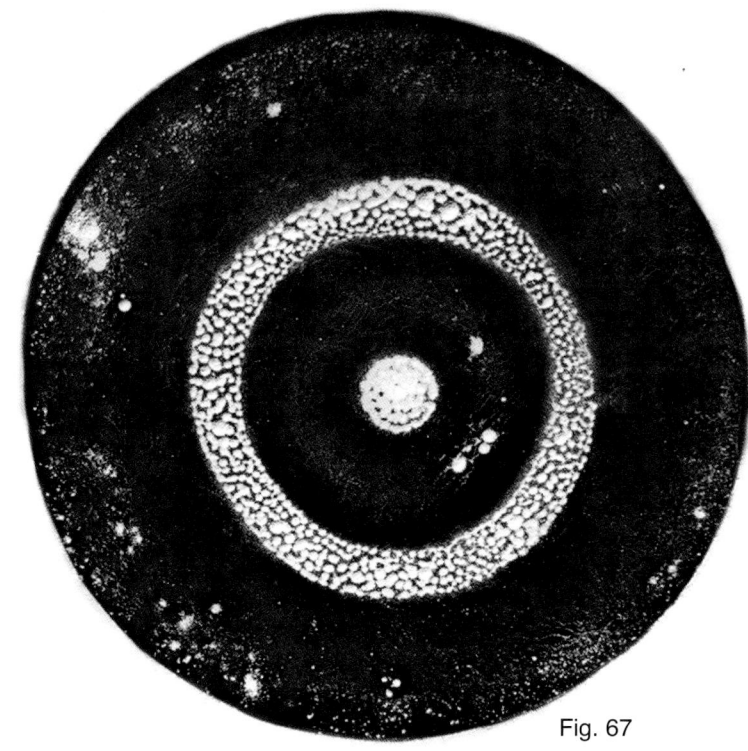

Fig. 67

Fig. 68

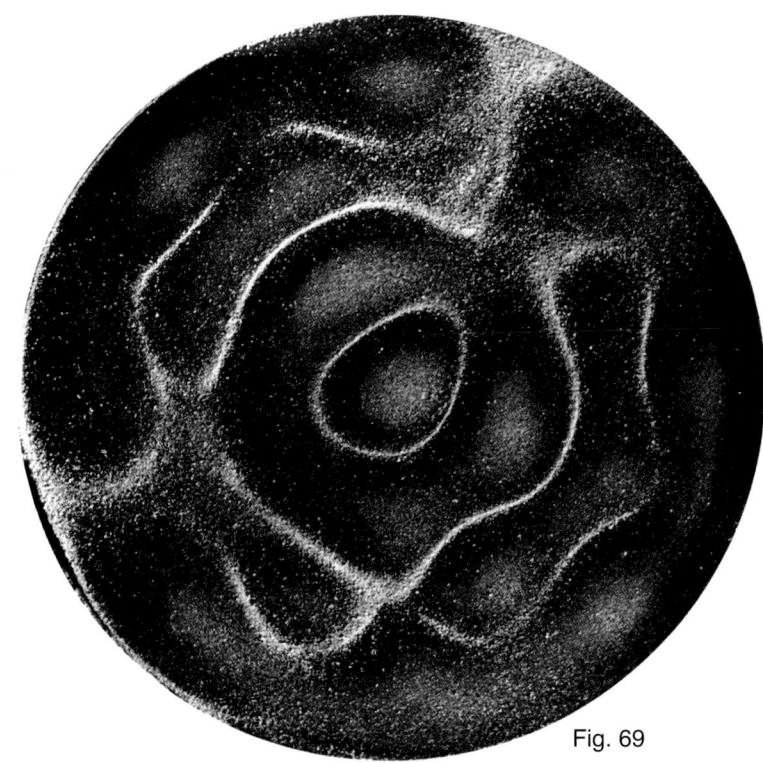

Fig. 69

Figs. 70-73
Like the human voice, music can also be made visible by means of the tonoscope. For this purpose an electroacoustic version of the tonoscope is used. Liquid was used as a reagent in the pictures shown here. The vibrational figures appear indirectly in the layer of fluid. They are entirely characteristic of the music played. However, the eye is unaccustomed to *seeing* music and is at first lost without the guidance of the ear. But with the ear to prompt it, the eye experiences these tonal events in full visual detail.

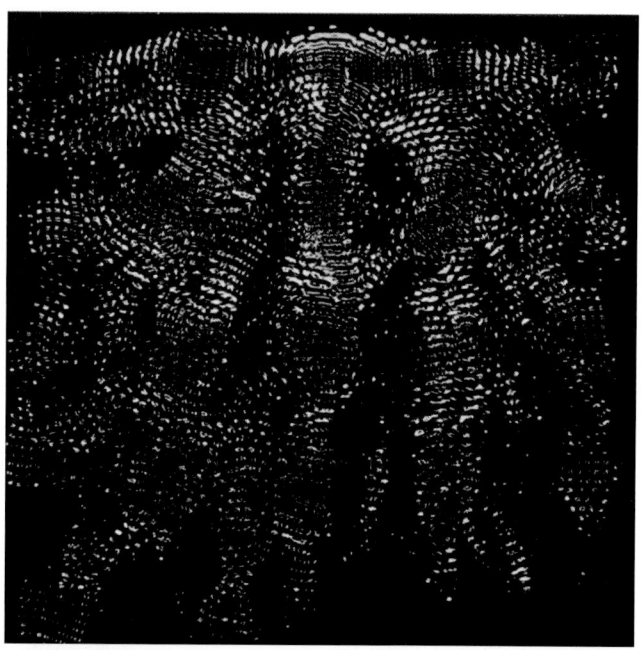

Fig. 70
Bach, Toccata and Fugue in D minor, 1st movement, bar 30, before start of andante.

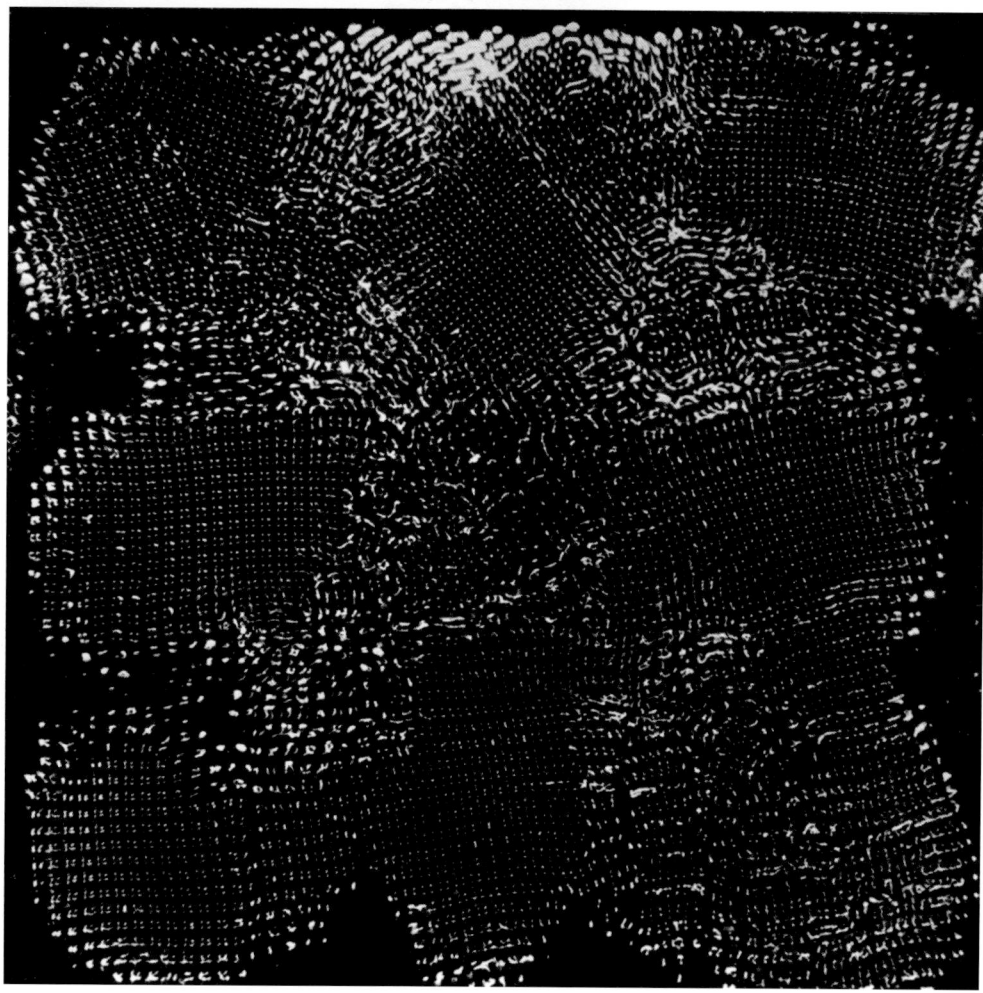

Fig. 71
Bach, Toccata and Fugue in D minor, 1st movement, bar 29, start of *fortissimo* passage.

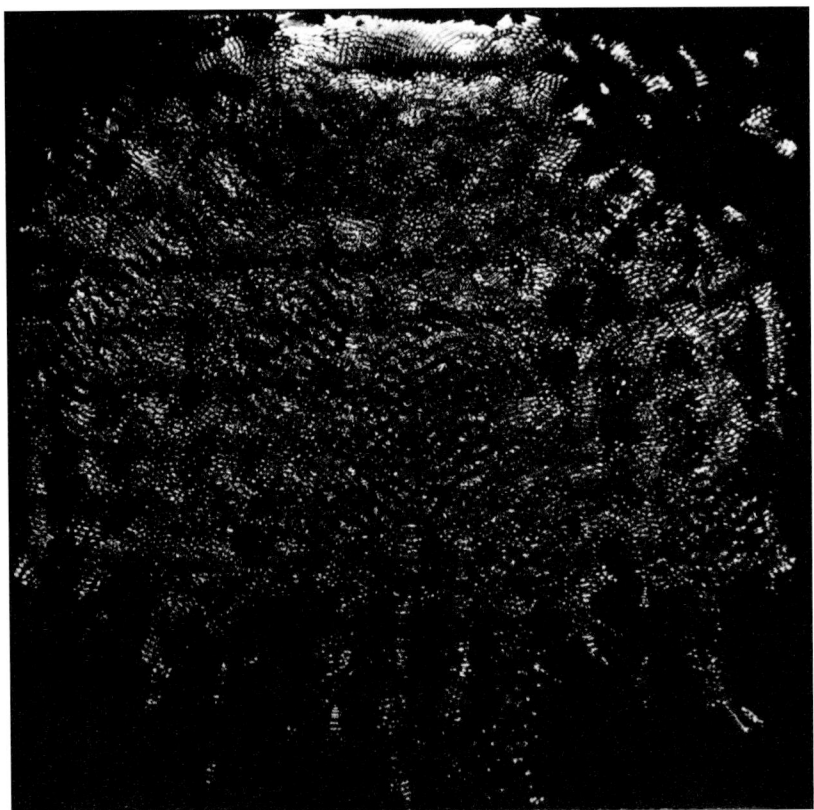

Fig. 72
Mozart, Jupiter Symphony, 1st movement, bar 173, beat 3.

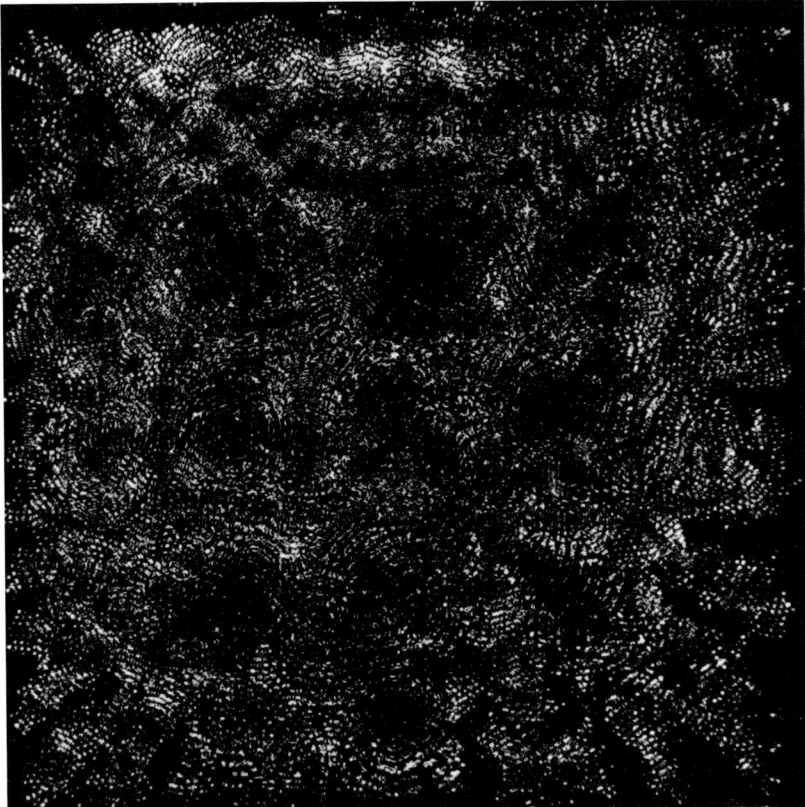

Fig. 73
Mozart, Jupiter Symphony, 1st movement, bar 59, beat 1 (trill).

5

The Action of Vibration on Lycopodium Powder

To provide a fuller account of the way in which a cymatic process may be observed, an example will be discussed at greater length. We shall describe a series of effects which appear when lycopodium powder is subjected to vibration. One can affect changes almost endlessly in experiments of this kind. There seems to be a risk of losing oneself in the infinite. All the same, the phenomenon must be made to yield as much specific knowledge as possible. There is no point in merely reading from the phenomenon what is immediately apparent. Time and again one is reminded of Goethe's words: "All have a similar form yet none is the same as the other, and thus the choir shows a secret law, a sacred mystery."

Of course, many processes can be analyzed, measured, and expressed in formulae. Of course, formulae, graphs, and probability relations are also derived from reality. They are images of reality and as such are also real. With their aid one can influence this reality, and indeed create new realities (e.g. in technology). But they are *derived* from reality; they are outside that which really exists; and the latter is more, much more, than the formula, than the quantitative determination, can indicate. Only certain aspects of the phenomenon are grasped. But how is it possible to grasp the complete phenomenon, the *really real*, at all? The creation of purely philosophical ideas, which paints Nature in mental images, is likewise incapable of grasping existence in its vital plenitude. It is "above" the really real. Even this speculative philosophy cannot penetrate the mystery of existence in all its fullness. This will only reveal itself progressively if we do not merely analyze it and anatomize it to a skeleton; if we do not merely try to take mental possession of it but instead patiently attend upon it, neither raising ourselves above it nor killing it. However much it may seem that "nothing" is thus achieved, this close observation is nevertheless the way that renders sources of knowledge accessible to research, enables the seeker to stay the course, and confers vitality. It will inevitably become apparent in the course of our descriptions whether this method of observation and more observation, whether this preservation intact of what is observed, this non-interference with the phenomenon, will elicit the outlines of basic phenomena and reveal something intrinsic and essential.

The lycopodium powder is strewn evenly on a diaphragm (Fig. 74). The diaphragm is excited by a tone. Immediately a number of small round shapes are formed (Fig. 82). If the amplitude is increased, these shapes begin to migrate to the places where the oscillatory movement is more pronounced (Fig. 83). As this happens, many of them unite to form new piles of spherules. If this process is continued, more and more of these formations join up. One large overall form might appear. Fig. 83 shows the migrating piles of spherules and the paths they follow. All these round piles are in a state of constant circulation; that is, the particles on the upper surface move from the center towards the periphery and those at the bottom in the contrary direction. They rise upwards at the center and then proceed towards the periphery, etc. There

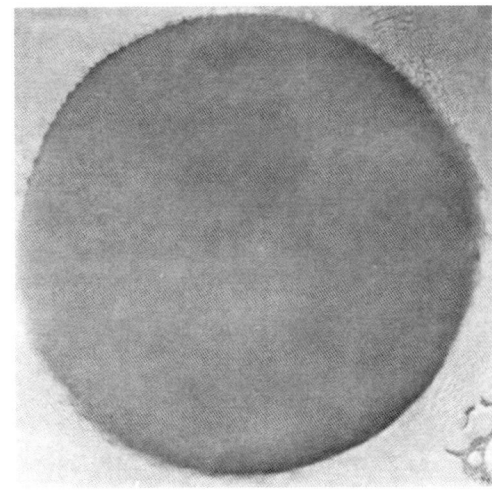

Fig. 74
Before any acoustic irradiation, the lycopodium powder (spores of the club moss) is spread evenly over the diaphragm. The diameter of the diaphragm is 28 cm.

Fig. 75
When the tone is loud (large amplitude) the powder is thrown up in fountains or even ejected. Yet by means of the stroboscope, which renders rapid phase sequences visible, it can be seen that vibration imposes its pattern on both the form and the dynamics of these eruptions.

is thus a kind of radial circulation. Marking with colored grains enables these circulations to be observed. The same circulation also takes place in the large forms; however, these processes can be seen taking place in the areas where organized patterns have been formed by the vibration. Fig. 84 shows a large complex of this kind. We can see the patterns whose configuration changes with shifts of frequency. Fig. 84 is formed by a soft tone. Fig. 85 shows a large circular shape about the size of a small chicken's egg at a frequency of 300 cps. If the amplitude is further increased corresponding to a crescendo for the ear, the movements become increasingly violent. The circulation becomes faster and proceeds on a larger scale. Eventually masses are thrown up and ejected. There are eruptions and fountains of powder which swirl high into the air (Fig. 75). A vortical movement can be seen in one of these powder fountains (Fig. 76). But while masses are ejected at one moment, they fall and are returned to the system at another. Thus, for all the changes and transpositions, a certain constancy of mass is curiously retained. The forms move to-and-fro both when the frequency is changed and the amplitude. In doing so they move as wholes, i.e. when they move in one direction by putting out an arm, the whole follows, and there is retraction at another place. Thus locomotion is truly correlated throughout. Instead of parcelation, there is a uniform

Fig. 76
Here again a lycopodium fountain can be seen. A spiral formation can be discerned in the column of powder. 130 cps., large amplitude.

streaming or flowing. This character is, of course, determined simply and solely by the vibrational pattern. Once the tone is discontinued, the powder lies undisturbed as an accumulation of particles. Even the separation and rejoining of the formations take place in this uniform manner. Whether they engulf each other or disjoin from each other, there are no indications anywhere of a disintegration or disruption. It can sometimes be observed by stroboscope that the conglomerations are pulsating.

If the amplitude is greatly increased (*fortissimo*), the powder changes into a cloud of dust. The phenomenon enters the spatial or three-dimensional. Even so, the stroboscope still reveals in these turbulent, eddying masses certain arrangements which can be traced back to the vibrational pattern (Fig. 77). Where the turbulence is most marked, vacant zones, spatial eddies, "empty spaces" alternate in conformity with the patterns imposed by the vibration. If at the greatest amplitude, when the powder is thrown up into a veritable cloud storm, the sound is switched off abruptly, the powder falls back onto the motionless diaphragm more or less evenly. At this maximum modulation of amplitude, then, what might be described as an integrating effect is possible. The configuration of particles, for all its multiformity and individualization, can be integrated in an instant, and we are back at the initial position.

How precisely the spherules are concentrated at the sites of movement

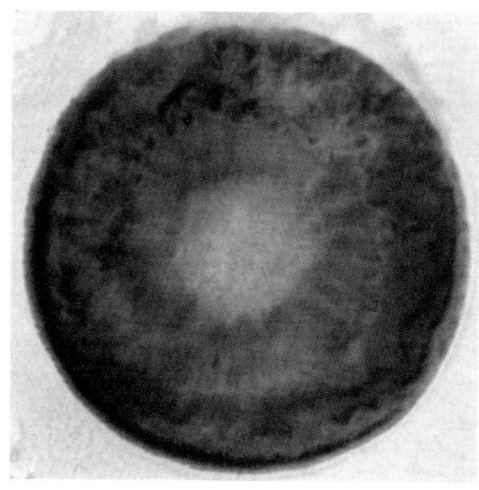

Fig. 77
A still larger amplitude throws up a cloud of lycopodium. The process actually takes place in the space above the diaphragm. Vortices, turbulences and also patterned elements can be seen. There are areas of greater or lesser density. Here again the stroboscope reveals details of the vibrational topography. 300 cps.

Fig. 78
This figure shows how characteristically the various substances behave. Hexagonal steel plate, 3600 cps. The strewn material is quartz sand and lycopodium. The sand moves into the nodal lines and the powder into the antinodes, where it circulates in the center as a round shape. The behavior of the two substances under these experimental conditions is specific.

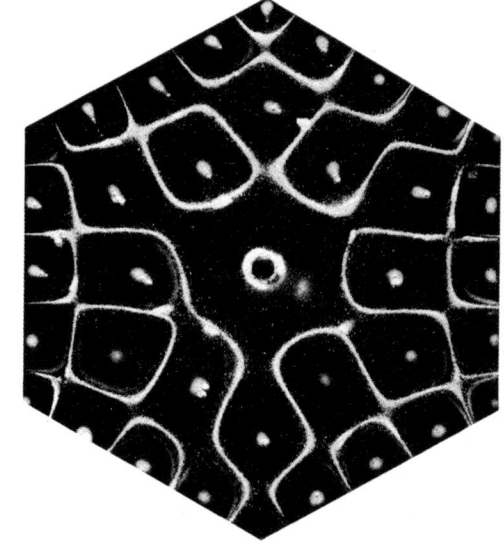

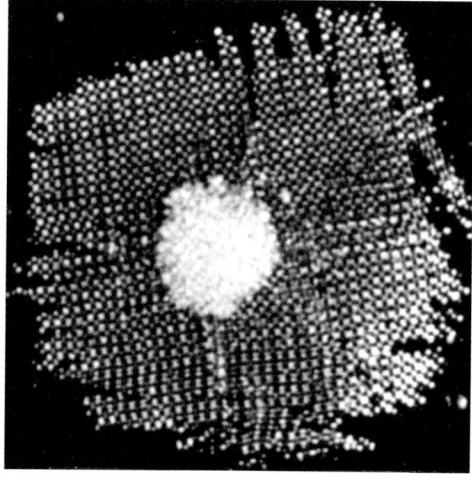

Fig. 79
Under different conditions the quartz sand also forms into circulating round shapes. Fig. 79 shows a wave field of vibrating water. The sand heaps up in the center and circulates. Frequency 1060 cps.

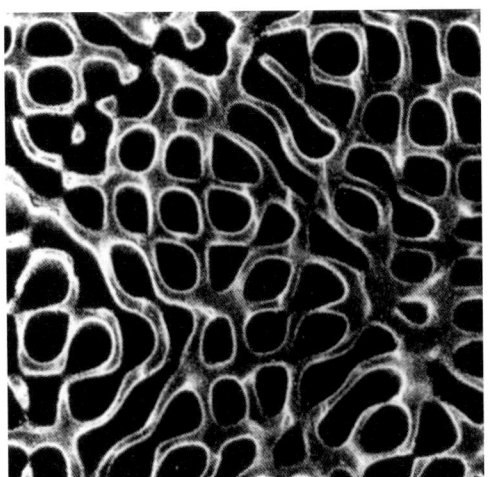

Fig. 80
When conditions are different, the lycopodium powder forms sonorous figures. Fig. 80 is a detail of a sonorous figure formed of lycopodium powder on a steel plate, 31 × 31 cm. Frequency 12,900 cps. This picture shows very strikingly how the powder is transported from the zones of movement towards and into the nodal zones. The precise way in which the double contours mark out the nodal zones is very conspicuous.

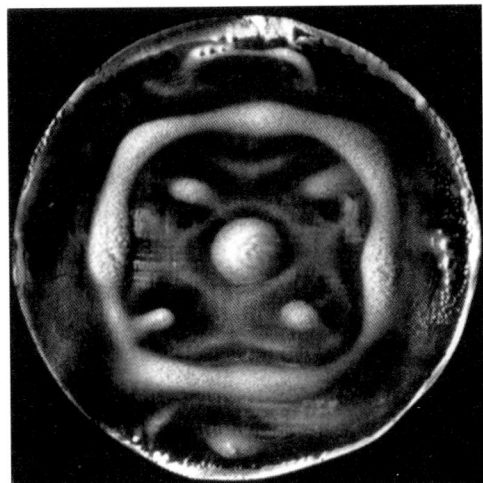

Fig. 81
Sonorous figure appearing in lycopodium powder on a diaphragm, 250 cps., large amplitude. The neatness with which the particles are transported and deposited is a recurrent feature. Before acoustic irradiation the powder was spread out evenly.

can be clearly shown if, say, a steel plate is evenly strewn with quartz sand and lycopodium powder. When vibration is switched on, the grains of sand move into the nodal lines and the lycopodium spores into the antinodes, in the center of which they join together and start to circulate. In Fig. 78 (steel plate, frequency 3600 cps) the quartz sand is in the nodal lines and the central white spots are lycopodium. The round circulating heaps are not to be found with lycopodium alone. If quartz sand is poured into a vibrating sheet of water, round heaps of circulating sand can be discerned in the wave fields of the liquid, i.e. in the areas of motion. Fig. 79 clearly demonstrates this.

Needless to say, the processes of interference referred to above can also be generated with lycopodium powder. The forms then move to-and-fro between two positions, or phases of rest alternate with phases of movement.

Since a vibrational system contains certain heterogeneous elements within itself, it is possible for interference phenomena to be generated by *one* tone and shown by the indicating material.

Rotational effects can also be seen here. In the circular heap, the movement created follows a circular path. If this process takes place in two directions, a seesaw effect results. The phenomenon occurs like this: a stream of powder follows a circular path through the round form, moving forward in what looks like minor eruptions; there is a period of rest, and then the same movement starts again, but in the opposite direction; another period of rest; and then the first movement is repeated, and so on. This seesaw effect has, of course, been filmed (as has all the phenomenology shown here).* It might also be noted that it can be produced with an impulse at only *one* frequency. Thus a system, if it is of such a type as to contain several oscillatory responses, can be made to produce interference by *one* frequency or even produce a seesaw effect (i.e. without frequency modulation in the generating tone).

This series of experiments is exceptionally diversified. Well-known sonorous figures may appear with lycopodium under certain conditions. Fig. 80 shows a sonorous figure rendered visible on a steel plate with lycopodium, at a frequency of 12,900 cps. By using lycopodium, patterns can be obtained in which the powder is driven into certain zones while others remain entirely free from particles, e.g. in Fig. 81. There are empty spaces on the one hand and powder erupting into clouds on the other. It might also be noted that a few isolated round shapes have formed where the vibration is less violent. Kinetic-dynamic processes can be instigated by frequency modulation. Here the situation is not static, as in Fig. 81; instead currents race around the empty spaces, partly in opposed directions. Figs. 86-88 are photographs of

*Editor's note: Dr. Jenny did make several films of his experiments, highlights of which are now available on video. See promo pages at the end of this edition.

this kinetic-dynamic phenomenon. These proceedings must be visualized as "current storms" or storm currents. However, this is not an unregulated chaos; it is a dynamic but ordered pattern. Fig. 88 gives a detail of such a raging mass of spherules. The laminar flow of these rapidly moving masses is clearly visible.

Let us summarize some of the effects we have observed in lycopodium powder under the influence of vibration.

> These vibratory effects are:
> > The creation of forms, formations
> > The creation of figures
> > Patterned areas
> > Circulation
>
> Constancy of the material in a system
> > Pulsation
> > Rotation
> > Interference
> > Seesaw effect
> > Correlation
> > Integration effect
> > Individuation
> > Conjoining and disjoining of a single mass
> > Dynamics of eruption
> > Dynamics of current flow, etc.

This list shows that vibration produces a great diversity of effects. Vibration is polyergic and many of its effects are specific.

It is not our intention to order or analyze these categories but rather— in accordance with the empirical method adopted— to leave the whole complex of phenomena as it is, with *this* category, now *that* category dominating. But it is through this generative and sustaining vibrational field that the entire complex comes into being, and this complex whole is omnipresent. There is no parcelation, no patchwork; on the contrary what appears to be a detail is utterly integrated with the generative action and merely acquires the semblance of an individual, of an individualized quasi-existence, the semblance of individuality.

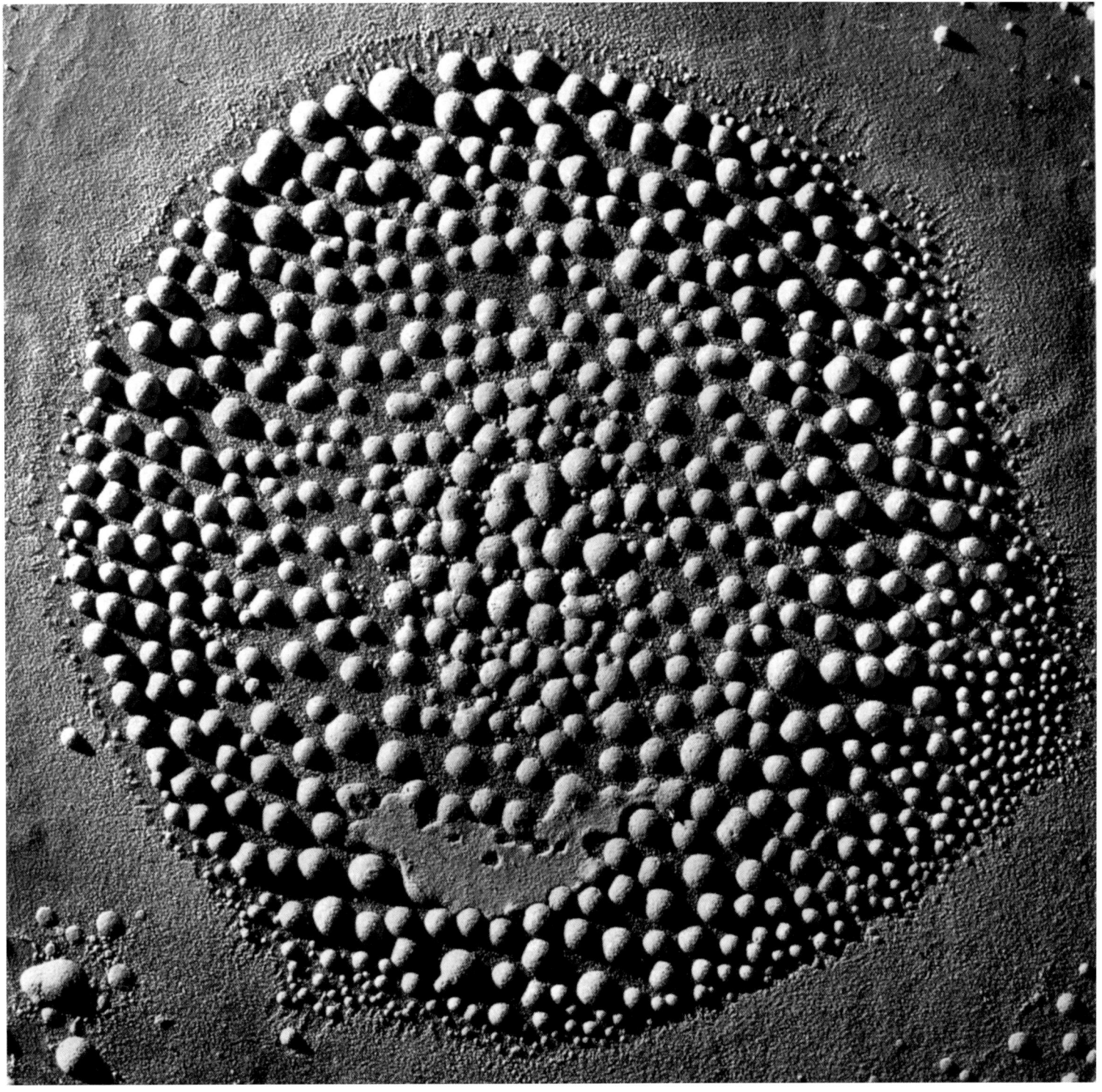

Fig. 82
Acoustic irradiation transforms the uniform layer of lycopodium powder into a number of round shapes. Each of these rotates on its own axis and at the same time circulates in a constant manner around the whole figure.

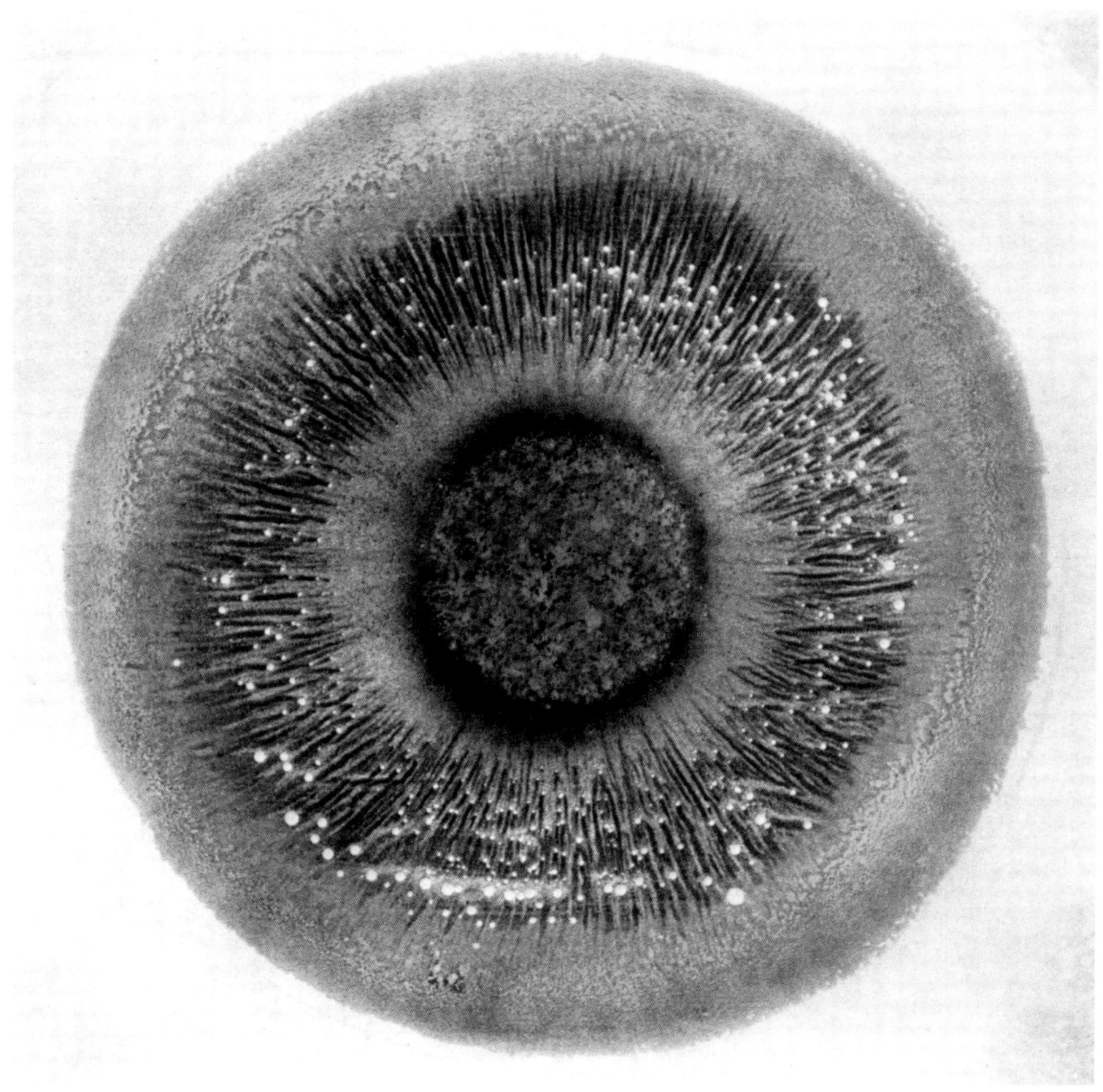

Fig. 83
On a crescendo, the round heaps migrate to the center following the topography imposed by the vibration. The numerous radial pathways can be seen.

Fig. 84
The individual shapes unite to form a large pile, which, however, continues to rotate on its own axis. At the same time patterned markings can also be seen. Figure 84 was produced with a soft tone. Frequency 50 cps.

Fig. 85
Again a large round form. This time the frequency was 300 cps. and the pattern is accordingly of a more delicate character.

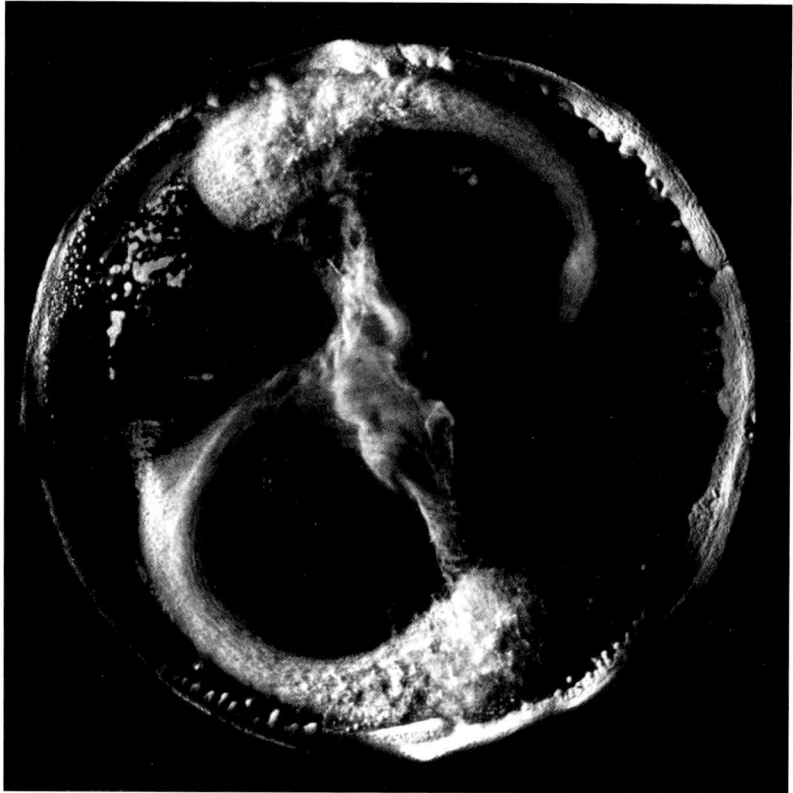

Fig. 86

Figs. 86-88
Since the apparatus used for these experiments allows frequency to be increased systematically, it is possible to produce not only sonorous figures (as in Fig. 81), but also flow patterns in which the streams of powder move in a significant manner. In Figs. 86 and 87 the powder must be imagined to be streaming along at a great rate in the visible forms, but exactly in the pattern and direction imposed by the vibration. The process is thus not chaotic. Fig. 88 shows a detail. The laminar configuration of these storm currents or "current storms" should be noted.

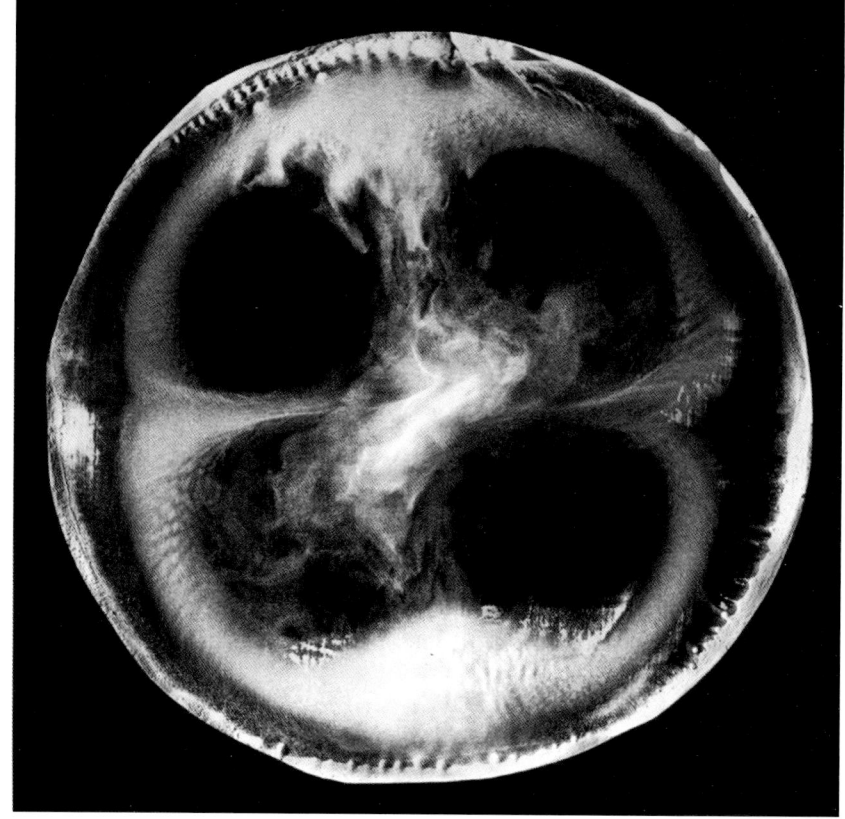

Fig. 87

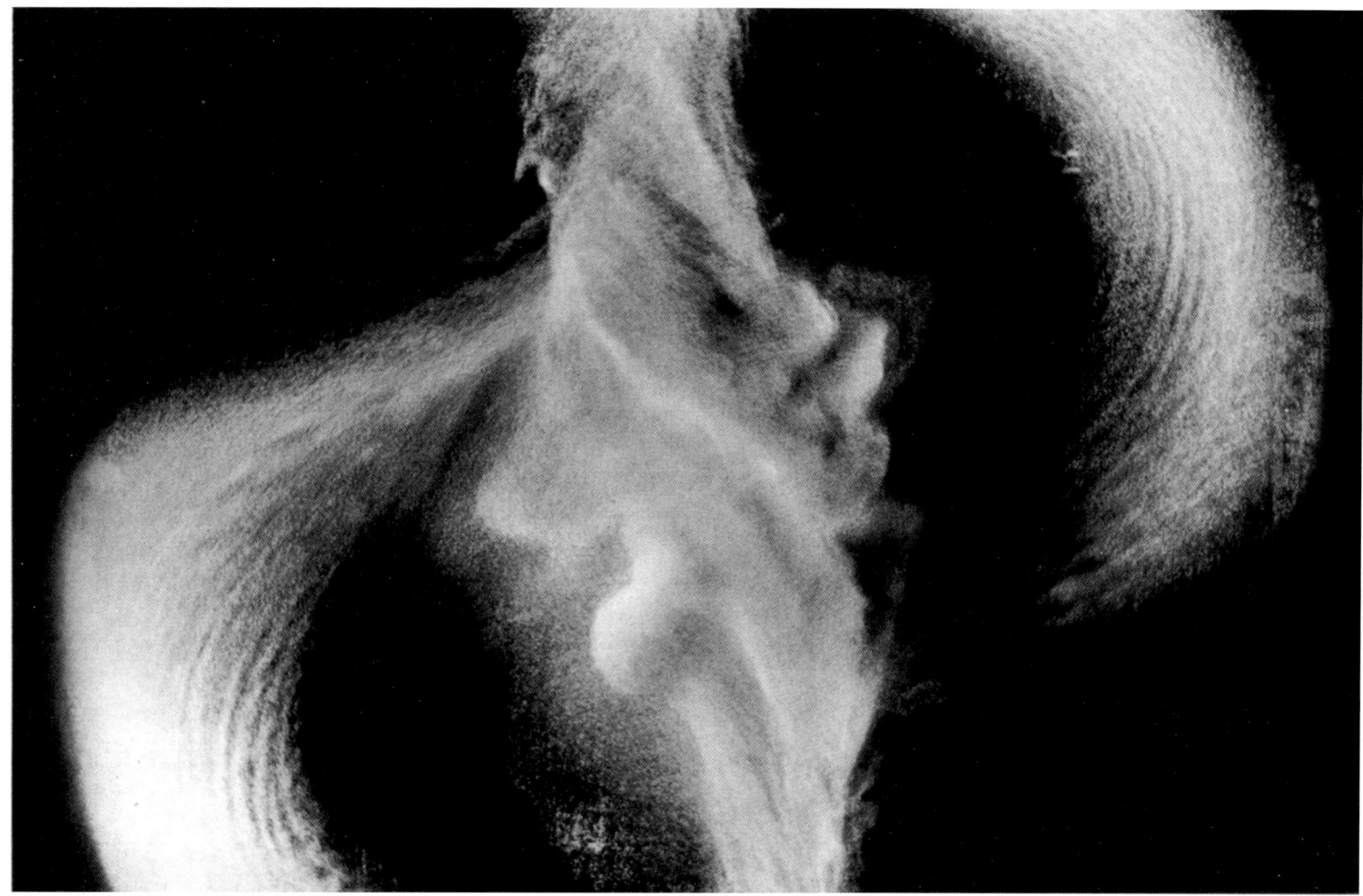
Fig. 88

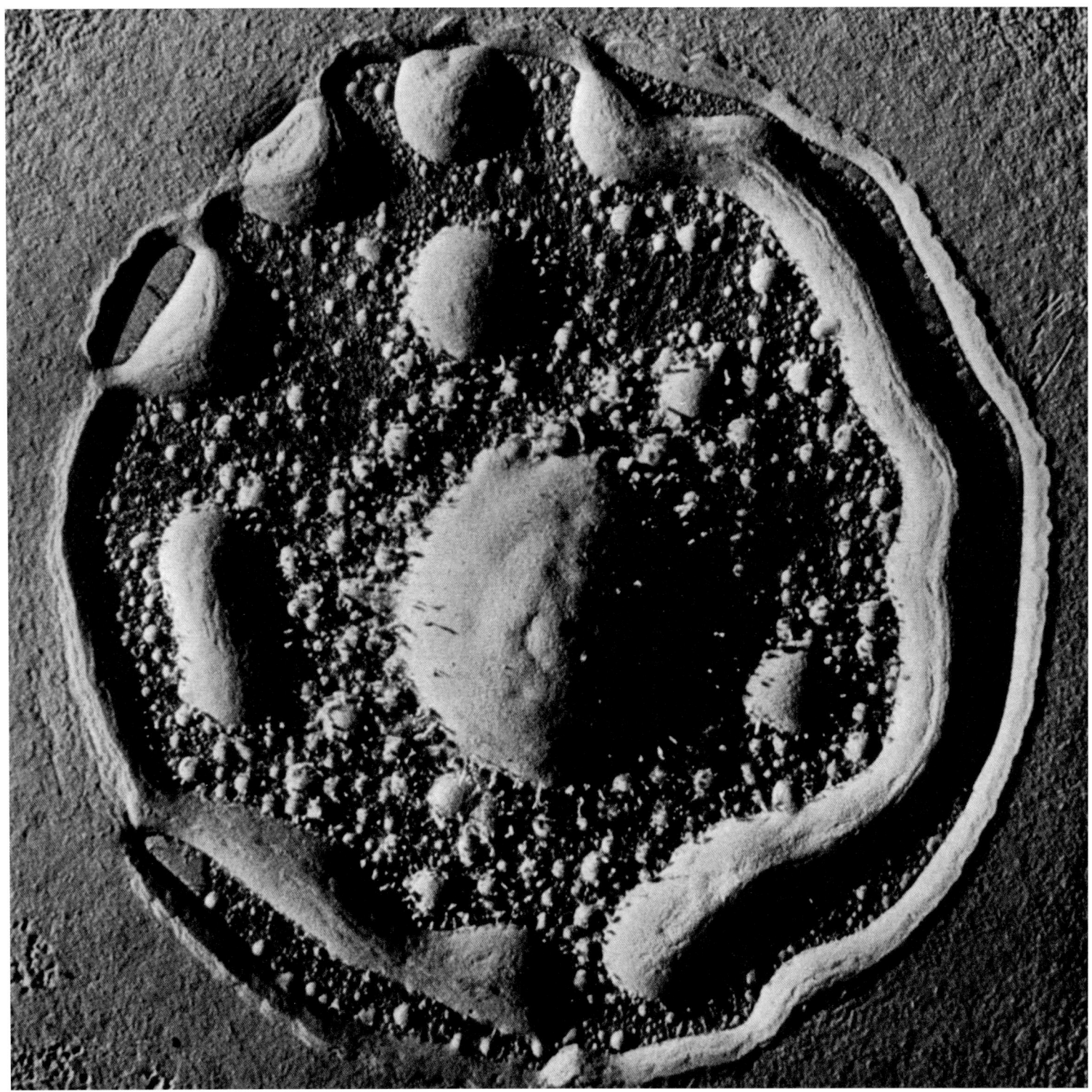

Fig. 89

Fig. 90

Figs. 89-91
As the amplitude is varied, the pattern goes through a number of changes. The dynamics of the moving mass of lycopodium particles alter, depending on whether the tone is loud or soft.

Fig. 91

Fig. 92
This landscape of lycopodium powder under the influence of vibration is a synoptic view in which the various phenomena can be recognized in all their diversity. Everything must be imagined as circulating, moving, pulsating etc., and all this is caused by the vibrations of a tone which can be heard the whole time the experiment is in progress.

6

Periodic Phenomena Without an Actual Vibrational Field

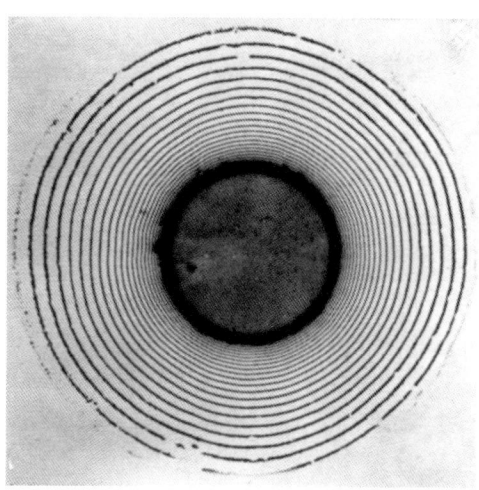

Fig. 93
Liesegang rings or periodic precipitation. This phenomenon is a periodic process taking place in a field of chemical reaction without vibration. The salt, silver chromate, is precipitated in regular concentric zones which proceed from the center to the periphery.

From the very large number of periodic phenomena in which no actual vibration is involved we have singled out three examples and sketched them in their phenomenal form.

The first example concerns Liesegang rings or periodic precipitations. They were discovered by Raphael Eduard Liesegang (1869-1947), and they involve processes in the field of chemical reactions. The experiments described in Figs. 93-95 and 102 were performed essentially according to Liesegang's method. A layer of gelatin containing the chromium salt potassium bichromate, is poured onto a glass plate: a drop of silver nitrate is then placed on this gelatin layer. A chemical reaction takes place between the two salts and silver chromate is formed. This precipitated salt however, is not diffused evenly, but periodically or rhythmically; zones of precipitation alternate regularly with zones free from precipitation.

In this way the concentric rings seen in Fig. 93 are formed. The process proceeds from the center towards the outside, pace for pace with the diffusion of the substance. Fig. 94 shows the same process in a vertical direction. The layers of precipitation follow from top to bottom. Hence these are not crystalline configurations, but the product of periodic precipitation, the result of a process which is characterized by rhythmicity. In Fig. 95 Liesegang precipitations appear again. The innermost of the zones enclosed by the curved lines, however, displays crystallization. Thus the features of the two patterns can be seen side by side and compared. If the experiment is on a larger scale, whole "precipitation landscapes" are created as seen in Fig. 102. However the drops of silver nitrate are placed, the same picture of periodic precipitation is obtained everywhere.

The curvilinear patterns of periodicity are brought about by these precipitations and processes of diffusion. In every pool one can see the first beginnings of a rhythmic pattern.

The various theories advanced to explain this phenomenon will not be discussed here. Even the many analogies drawn with patterns in the animal and vegetable kingdom (as for example the arrangement of pigment on butterfly wings, comparisons with bone lamellae, etc.) cannot be verified. From 1913 to 1935 Liesegang himself probed deeply into the true nature

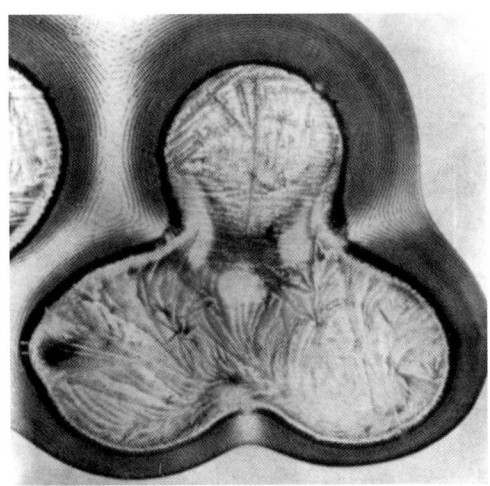

Fig. 95
Liesegang rings. The salt has crystallized out in the interior areas. Rhythmic precipitations which are not themselves a crystallizing process can be compared with crystalline formations.

Fig. 94
Liesegang precipitations proceeding vertically from top to bottom. The layers follow each other in rhythmic sequence reflecting the periodic nature of the chemical process.

of these reactions. He studied agate, the banding of rock formations, the formation of layers in salt deposits, and numerous other geological examples, and also the concretions of the body, such as gallstones and calculi. In these studies he made a clear distinction between internal and external rhythm. He assumed the first to be present only when the rhythmicity necessarily arose from the system in question.

We chose the example of Liesegang rings in order to direct attention to the wide field of periodicity without vibration in chemical reactions. This opens up a whole realm of phenomenology, the basic outlines of which were already sketched by Liesegang; simple and multiple periodic precipitation, periodic formation of semi-permeable membranes, rhythmic precipitation without gelatin, rhythmic crystallizations, stratification in drying colloidal layers, rhythmic deposition of precipitates, rhythmic precipitation of vapors. The problems of "chemical periodicity" take us far into the fields of biology and physiology. Indeed, it is hardly surprising that these phenomena bring us to the question of how catalysis functions.

Figure 103 portrays another example of a periodic process taking place without vibration. Drops of an emulsion dyed red are placed in a weak solution of Indian ink. The diameter of the area of ink is about 30 cm. First of all a delicate film forms over the entire surface. Then the emulsion slowly diffuses into the fluid according to the concentration gradient. The focus of interest here is on this diffusion: it proceeds in a periodic, rhythmic, to-and-fro manner. The flow of the diffusion is serpentine in character. Long winding processes are sent out and tail off to nothing as they become diluted. If drops are placed in several positions, the pressure of diffusion causes the circumambient areas to become displaced in relation to each other; some expand while others contract. In this way cell-like zones are formed whose boundaries consist of films of constantly increasing thickness. Since there are no walls or "banks" within each individual region and liquid can thus move in liquid, the emulsion can diffuse away without impediment, and the process of diffusion is as free as can be imagined. This is what happens on the one hand. On the other hand, this spreading does not proceed on a uniform linear front, but according to a hydrodynamic pattern displaying periodicity in great variety. If the quantity and the concentration are altered, a whole array of phenomena make their appearance: thick serpentine spurts of fluid; then again, delicate formations vanishing like wisps of mist (Fig. 96); centers flowing out radially in all directions; in several places tiny streams crisscrossing and forming a lattice; and then again hair-like threads of fluid anatomizing (Fig. 97). In Fig. 103 a whole series of these processes can be observed. However, it must be remembered that all the red structures are in a state of flux, that the cells are shifting to-and-fro, and the liquid system is constantly diffusing away. As we mentioned before, all this takes place

Figs. 96-97
Dispersion at a concentration gradient similarly takes place in an oscillating, periodic manner. A whole hydrodynamic phenomenology makes its appearance. In Fig. 96 the concentrated emulsion disperses by, as it were, trailing off on every side into delicate rivulets. In Fig. 97, on the other hand, we have a picture in which the delicate, flowing rivulets anastomize and form a network.

Fig. 98
Here we have an entirely different picture. The whole has been made to oscillate and vortices have appeared at once. As soon as the tone is discontinued, dispersion is resumed in the manner seen in Figs. 96, 97 and 103, and this continues until the gradient has ceased to exist.

Figs. 99-100
If a doughy mass is pulled apart, the surfaces do not reveal a random but rather a characteristically organized pattern. This phenomenon of dehiscence produces tree and branch-like patterns, delicate filigrees and fine networks. The branch work in Fig. 99 is sometimes thick, but in Fig. 100 the dendritic pattern is more delicate.

Fig. 101
This cell-like pattern is again the result of dehiscence.

Fig. 96

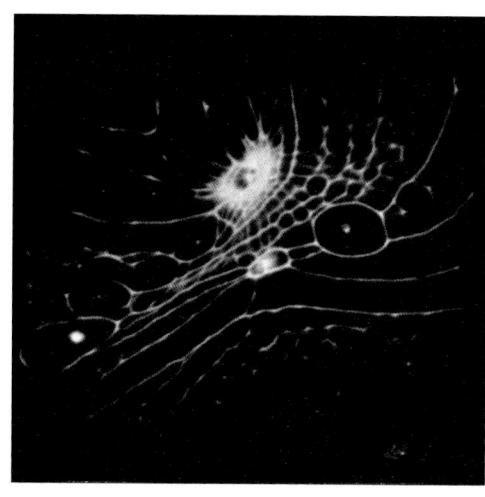
Fig. 97

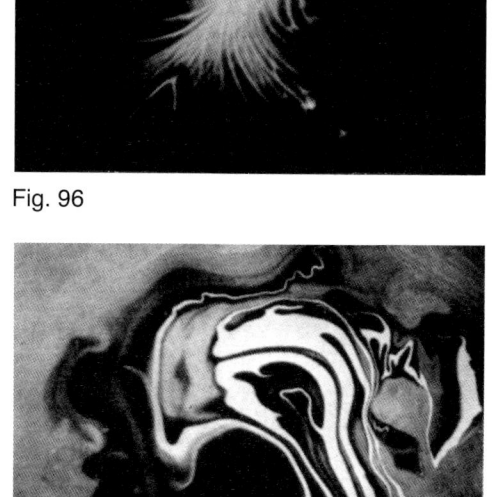
Fig. 98

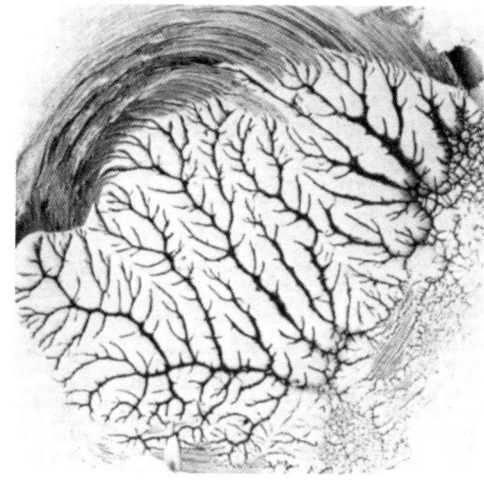
Fig. 99

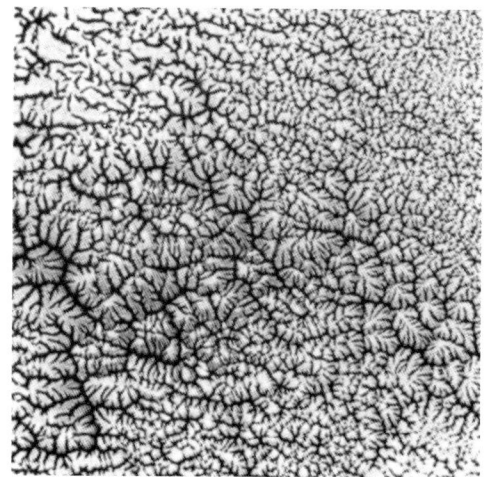
Fig. 100

Fig. 101

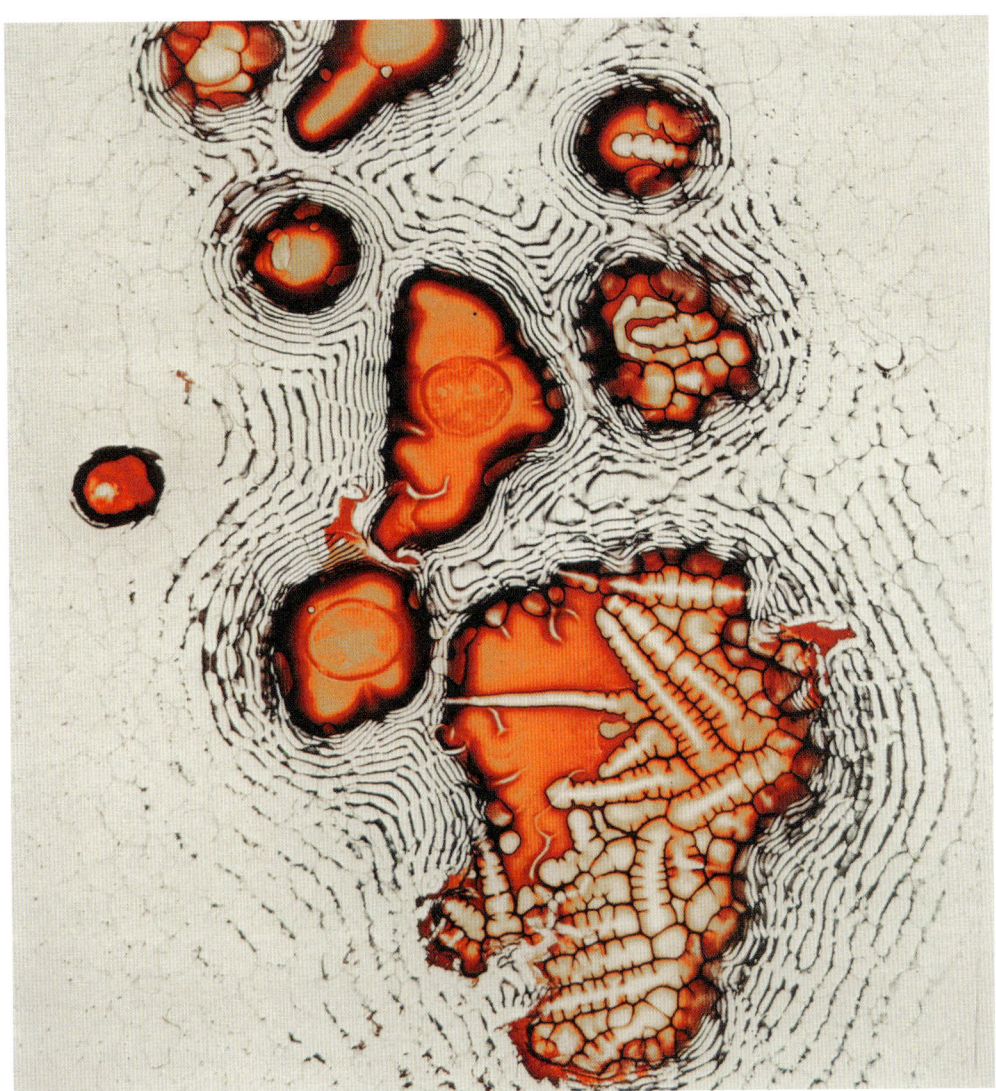

Fig. 102
Precipitations with Liesegang rings have been started by dropping the silver salt at random on the plate. The zones of periodic precipitation reveal how the precipitates are arranged. Everywhere there is a tendency towards a rhythmic process. This tendency can be seen in each pool.

without vibration in the true sense of the word. If the fluid is energized by vibration there is an instantaneous change in the picture (Fig. 98). The masses are transformed into vibration patterns: turbulences, vortices, wave fields appear. Upon the sound being discontinued, the weaving, wandering "red structures" reappear: centers, cells, flows and current networks emerge again. Once again the interplay of this hydrodynamic pattern with its rhythms, to-and-fro motions, and periodic floating movements is dominant. Gradually the concentration gradient disappears.

One more phenomenon demands our attention and that is the dehiscence of doughy masses. If such a mass is torn apart, curious structures are revealed. These forms can be accurately described as dendritic (Figs. 99, 100, 101, 104 and 105). This dehiscence can be most simply produced in the following manner. A layer of the doughy plastic mass is placed between two glass plates. The glass plates are then drawn apart and the cohesion of the mass is overcome. The result is that the mass is torn apart. The appearance of the two surfaces is dominated by structures which interbranch and interdigitate in conformity with a law (Figs. 99-101, 104 and 105). These formations change depending on the consistency of the material and the quantity used. "Trees" can be produced, then fine "bushes", filigree patterns, the most delicate network, and cellular textures

(Fig. 101). What the eye perceives here is the regularly patterned tear of the mass. Dehiscence proceeds in such a way that true periodicity is revealed in its appearance.

Any number of similar examples could be adduced, but these three must suffice. A number of analogies immediately spring to mind in this connection. The Liesegang experiment invokes the picture of a zebra, the systematic banding in the plumage of many birds, and the regular banded and criss-cross patterns of pigmentation in many species of animals. The to-and-fro hydrodynamic pattern is reminiscent of the circulatory systems in the lower animals or in the development of the embryo. And there is no end of organizational forms which are brought to mind by the dendritic structure of dehiscence. Nevertheless such similarities and interpretations as these, derived as they are from external appearances, invariably and inevitably peter out in generalized statements which are devoid of authentic meanings and never reach the real unity of things. Each phenomenon must speak for itself and must be studied in its own terms. If the examples adduced here are studied along these lines, they not only display periodicity as a general feature, but they also reveal that this periodicity has various aspects. Whatever comes to light by way of kinetics, dynamics, chemical reactions, and results in connection with Liesegang rings, whatever elements of hydrodynamics and hydromorphology

Fig. 103
A red emulsion is introduced into a liquid which is stained black. The red emulsion begins to disperse according to the concentration gradient. Fig. 103 shows some of the processes involving periodic hydrodynamics by which the dispersion takes place in an oscillating meandering pattern. It must be imagined that "everything" is not only flowing, but actually flowing in patterns and rhythms.

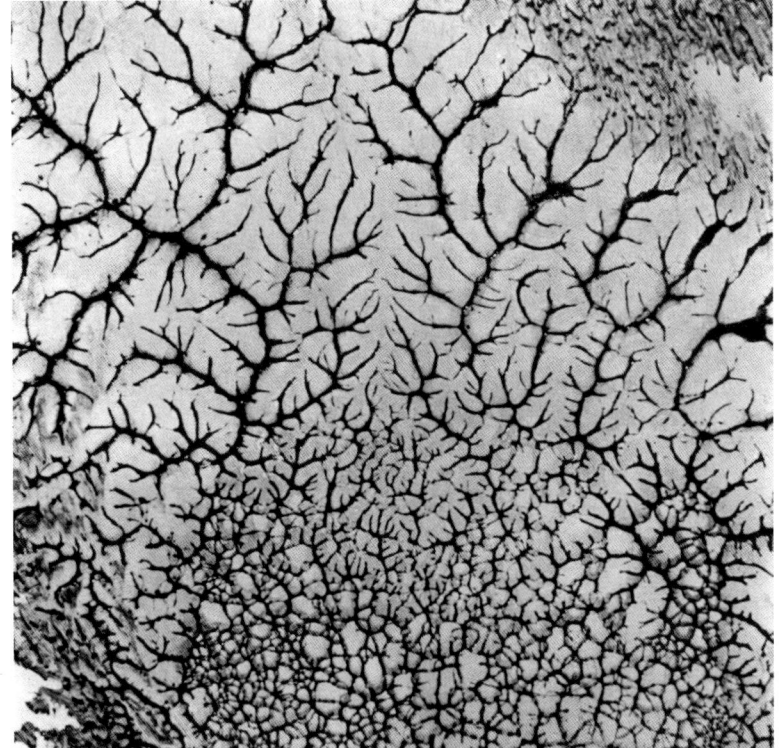

Fig. 104

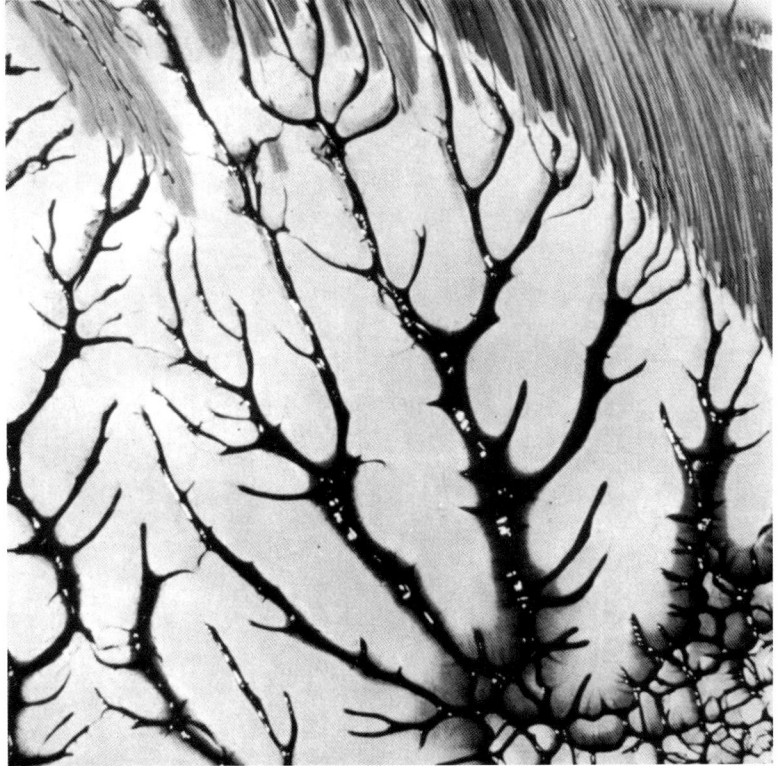

Fig. 105

are revealed in the experiments with concentration gradients, whatever mechanical forces appear in the mass system in dehiscence — all these are inseparably and essentially associated with regular recurrences in the movement, dynamics, configuration, patterns, etc. Rhythm dominates the outward form. The systems and their phenomenal classifications are essentially periodic through and through. To trace these elements down to their minutest manifestation, to grasp what is really happening, to observe the pulse and formative rhythm in milieu and medium, to recognize those aspects of the phenomenal world which are permeated by the essential nature of vibration, waves, and periodicity — these are the activities characteristic of the cymatic process.

Thus vibrational phenomena may be said to represent a prototypal chapter of cymatics; but, in the light of the three examples adduced, it may be said that there is also a general cymatics in the very widest sense.

Figs. 104-105
Examples of dehiscence. If doughy masses are torn apart, branched, ramified patterns are revealed. The formation is dendritic throughout. The pattern ranges from bush-like growths (Fig. 104) to the most delicate tracery (Fig. 105) with every variation in between. In the lower part of Fig. 104 there is even a cellular pattern. These phenomena are entirely periodic in character. (Fig. 105 is a detail of Fig. 99.)

Figs. 106-107
When a plate vibrates, adhesion to its surface is reduced by the movement of the plate and of the layer of air immediately above it. Iron filings placed on a vibrating diaphragm and at the same time in a magnetic field are in a constant state of motion. As a result of the reduced adhesion the filings dispose themselves in the magnetic field with certain degrees of freedom. The configurations in Figs. 106 and 107 move, incline, and migrate as the magnetic field is changed.

Fig. 106

Fig. 107

7

Sound Effects in Space; Spacial Sonorous Patterns

If a vibrating surface is not quite horizontal, objects resting upon it easily slide off. They start to move even when the angle of gradient is very small. The stroboscope reveals that some objects are, as it were, hovering, or at least that their adhesion to the surface under them is reduced. When iron filings are distributed on a vibrating diaphragm in a magnetic field, the reduced adhesion is revealed by the fact that iron particles acquire certain degrees of freedom and their arrangement in the magnetic field is a mobile one. Large polar figures are formed which move as the density of the magnetic field changes, but in addition to these, the iron particles also form figures which migrate together and, if the direction of the magnetic field changes, take up an appropriate disposition. Figs. 106 and 107 represent such patterns of iron filings. If the direction of the magnetic field in Figs. 106 and 107 is changed, the erect piles of filings are mobile enough to respond by a change of inclination. These processes and others like them, direct the attention to the environment of vibrating objects, to the space surrounding them. What happens when the sound impinges not on flat layers, but on more substantial masses? If a readily plasticized mass is placed on the diaphragm, the material is set in motion. The masses are pushed to-and-fro. Whereas the field empties in various places, in others the substance forms lumps. Whole landscapes are plastically molded; a kind of relief which is a reflection of the vibrational processes is shaped out of the material. The lumps formed frequently sidle to-and-fro, and also show rotational movements within themselves. This rotational plasticization leads to the formation of balls if the consistency of the material permits easy shaping. Material which is applied horizontally at the beginning begins to rotate around local centers and clumps into balls, so that finally there are a number of spherical formations which mass together and go on producing larger and larger globular shapes. Figs. 108 and 109 are photographs of these processes. The largest of the globes attained the size of a pigeon's egg.

The next step was obvious. It was to study the action of sound on a three-dimensional element. Because of their high surface tension, drops of mercury are globular elements which have a very strong tendency always to return to this ball shape and to maintain it. Under the influence of sound, large quantities of mercury show wave patterns, vortices, etc.; that is, hydrodynamic phenomena like other liquids. Drops of mercury on the other hand, always behave as homogeneous elements within certain limits: they only fly apart when the amplitude is high. What happens then to such a drop? It must be remembered that reflections make photography very difficult. The photographer had to take appropriate measures in order to photograph the whole phenomenon without dazzling

Fig. 108

Fig. 109

Figs. 108–109
A plasticizable mass is formed into balls by oscillation. The material is first spread uniformly on the plate and then forms into balls when exposed to the effects of sound. Fig. 108 shows a general view of this process and Fig. 109 a detail. When the plasticity of the material is exhausted, the balls (about the size of a pigeon's egg) roll about on the diaphragm.

Fig. 110
A drop of mercury before acoustic irradiation. Mercury is difficult to photograph because of the reflections. In these photographs they were avoided by using diffused light. The black circle is the reflection of the lens. Changes in the surface were rendered visible by the resulting deformations in this mirror image.

Figs. 111-113
The surface upon which the drop of mercury is resting is made to vibrate. Concentric waves are the first effects to be produced by a small amplitude (low tone). Fig. 111 was excited by a frequency of 150 cps., and Fig. 112 by a frequency of 300 cps. A figure (triangle, hexagon) is already apparent in Fig. 113. At the same time, currents become evident in the drop, often in the form of twin vortices.

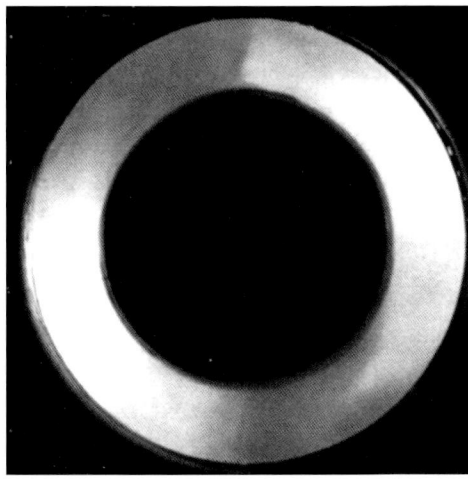

Fig. 110

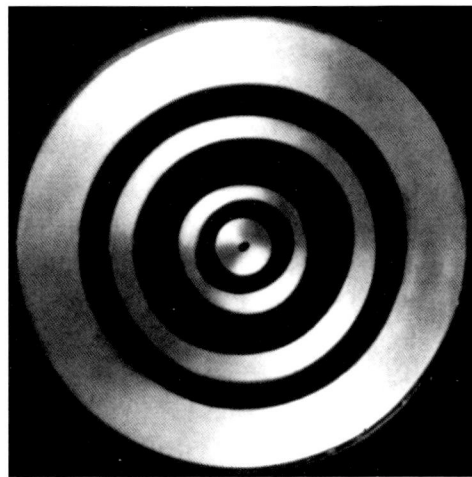

Fig. 111

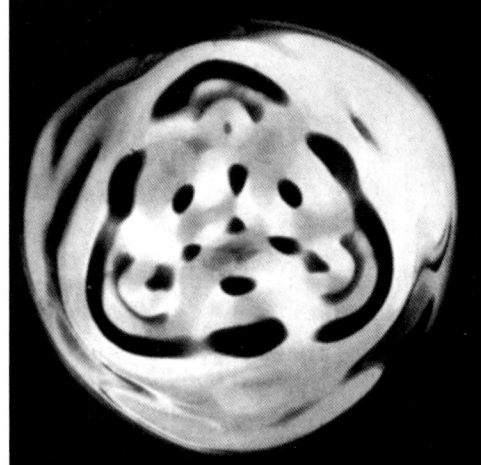

Fig. 113

Fig. 112

reflections. In Figs. 110-113, the lens is reflected in the drop as a dark, round circle. This is actually an advantage in that curvatures on the surface of the mercury are rendered visible by the distortion of the mirrored image of the lens itself. In Fig. 110 the resting drop of mercury can be seen. As sound is applied, concentric rings are formed (Figs. 111, 112). Then a carefully managed crescendo causes vibrational patterns to emerge (Fig. 113). If the action of the sound is more intense, the whole globe of mercury is affected, and since the whole coheres and remains a unit, the oscillation can be revealed as a formative factor. The series of figures 118-123 illustrates this phenomenon. We see spatial sonorous figures which are tetragonal, pentagonal, hexagonal, heptagonal and octagonal in shape, and there are many other forms. The production of these shapes is very accurate, but not rigid throughout. Here again the stroboscope reveals pulsations. For instance, a drop can move in such a way that two triangular figures pulsate through each other: the figure oscillates in these triangular forms. In addition to these pulsations there are also currents within the mass of mercury. A few lycopodium spores strewn on the surface make these visible. Here again we find ourselves concerned with a whole series of oscillatory activities within a system. Figs. 114 and 124 show a lateral view; the three-dimensional character is manifest. One might speak roughly of edge

Fig. 124
This lateral view clearly shows that Figs.118-123 are really sonorous configurations in space.
Here, they appear, as it were, in elevation. They tower up in their numerical progressions.
One might speak of a law of numerical configuration.

8

The Spectrum of Cymatics

We shall now describe two examples of cymatic processes. In our account we shall record the various aspects of their outward appearance and make a compilation of them. It will then be seen whether there are any features of regularity between the various categories of phenomena revealed by this compilation and what the relationships between them may be. It is not a scheme we are looking for; the processes will be allowed to speak for themselves. They will show how they form into series and whether there emerges from the empirical field a spectrum of phenomena which stems entirely from their own essential nature.

From the numerous series of experiments, we will first single out "solidescence." What is involved here is the change in the state of matter wrought by vibration. In the present experiments the cooling process was used in order to observe the solidification of a substance. A kaolin paste is heated; it liquefies and is then poured onto a vibrating diaphragm. Wave fields are created and typically, standing, traveling and interfering waves appear. Figs. 134 and 135 afford views of these "wavescapes." During the cooling process, the picture begins to change. The mass becomes semi-solid. Depending on the vibrational topography of the diaphragm, the material masses together and forms round shapes (Figs. 136, 137). There is marked circulation in the sense that the material at the top is transported from outside to the center and at the bottom from the center to the outside. There is thus radial circulation within these round shapes themselves. In addition organized patterns appear, so that the whole presents a relief of ribs in radial movement (Figs. 136, 137). At the same time the stroboscope reveals pulsations throughout the formation. It need hardly be stressed that these are the results of vibration. If the generating sound is discontinued at this stage, everything again becomes a semi-solid paste and cools to form an evenly distributed mass. But if vibration continues while cooling proceeds, the picture changes yet again. The material assumes the consistency of a tough dough. It begins to rotate around and around. It rolls, turns on itself like a ram's horn and is plasticized in volutes (Fig. 138). These processes and their formations are reflected in the figures which have formed on the diaphragm and are in the process of solidifying. Fig. 139 is a photograph of an area which has solidified completely. The areas where the substance has solidified under the action of vibration can be recognized. Fig. 140 shows a whole strip of solidification. Fig. 141 a detail of such a patterned field. Finally the masses solidify completely. The product is now in a solid state and is preserved. The patterns are of a peculiar nature. In the experiment performed with this paste, they are not in any way crystalline. Instead they branch

Fig. 125

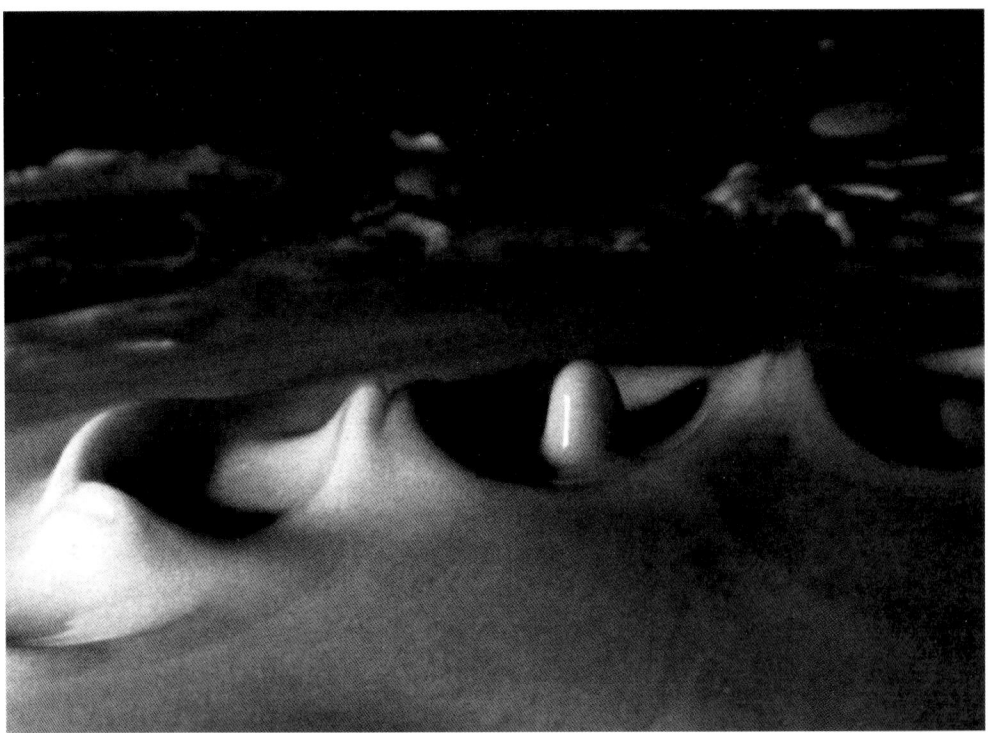

Fig. 126

Fig. 125
A viscous fluid is poured onto a vibrating membrane. An annular wave is formed surrounding a hollow.

Fig. 126
Continuation of experiment 125 results in the formation seen in Fig. 126. A sequence of annular waves appears with a protuberance in the center. Two trends can be seen: one tending to more actual wave trains, and the other to round forms.

Fig. 127

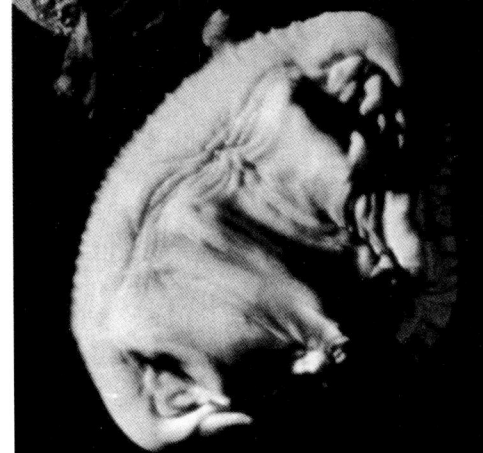

Fig. 128

Fig. 129

Fig. 130

Fig. 127
The wave principle predominates. It might be described as a wave sculpture. The photograph is taken looking down.

Fig. 128
Here a sort of rampart is added to the wave form: this is a nascent round form.

Fig. 129
The round form predominates. There is nevertheless a concentric wave train as well. A cone protrudes in the center. The whole configuration is in a state of radial circulation.

Fig. 130
This configuration combines the round form with marked wall-like waves in rings. The whole must be imagined as moving, flowing, circulating and pulsating, etc. During the experiment the tone producing these forms is, of course, audible.

and ramify. The pattern runs from large thick branches to increasingly more delicate ones until finally there appears a kind of filigree. They are true dendritic formations which can be produced again and again in this typically serial configuration.

What exactly is revealed in this series of experiments with solidification? If we follow the course of events, we find first of all wave phenomena, which are the prototypes of periodicity. These are followed by formations and organized patterns. At the same time different forms of movement appear: rotations, circulations, to-and-fro motions, pulsations. But these processes are caused purely and simply by vibration and nothing else. Periodicity is inherent in them, it lies in their nature to be rhythmic, whether in form, in configuration, in movement or as a play of forces. Sculptural shapes are actually formed. It sometimes seems as if one is vouchsafed a glimpse into the origins of Baroque. All these phenomena take their rise in the field of vibration and owe their existence to vibration. They have therefore originated in vibration and are in some measure its specific effects; vibration produces a multiplicity of effects or is polyergic.

In the course of these experiments pure periodicities appear which coact and interact; sometimes form is dominant, sometimes movement; sometimes the picture becomes organized as a pattern and there is real configuration, sometimes

Fig. 131
Minor changes in viscosity bring about changes in the forms seen. The liquid has been rendered more fluid. When greater amplitudes are used, the masses are flung high and ejected. The experiment can be continued in this direction until the liquid forms a spray. The waves also increase in height and look like cups or pots although they are also in a state of flow.

Fig. 132
These fluid, flowing sculptures assume any number of different forms. Wall-like waves rise in some places. Where trains of waves interpenetrate in lattices, the waves rise in columns. Even these phenomena which persist for some time are "living." The mass flows and pulsates within itself. If the tone is stopped, the liquid returns to its uniform state.

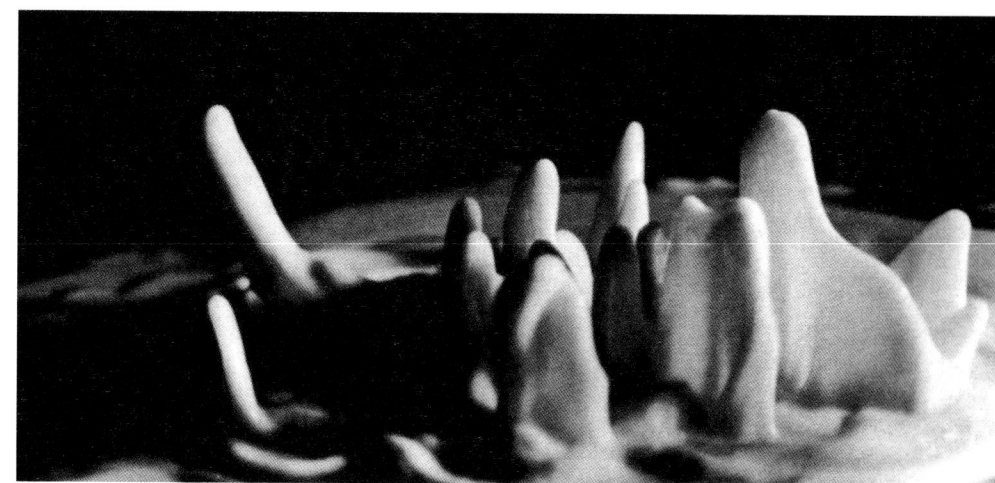

Fig. 133
If the liquid is of a more viscous character, figures of the most varied kind take shape. The club-shaped configuration seen here has actually been raised out of the mass by the vibration. It is not a finished sculpture but a configuration in a state of flux. The substance flows up the stem of the club and circulates. These figurines also pulsate in themselves. They may also move around depending on the topography of the vibration. Such processes are not purely adventitious but can be reproduced systematically.

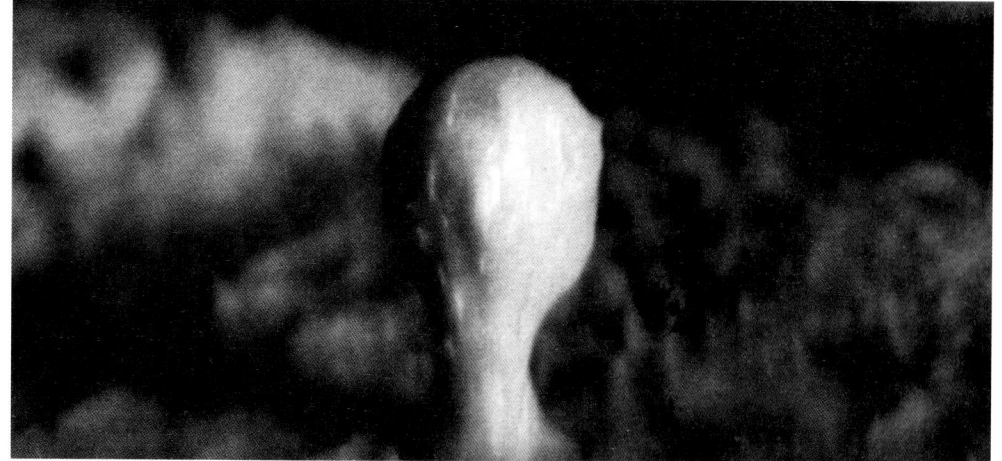

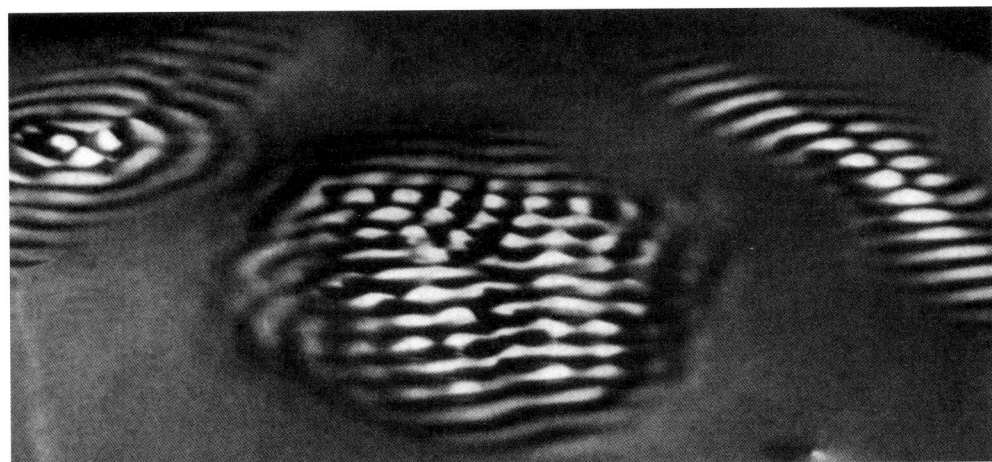

Fig. 134

Fig. 135

Figs. 134-135
The transition from the liquid to the solid state while under the affect of vibration can be seen in this and the following figures. The process can be exemplified by cooling, evaporation, chemical rearrangement, etc. The liquid state is invariably taken as the starting point. Oscillation causes the substance to form waves. Figs. 134 and 135 show wave fields of this kind. What can still be seen here as static liquid becomes solid and rigidifies during the experiment.

everything is in a flux, in a plastic flow. A certain polarity becomes evident; for it is not inappropriate to speak of polarity when on the one hand the object assumes form, even a rigid form, and on the other hand, moves, finds itself in constant change, and appears as fluid form and as formative flux. Since all these aspects are essential and necessary parts of the picture, since they all represent exclusively the effects of vibration and are caused by the same vibration, they fall, as it were, into a series of their own accord. Taking this idea of polarity as basic, we have a spectrum of phenomena running at one end into the figurate, into shapes, into organized patterns and textures, and at the other end into movement, circulation, into streaming and flowing, into kinetics and dynamics. The whole is created and sustained by oscillation, the most explicit and typical expression of which is the wave in the widest sense.

What it all boils down to is this, we must keep on asking ourselves as Goethe did: "Is it you or is it the object which is speaking here?" If we were to establish rigid definitions and split up the various manifestations into sections, we should be artificially dismembering the phenomenon by applying the analytical instrument of the intellect. If the phenomenon is to remain vital, its spectrum must be grasped as a fluctuating entity. True, there are significant forms there; but what we have to evolve is the concept of "moving form" and "formative movement."

Sharply defined patterns emerge, but they flow away into nothing. Flowing patterns and patterned flux appear before us. Thus, the problem of cymatics exists not only in observing in the experimental field but also in formulating concepts with which to press towards comprehension of the actual realities. In attempting to leave the cymatic phenomenon intact and unharmed in our intuitive vision, we can derive from it the following spectrum with form at one end and movement at the other: figurate, patterned and textural on the one hand, turbulent, circulating, kinetic and dynamic on the other, and in the center, acting in either direction, creating and forming everything, the wave field, and thus as *causa prima*, creating and sustaining the whole, the *causa prima creans* of all — vibration.

So that we can describe and define conditions from many different points of view, we will illustrate and explain a further example. This time it concerns a viscous liquid. Here again the material is irradiated with sound. In this instance not only the frequencies and amplitude are modulated but the liquid itself is changed by modification of its viscosity. First of all wave formations appear.

Fig. 136

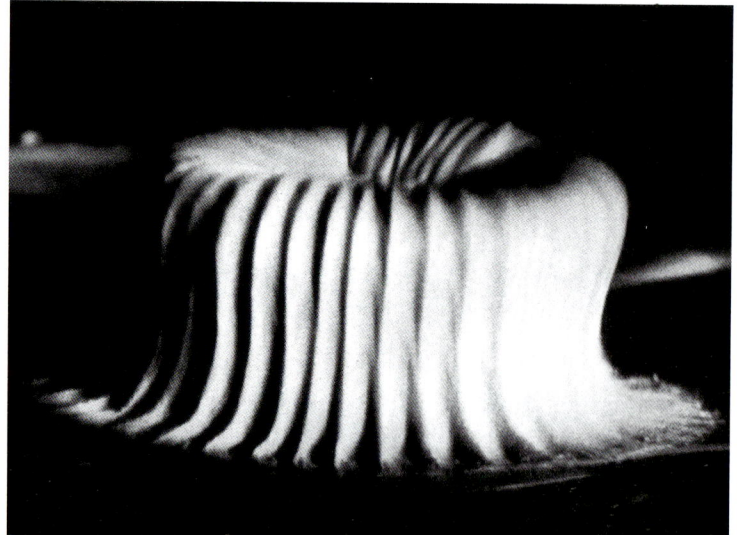

Fig. 137

Figs. 136-137
The consistency soon changes. The substance becomes plastic and dough-like. The round configurations which appear are in a state of circulation and also display radial patterns. Fig. 136 is taken from above and Fig. 137 from the side.

Fig. 138
The process of solidification proceeds further. The mass becomes a viscous paste. It is rotated into volutes and rolled into a shape like a ram's horn.

Figs. 125, 126 and 143 are photos of imposing wave crests. Fig. 144 shows a series of waves seen from above. Then round forms appear. Interesting transitions can also be found. Fig. 128 shows a picture in which the sequence of undulation is still clear, but already there are intimations of an encircling rampart which, as this process begins to dominate, flows together to form a round shape. These round shapes reveal a radial circulation within themselves. Yet, also inherent in them, is the wave principle which now however, appears as an annular formation (Fig. 145). The ring form may appear with a high degree of perfection (Fig. 146). This double-ring form looks as if it has been turned on a lathe or a potter's wheel. In spite of its perfect shape, it must be imagined as constantly rotating. In other places on the vibrating diaphragm, movement processes arise as amplitude and frequency are changed. The mass creeps around in vermicular forms. In Fig. 147 one can see such a formation which is readily derivable from the round shape (Fig. 146). If the vibration is now increased (corresponding to a crescendo for the ear), the round configurations rotate more rapidly. The processes are intensified. The wave rings rise up to form walls (Fig. 130). The masses lift up into peaks (Fig. 148). Crypts like honey-combs are formed while, close by, the waves rise high. There are pillar-like prominences: column waves and wave columns. The mass shoots up, is hurled away in the form of spiculae (Figs. 149, 150). Protuberances of every kind are thrust up, some with explosive dynamic force, others in apparent tranquillity. We say "apparent" advisedly, because in all these objects, in the eruptive as well as the persistent, the stroboscope reveals pulsations, turbulences and currents. Figs. 131, 132 and 148-151 are a series of photographs showing these processes. The round shapes and their ring formations can be recognized; the process of intensification leading to wave walls and wave columns can be followed. The enormous dynamic force which ultimately causes the mass to atomize can also be seen. In everything we find the effects of vibration undergoing a series of changes. The highly specific character of the relationships between the factors involved is recognizable in the following pictures. Viscosity has been increased compared with the preceding series of experiments. The mass is now more gluey. Now organized patterns imprint themselves. The round shapes display radial ribs; so do the "worms"; they even undergo actual segmentation (Figs. 152, 153). If the amplitude is now increased, a process emerges which is entirely different from that witnessed when amplitude was increased in the experiment with highly fluid substances. The mass rears up; it assumes a shape like a figurine. Fig. 133 shows a club-like excrescence. What must be realized is that the mass flows up into the figurine, circulates, flows down and pulsates.

Fig. 139

Fig. 140

Figs. 139-141
The mass has solidified. The patterns in the substance on the diaphragm are also rigid. They reflect those of the rotating round shapes (Figs. 136, 137) and reveal a dendritic formation; they are not crystalline in character.

Fig. 141

But the shape of the figurine persists in spite of the turnover of material. At certain places on the diaphragm, depending on the different vibratory topography, they move around, and run to-and-fro in the vibrational field.

These descriptions, which take account only of the most important aspects, have a bewildering effect on many people at first. The "profusion of appearances" is difficult for the mind to grasp. But this is a stage which must be passed through. One can then, of course, proceed to analyze this or that category; but always one must return to the whole, otherwise one is left "holding the pieces in one's hand." Analysis is essential, but the eye and the brain must restore what they have dissected to the active phenomenon and reintegrate it with the complex reality so as to see it again in the nexus in which it is alone existent.

In order to get one's bearing, the processes here can once again be tentatively categorized without injury to their fluctuating character. Once again we have, on the one hand, figures, configurations, organized patterns and formations, textures and webs; on the other we have turbulences, currents, movements, and the play

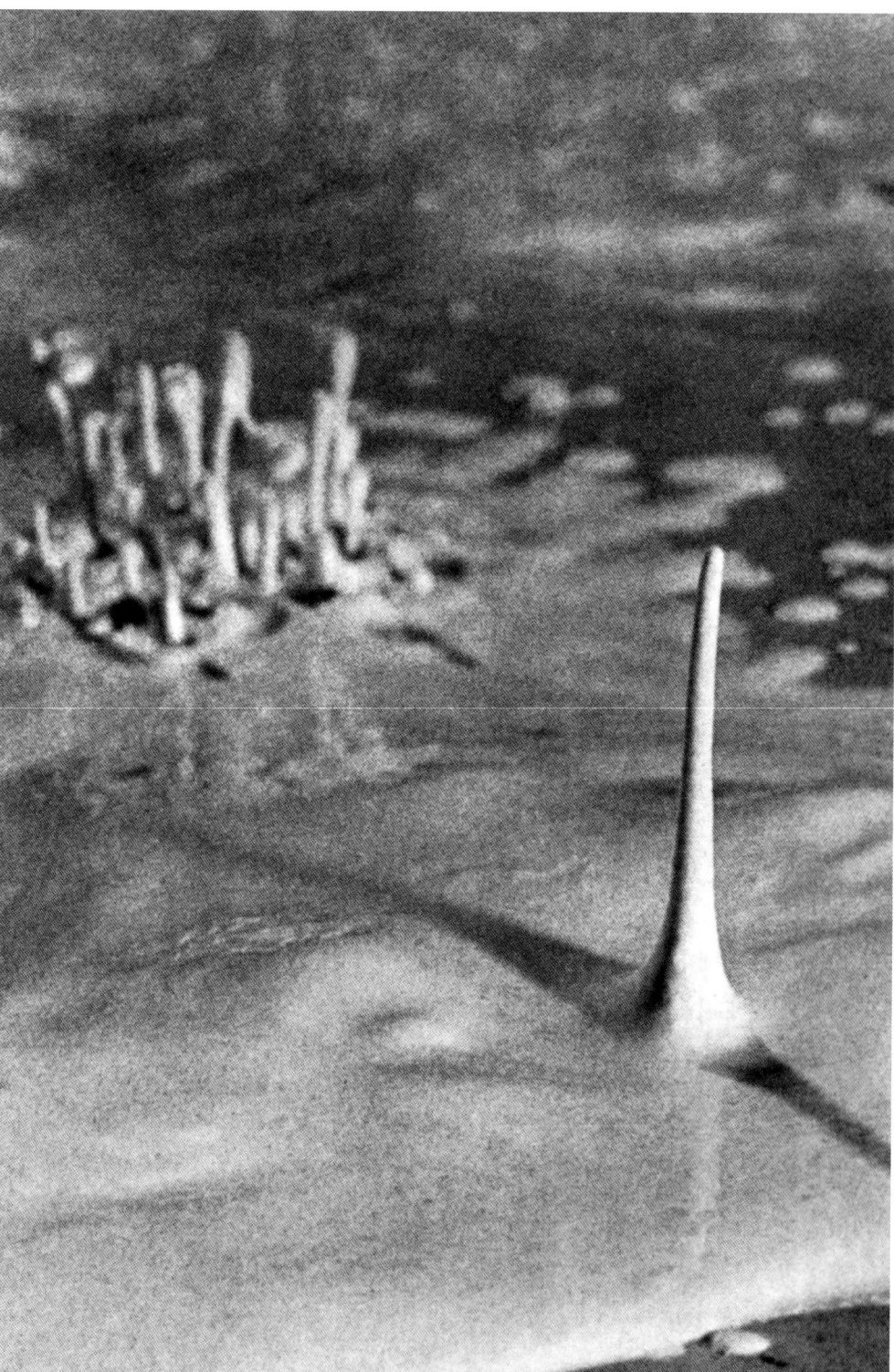

Fig. 142
In experiments with "solidescence," curious effects resulting in figurines are repeatedly observed. The tall, columnar figure is driven up by vibration. At the time the photograph was taken, the mass was still viscous. Although the figure persists for some time it will vanish in the course of the experiment as the material enters the solid state. Other morphological elements will also appear as the result of vibration.

Fig. 142

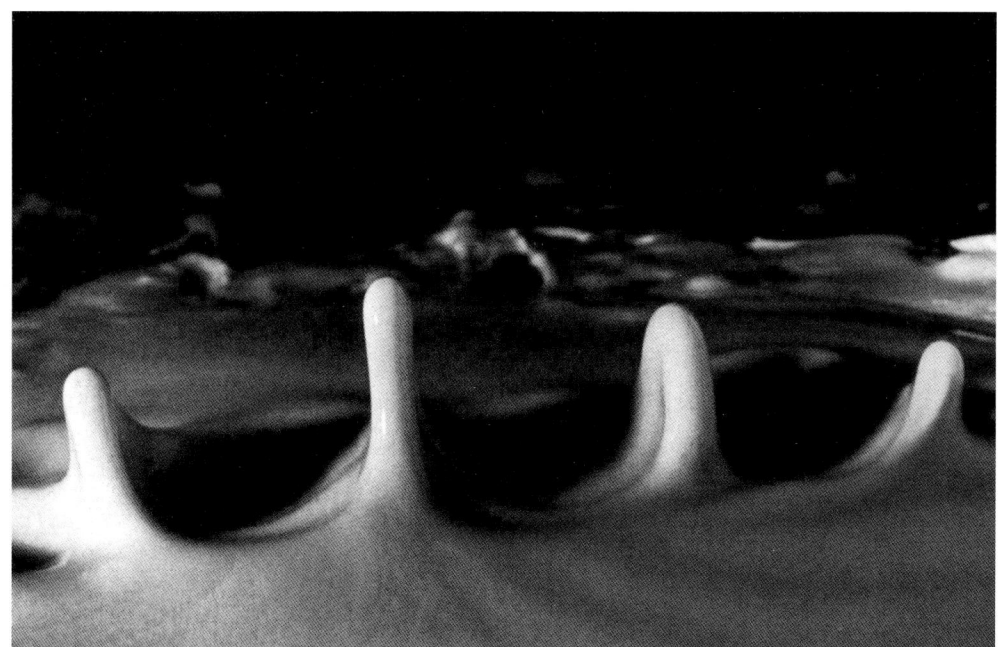

Fig. 143

Fig. 144

of forces on and in the masses. Once again the waves are a kind of middle category; they are the periodic element *par excellence*. But here again it must be noted that it is the essential character of the waves, their characteristic periodicity, that is the basic causative factor throughout. For the whole drama of these phenomena is played by the vibrational orchestra of an oscillating diaphragm.

The phase we have reached in our study is characterized by the discovery of a spectrum of phenomena. It can be approximately described in the following terms: figures; organized patterns; texture; wave processes in the narrow sense; turbulences; kinetics; dynamics. These terms are not conceptual pigeon-holes: they do not package reality. They are derived from empirical perception. We will therefore leave the spectrum in this form for the time being. Whether it can be developed further, and if so, how, are questions to be dealt with in the next chapter.

Fig. 143
In the following series of experiments the remarkable diversity of the phenomenology of vibration will be shown. Fundamental changes are produced not merely by variations of frequency and amplitude, the character of the material is also a factor. Very small differences in the viscosity produce entirely different formations. In Fig. 143 huge wave crests appear as a viscous liquid is poured onto the plate and vibration begins.

Fig. 144
Photograph of the "wave trains" seen from above.

121

Fig. 145
Round shapes appear in addition to the wave trains, all depending on the topography of the vibrating diaphragm. In Fig. 145, these waves trains appear in a concentric annular pattern.

Fig. 146
Round shapes of the greatest regularity appear. The one seen here is absolutely perfect in its form. It is not some finished design in porcelain but quite literally a "fluid configuration." Within this form everything is in a state of circulation. The material flows from the periphery to the center and from the center to the periphery. Everything is generated and sustained by vibration. Without vibration everything would be simply a uniform paste.

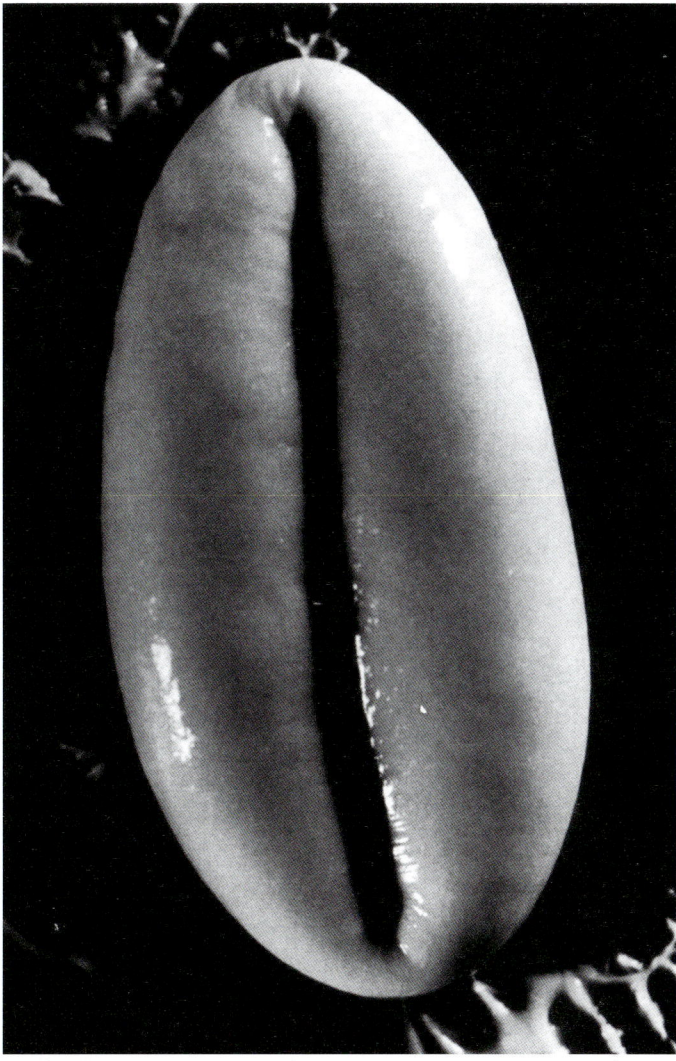

Fig. 147
In places of continuous movement these round shapes become elongated. This figure shows such a stage in the nascent state. Even this long oval shape is still circulating within itself. The longitudinal fissure is the zone in which the mass flows in. If the amplitude is increased (a crescendo for the ear) the form creeps hither and thither like a worm. (See also Fig. 153.)

Fig. 148
If the mass is made more fluid and a greater amplitude is used, a dynamic element appears. Waves rise and eddy up to form walls, plates, columns.

Figs. 149-151
Figs. 149 to 151 also show kinetic dynamics due to vibration. While on the one hand there is a moving world of forms in which the mass may even be ejected in the form of spiculae (Figs. 149, 150), we may also find wave sculptures (Fig. 151) of a more persistent nature. But since shapes and patterns appear simultaneously, whatever the kinetics and dynamics, the cymatic spectrum is omnipresent and manifests itself consistently everywhere as the basic triadic phenomenon.

Fig. 148

Fig. 149

Fig. 150

Fig. 151

Fig. 152

Fig. 153

Figs. 152, 153
We can again vary the experiment by making the mass more viscous. Again round and vermicular forms appear in a state of circulation and pulsation, but now the patterns are more marked with radial ribs in the round configurations, and a kind of segmentation in the elongated forms. While the round shape remains stationary if the amplitude is unchanged, the elongated form creeps around, but always in conformity with the vibrational pattern.

Fig. 154
Viscosity changes again. With increased amplitude the masses are driven up and shaped like figures. Compare Figs. 133 and 154. There is also a figurine there, but its stature is quite different. Thus slight changes of viscosity produce entirely different forms. The figurine seen here must be imagined to be in a state of flow and pulsation.

9

The Basic Triadic Phenomenon

Since the various aspects of these phenomena are due to vibration, we are confronted with a spectrum which reveals patterned, figurate formations at one pole, and kinetic-dynamic processes at the other—the whole being generated and sustained by its essential periodicity. These aspects however, are not separate entities but are derived from the vibrational phenomenon in which they appear in their "unitariness." Even though one or the other may predominate in this or that phenomenon, we invariably find these three elements present. In other words, the series we have formulated is in reality confluent in homogeneous activity. It is not that we have configuration here and organized pattern there, but that every effect of vibration bears the signature of configuration, movement and a play of forces. We can, so to speak, melt down our spectrum and observe the action of its various categories as a continuous play in one and the same entity. If we wish to describe this single entity, we can say this: there are always figurate and patterned elements in a vibrational process and a vibrational effect, but there are also kinetic and dynamic elements; the whole is of a periodic nature and it is this periodicity which generates and sustains everything. The three fields—the periodic as the fundamental field with the two poles of figure and dynamics—invariably appear as one. They are inconceivable without each other. It is quite out of the question to take away the one or the other; nothing can be abstracted without the whole ceasing to exist. We cannot therefore number them one... two... three, but can only say they are threefold in appearance and yet unitary; that they appear as one and yet are threefold. All the examples in this book can be considered from this point of view. It will be seen from the selection shown in these pages that in every case there are formations, textures, and forms, we can see movements, currents, circulations, rotations, etc. and yet all these display throughout a rhythmic, serial, vibrational character. Hence we cannot say that we have a morphology and a dynamics generated by vibration, or more broadly by periodicity, but that all these exist together in a true unity. This can be seen from all the experiments described here; all the examples adduced— whatever their variations, are recognizable as this unitary element. It is therefore warrantable to speak of a basic or primal phenomenon which exhibits this threefold mode of appearance. It must be stressed that this is an inference made from appearances. The basic threefold or triadic phenomenon is not a preconceived conceptual form which is forced on the nature of things: these things themselves *are* the basic triadic phenomenon.

It might be argued and discussed that this is not really a true morphology, but only a vibrational form; not an inherent dynamics, but a vibrational dynamics, etc. If, however, we restrict ourselves to experience and speak its language, we shall find that we speak of every metamorphosis and variation in terms of the basic triadic phenomenon. But the few examples given here do not exhaust the phenotypes of this phenomenon. Using this basic phenomenon as a perceptive organ (not as a dogmatic formula) we can watch and observe the most varied fields to see whether the language of periodic triadism is also current there. As far as research has proceeded along these lines, it has been found that this basic model is fundamental to the most varied fields and constitutes an essential part of their nature. Contact with scientists and research workers in the most diverse areas of study has afforded wide vistas of this kind. We are thus confronted with the concrete research task of starting a monograph on periodic phenomenology which would cover many different fields. The basic triadic phenomenon is an empirical notion which comes to mind in the study of histology, cell physiology, morphology, biology and functional science; likewise in the study of geology and mineralogy, and atomic physics, astronomy, etc. This does not mean that this basic theme should be inflated to be a model of the universe; the interpretation of phenomena calls for subtlety of mind and love of research. Rather than blunt the reader's mind with countless references, we shall describe in detail one example to show how this basic phenomenon appears in a specific case and how appropriate it is as an instrument of knowledge for probing into fields of research. This one example will serve for many. It is concerned with what is now known as fluid dynamics. In order to give the reader some idea of the subject we shall quote some appropriate passages on hydrophysics. First we quote from Wolfgang Finkelnburg's <u>Einführung in die Atomphysik</u>, (1954): "Since it is at first sight surprising to claim that liquids have a quasi-crystalline structure, we will look first of all at the most important evidence supporting this contention. Direct evidence is provided by Debye's method of X-ray diffraction in liquids. If the molecules and the distances between them were completely irregular, the intensity of scattering would decrease uniformly as the scattering angle increased; if the molecules are in a regular arrangement, the X-rays scattered by the various molecules would show peaks and dips in the intensity plotted against the angle of scattering, and these have in fact been observed. If the scattering curves are obtained for the various geometrical arrangements (i.e. liquid structures) which might be considered possible and compared with those found empirically for a liquid, the particular molecular arrangement in the liquid can be ascertained with some accuracy provided the case is not too complex. Recently the distribution of atoms in liquid mercury has been accurately studied by Hendus by means of monochromatic X-ray irradiation. At a temperature of 18°C he found an atomic arrangement which was almost exactly the same as that of the crystalline state and substantially different from the densest packing of particles expected in liquids."

Further evidence for a quasi-crystalline arrangement in liquids can, according to Sauter, be found in the fact that the specific electric resistance of pure metals increases surprisingly little on the melting point being exceeded. The measurements appear to be consistent with the existence of crystalline groups of 50 to 150 atoms.

"Another no less cogent proof of the semi-crystalline structure of liquids is to be found in the fact that a value of 6 cal./Mol. degrees has been obtained for the atomic heat of monatomic liquids such as mercury and liquid argon, which is twice the value to be expected for freely movable atoms whose three degrees of translational freedom each contribute R/2 to atomic heat. In crystals, on the other hand, the building blocks vibrate around a position of equilibrium so that there is added to the contribution of the kinetic vibration energy of 3R/2, the same amount again for the potential energy of mean equal magnitude generated by harmonic oscillation. The atomic heat of solid bodies is consequently (on all degrees of freedom being excited) 2x3 R/2 = 6 cal./Mol. degrees. The fact that this value is found for monatomic liquids can be explained only by assuming that here again there is a three-dimensional oscillation of the atoms around a position of equilibrium with the difference that, in contrast to solid crystals, these centers of oscillation themselves perform a translational movement which depends on the temperature.

"There are a number of other optical and electrical measurements which afford equally cogent evidence in favor of a crystalline-like arrangement of the molecules of a fluid." In order to make the example more specific, we will give further extracts from Finkelnburg's Einführung in die Atomphysik dealing with the constitution of water. "In general the existence of strong dipolar or quadrupolar factors brings a complication into the normal structures of liquids by reason of the formation of chains of molecules (e.g. in the alcohols) or clusters of molecules in the 'associated' fluids, whose anomalous behavior is due to the formation of these molecular clusters. By far the most important associated liquid is water, whose anomalous behavior has long been attributed to association. Whereas ideas have hitherto been focused on polymeric molecules of the type $(H_2O)n$, whose degree n has been held to be constant and impossible to determine, it has recently been concluded on good grounds that what is actually involved is molecular clusters of indeterminate size. The precise crystalline structure of these molecular clusters was ascertained by comparison of theoretical X-ray scattering curves and those actually observed: they were found to be tridymite-like in structure with each 'O' atom surrounded in tetrahedric configuration by four 'H' atoms. However, as might be expected for bonding reasons, two of the 'H' atoms were bound rather more firmly to an 'O' atom and hence more closely to it than the two others which were bound to it only by hydrogen bridges. This special geometrical arrangement of the H_2O molecules, i.e. the semi-crystalline structure, is the reason for the special position water is generally known to occupy. Similarly the change in the structure, and thus the characteristics, of water, by the introduction of relatively few ions (or the addition of a little alcohol) is now understandable: the tridymite structure of water, which is determined by typical secondary valency forces (van der Waals forces) is seriously disturbed by the electrostatic forces of the ions or the addition of even a few large foreign molecules, and at least the size of the molecular cluster is influenced. Conversely, the addition of a few H_2O molecules to pure alcohol has no noticeable effect on its chain structure, and indeed the properties of alcohols are scarcely affected by the addition of small amounts of water. Other more subtle properties of liquids ascertained empirically can thus be understood by reference to their crystalline structure in terms of atomic theory." From Pohl's Einführung in die Physik, (1959) we quote: "A liquid is a crystal in a state of turbulence with very small but still crystalline elements of turbulence. As 'individuals of a higher order' these elements are in a state of constant change and go through movements and rotations which proceed in common."

What impressive insights this affords us! One must realize what an epochal change this represents in our idea of liquids and particularly of water. The problem that concerns us here is the way in which physicists are striving to obtain a conceptual picture of water. Let us look at the facts: pattern (crystalline-like, quasi-crystalline, semi-crystalline) on the one hand, and on the other turbulence, movement, both in a flux of constant change. It would be wrong to state that what we have here is the basic triadic phenomenon. All we can say is this: the search for a conceptual picture moves along the lines suggested by this basic phenomenon (periodicity, pattern, kinetics). From the very nature of things this aspect must be continually recurring in research and constantly confronting the astonished eyes of the researcher.

How extensively the triadic nature of vibration is found, is brought home to us when we realize that the complex organizations of movement, of rhythmic systems (circulation and respiration), and of nerve physiology become evident to us as frequencies and modulations including amplitude modulations. We spoke at the beginning of striated muscle and its real vibration; cardiology is, of course, "rhythmicity" *par excellence*. Neurology is a field of frequencies and the laws to which they conform (cf. the wave bands of electroencephalography). These systems have patterns of a serial nature and a dynamic of rhythmic impulses. The electrograms are, of course, only the bioelectric expression of processes which are of a chemical, thermal, energetic, kinetic and structural kind. The dominant role of the periodic in other organs and their functions is merely mentioned in passing. (Protein synthesis, the model of genetic information in the living cell, respiratory enzyme chains, catalysis, etc.)

And now into the organization of the locomotive system, of the circulation and the respiration, and of nerve activity— all of which have their being in rhythmicity— living, experiencing man now implants himself. He lives in these fields in that he grasps them and acts

with them,

in them,

on them,

through them,

and only thus takes on a tangible appearance himself. We used the term "implant." In point of fact in, say, the vibrational field of skeletal muscle, man has developed to the point where he can now manifest himself in and through this medium (facial expression, gesture, gait, dance). The same holds true for breathing, the stream of breath, the formation of sound. The physiological periodicity of nature is raised to a higher plane by the development of rhythmic activities on the basis of physiological fields of rhythmicity. An example of this elevation to a higher plane is the way in which organic periodicity develops into speech. Speech is a pure field of rhythmic phenomena; here again we have kinetic dynamics, here again there is configuration (tonoscope).

By drawing attention to these relationships, the natural scientist is in no way straying from his proper domain; indeed it is in this way that the phenomenon of man can be grasped by the senses and by the intellect through the operation of the empirical method. For how could man develop and operate a speech organ unless he himself were a manifestation of the basic triadic phenomenon at a different level.* Series are revealed leading from the vibrating to the dancing muscle fiber, from respiration to speaking or singing, from the frequency modulation of the ganglion cell to the scientist's formation of ideas.

Again and again, and in ever new forms, the cymatic method reveals the basic triadic phenomenon which man can feel and conceive himself to be. If this method can fertilize the relationship between those who create and observe, between artists and scientists, and thus between everyone and the world in which they live, and inspire them to undertake their own cymatic research and creation, it will have fulfilled its purpose.

---

*In actual fact these studies lead by their very nature into fields which are beyond the scope of this work to discuss. Contact with artists, sociologists, psychologists, jurists and historians has shown us that not only the idea of general periodicity, but also the notion of a triadic world model (the trinity of configuration, wave, power) have validity in these fields. Rhythms in history; resonances, interferences, standing and traveling waves in human relations; the wave-like rise and fall of memories, thoughts and emotions in a periodic manner; poetry and music— all these are themes which have been illuminated by this concept of the basic triadic phenomenon during our conversations with numerous personalities. These views must be described elsewhere. But it must be stressed that these affinities are not merely metaphors or analogies, but involve the recognition of homologous systems.

Intermezzo:
A Closer Look at the BIG Picture
Imagining our Evolving Universe as an Elaborate Play
of Structure, Form, and Motion
by Jeff Volk, publisher

The dynamic pulsation of life permeates the entire cosmos, from the tiniest subatomic particle to the inconceivably slow and deliberate unfolding of the galaxies, and everything in between— contouring the rhythm of the cosmic heartbeat with astounding harmonic precision, each according to its unique qualities and composition. Everywhere we look, the eye perceives harmonic structures as patterns or prototypes of forms in nature, in every stage of their development— each stage a symbol of becoming— a letter, syllable, or word; a sentence, paragraph, or chapter, within the ever-expanding libretto, cast into the Book of Life. A constantly evolving dance of structure and motion… And where there's dance, there must be music! A melody so delicate, yet so compelling that not one particle can resist its siren song. Unfolding in perfectly measured steps, proceeding in graceful and distinctive stages— this dance of life plays out impeccably.

Just as surely as pollen carried on the summer breeze, in a few short seasons, will coalesce again into the seed— so too (though at a rather different scale of time and space) will stellar dust (ejected in some great cosmic sneeze) eventually again become a spherical celestial body— a galaxy, a sun, a planet… and through a few more iterations of this same dynamic process, even the very terrestrial bodies that you and I find ourselves inhabiting right now!

This intricate and astounding process unfolds in proportional stages, adhering to the same laws and principles as music, and the essence of language. How could it be any different? Can you imagine all that exists around you and within you— this multiplicity of forms— as innumerable manifestations of that same burning desire to express itself, just so we might enjoy the evidence of our own existence, and in so doing, eventually discover who and what we truly are!

CYMATICS
Volume II

Wave Phenomena
Vibrational Effects
Harmonic Oscillations
With Their Structure,
Kinetics and Dynamics

1

Cymatics

What effects do vibrations produce in a concrete medium? What effects appear in a system and its environment when wave phenomena are inherent in that system? An answer to these questions was sought first of all in the acoustic field, where experiments revealed a characteristic phenomenology of vibrational effects and wave phenomena with typical structural patterns and dynamics (cymatics). Serial phenomena of this kind were presented in our first volume *(Cymatics, Volume I)*. These studies are continued in this second volume.

Here again methods have been employed in which the phenomenon is treated as a whole and not dissected. Why is this? When we observe a phenomenon, it is natural to concentrate on one single factor and make it the focus of our attention. Now, if such a factor is abstracted from its context and allowed to dictate our procedure, the investigation tends to become biased and other characteristics of the object under study are easily missed. This is clearly reflected in the history of science in the way interest has alternated between opposed theories. Of course, certain special characteristics of things do at first appear more conspicuously than others in the field of vision. Immediately, our habitual way of thinking directs us to single out by analysis the various aspects of the phenomenon under observation. Even so, the attempt must be made to return time and again to the original phenomenon and time and again to look at it with fresh eyes.

For instance a system may contain structural patterns which are quite overlooked because the observer is concentrating on dynamic processes, but which show up when, as it were, he takes a fresh look at its features. It may be objected that with this method the end is never reached, that there is always something which may elude observation, and that there is no way of knowing whether one "has seen everything." It is a fact that research opens up one new vista after another and that natural science is a "never-ending business" (Goethe). It follows, then, that the investigator must remain continually alert within the field of observation. By failing to do so, he loses touch with reality and forfeits his chance of attaining the "really real." It is characteristic of this ultimate reality that the most diverse processes should be manifested in one and the same phenomenon: pulsation, circulation, pattern formation, changes of phase, etc. Between all these there is an intimate connection. Were one of these processes to be suppressed or eliminated, the phenomenon would cease to be. What we have then, is something of a unitary nature, something like a necessary connection between such diverse processes. Here we may be said to have a "whole", something that must be conceived and understood as a totality. The individual aspects of the phenomenon cannot be regarded as being self-existent. One must speak of totality and of a total phenomenon. Any attempt, however, to lay hold of this wholeness in conceptual terms raises epistemological difficulties. And, indeed, attempts to define "the whole", "the totality" often lead to abstractions which are incapable of comprehending the whole living phenomenon and, as it were, allowing it to take shape in the cognitive mind. What, then, is to be done? The answer is we must make bold use of our judgment based on concrete vision (Goethe). This brings us to a form of cognition which analyses *and* synthesizes, separates *and* unites, establishes categories, notes their correlations, follows their appearance and disappearance, discovers their unity in diversity, and, at the same time, continuously keeps watch on itself, remains accountable to itself, and constantly asks itself: "Is it you or the object that is expressed here?" (Goethe) The process of comprehending the phenomenon as a whole calls for such a methodology. What this research really involves, then, is a method, a way— and, what is more, a way which must always first constitute itself.

The position, then, of empirical observation is a special one. It asks questions, as it were, of the object. Changes can be rung on these questions. One condition or another may be stressed. An interplay between the various factors comes to light. Phenomenological categories are revealed: the phenomenon acquires characteristics in time and space, assumes quantitative and qualitative features, becomes defined as material, as structure, or as movement, reveals its origins, becomes individualized in its transformations, metamorphoses and self-integrations, in the centering of its polarities, in its disposition, in terms of its numbers and symmetries, and in its intensifications; it appears as entelechy (as uninterrupted activity and effect), as the configuring and the configured, as entity, individuation, as essence, etc. And through such definitions and conceptual operations runs cognitive sentience and sentient cognition.

A very special feature of the study of vibrations is the way in which the observer penetrates to the genetic element. Before our eyes we have the creative and the created, the vibrating and the sounding, and also what is produced by vibration and sound. Now none of this can be simply and harmlessly dissected out for examination. The events of the wave sequence transpire under complex conditions, in interferences, resonances, turbulences, in harmony, consonance, in disharmony, in dissonance, in frequency spectra, amplitude relations, etc. It is in this sphere of multiple creation that the investigator must carry out his observations. He must find out whether amidst all this tumultuous activity there are basic or ultimate phenomena in terms of which "everything else" can be comprehended. It happens often enough that we have the parts in our hands but unfortunately lack the "mental ribbon" (Goethe) with which to bind them together. What is the status of the parts, the details, the single pieces, the fragments? In the vibrational field it can be shown that every part is, in the true sense, implicated in the whole. If we single out a detail, if we follow an individual part, it will be found on careful observation that the sum total of connections, albeit specifically transformed, is reflected in it. Or, vice versa, remove the part, its properties, the fact of its existence, and the whole will cease to be. So the whole is inherent in the part; the part exists in the whole in a special way; the multiplicity which we perceive as parts is at the same time the multiplicity of the whole. The whole is always present in its totality; the part, if we scrutinize it properly, reveals itself to be the whole. These general propositions must be verified in the course of our documentation.

One may object that it is not possible to talk in such simple terms of the "whole" since there is no general agreement that what we have in front of us actually represents the whole. Research may have failed to identify some essential constituent which still remains hidden. It is perfectly true, of course, that we are confronted by the largely unexplored. But what we have observed and comprehended is, under the conditions we recognize, stamped clearly enough with the character of totality. Something with the features of wholeness is to be seen "in the plan and elevation." Research draws closer to totality by successive approximations; that is the way research proceeds. The objection ceases to be valid, then, if the truth of what has previously been posited is substantiated. Namely, that the whole is present in the so-called part, transformed and particularized perhaps, but nevertheless reflecting the totality in all its basic functions. The further we delve into cymatic effects, the more apparent these relationships become to us. Let us take a look at biology in this context. If we remove all the properties displayed by a living cell, if we do away with everything affecting this "part", all organic life is extinguished.*

Now it might be said that all we have to do is simply to observe the parts, that it is quite unnecessary to concern ourselves about the whole; if the parts are already the whole, then they provide us with an appropriate object through which to attain to the whole. This would be true if we had the power of truly comprehensive observation; but this is exactly what must first be acquired. And observation of the real, immediate phenomenon is the best and most reliable way of acquiring it. For one aspect of the phenomenon points to the other, and in this way one gradually comes to grasp how the whole is inherent in the part. If our propositions concerning the part as a whole can be verified, they also imply an epistemological factor since the cognitive organs from which knowledge is derived are a part of the human organism. Now if such a part contains the whole, one may suppose that in this part, namely in the brain as the instrument of cognition, *the whole* can appear in its specific sphere in an ideational form— as thought. In thinking, then, the whole will be capable of manifestation in terms of thought. Thus we see that man is able to excogitate a phenomenology of the galaxies, quasars, pulsars, or of atomic nuclei, or of molecular biology, etc. How could we speak of such things at all if our cognitive faculties were incapable of approaching and comprehending the whole through the processes of thought?

To enable the reader to enter into the spirit of cymatic phenomenology, we shall first of all present some additional sets of experiments in this second volume, beginning with a simple exposition of certain phenomena. These will afford some idea of the great multiplicity of vibrational phenomena. At the same time the reader will be gradually introduced to the complex nature of periodic sequences. The phenomena occurring in capillary interspaces under the action of vibration convey some notion of the richness and diversity with which wave phenomena are manifested. Again, remarkable formations are obtained if ferromagnetic substances are exposed to cymatic influences in a magnetic field. The forces of adhesion and cohesion are reduced by vibration and the masses can flow plastically, giving rise to curious configurations.

The additional documentation presents symmetrical phenomena exhibiting features of number and regularity, which can also be demonstrated in space. To

---

*This aphoristic remark on the living cell is intended to imply that in the individual living cell *all* basic functions are present (respiration, metabolism, mitosis, plasticity, metamorphic potential, regulative processes, controls, sentient capabilities, contact potentialities, outside/inside relationships, correlations, etc.) Remove these processes, and life is no more. So the "whole of life" is present in the individual cell. If everything contained in the cell is known, then the whole of life is known.

throw light on these characteristic numbers and symmetries we investigate the vibrations of the electron beam and the mechanical pendulum. Observations are also made of harmonic phenomena in fluids. Here mathematical arrangement and symmetry are observable throughout the phenomenon, its structure, its dynamics and pulsation. An attempt will be made to explain these harmonic processes, which occur in all categories.

With this cymatic phenomenology for reference, we can turn our attention to other fields of natural science on the principle that, if vibrations are among their essential constituents, they will also exhibit cymatic effects. This will be illustrated by reference to certain examples. Above all, however, we shall be dealing with the problems raised by symmetry and mathematical arrangements in organisms. Out of these questions a real research program has taken shape.

The relations between analogy and homology must be discussed. Since really perfect harmonic relationships appear to the observer in terms of sound and intervals, anyone interested in history will wish to look at earlier philosophies which were largely modeled on number order, symmetry, and harmony in the cosmos. A few connecting points with such world pictures are intimated.

The book closes with mention of some additional aspects of cymatics and a glance ahead at the continuation of the experiments. It must be realized that the observations represented here are only, as it were, stages on our journey of exploration into cymatics. The action of acoustic vibrations has been traced— specifically the effects of tone, musical sound, and speech sound. These observations must be extended by investigating the continuous series of phenomena associated with tones, speech sounds, and musical sounds. We have, of course, visualized, photographed and filmed sequences of tones and speech sounds, and recorded the vibrational effects produced by music. But the real work on what might be called *melos*, or speech, is still to be done. This brings the larynx and its action into the scope of our studies. And at the same time we are confronted with the origination of vibrational effects, the generative element; we must learn about the larynx as a creative organ which displays a kind of omnipotent nature in its field.

In this second volume we have once more deliberately dispensed with descriptions of experimental design and particularly of the quantitative analysis of parameters. We have been primarily concerned with bringing the phenomenon into the field of observation. It is of secondary importance at this juncture how we excite the diaphragms, plates, films, etc., whether we use schlieren or photo-elastic methods, and whether we use mechanical or piezoelectric means of producing vibration. The phenomena of vibrational effects and waves can be visualized in a large variety of ways. Indeed, most of the phenomena recorded can be produced directly by the human voice in the tonoscope without recourse to electroacoustics. What is of prime importance is that the peculiar spectrum of cymatics— made up of patterns and figures on the one hand and dynamics and kinetics on the other— should be witnessed and recognized in its uniformity and totality, that the eye should be alerted to interpret the true nature of periodicities and rhythmicities, and that the features of cymatics should be discerned in the various fields of knowledge in the form and manner peculiar to any specific one of them. First and foremost, then, it is a question of developing a special sense for perceiving and observing rhythmic and periodic systems. Again, it is our concern in this second volume to train this ability and it is hoped that this documented excursion into the field of cymatics will serve this end.

2

The Wave Lattice as a Configuring Field

The wave lattice generated in a liquid by the action of sound (Fig.1) imposes a spatial pattern on a diffusion process occurring there. Into the vibrating liquid we drip some of the same liquid which has been colored with a marker dye, expecting that we shall see it mix intimately with the outspread film. However, instead of diffusing uniformly, the colored liquid first shoots, as it were, in jets through the meshes of the lattice. If we greatly intensify the process, say, by turning up the volume of the tone (increase of amplitude), we see how the colored liquid jets forth but always in a particular direction. If we examine one of these jets more closely (Fig. 2), we can see that the liquid moves through the lattice in a complicated manner. While the main jet winds along its course, there are suggestions of lateral branches with a curious rotary formation. The liquid appears to be streaming through the pattern in vortices. Gradually a homogeneous distribution is achieved by diffusion. This apparently unremarkable process is singled out for particular mention because it shows how, in this simple experiment, observation is guided forward step by step by the phenomenon itself. We might note that structural patterns generated in a particular medium by vibration do in actual fact exercise a spatially directive function and, having described the facts of the matter, leave it at that. But, if we leave it simply at that, there is basically a great deal which remains unexplained. First we speak in general terms of a wave lattice, and then of a jet-like vortical movement, and so on. Now, we have taken just such an elementary experiment as this to show that, by looking further, we can make some entirely new discoveries; for when we reach the chapter on harmonic vibrations, we shall refer back to this experiment (Figs. 1 and 2) and be able to see and appreciate everything that happens in the liquids there, indeed everything that is transpiring in the "simple" wave lattice before we add the drops, the structural and circulatory conditions in which the drops are involved on being added, etc. Let us take this experiment as an example of how observation is tutored by the phenomenon itself; how in this instance our eye is caught by the curious lateral formations of the diffusing jet and we are induced to investigate the matter further by, say, the schlieren method to ascertain exactly what is happening. As we have said, we shall return to this experiment when we have obtained a wider view of the subject as a whole.

Figs. 1-2
The wave lattice generated by vibration guides the liquid diffusing there (dark areas) in a particular direction. Within its "meshes" it functions as a true structure. (Actual size.)

Fig. 2
The detail shows, in addition, curious "lateral branches" which can be investigated by schlieren methods. (See Chapter 15, Harmonic Vibrations in a Concrete Medium.)

Fig. 1

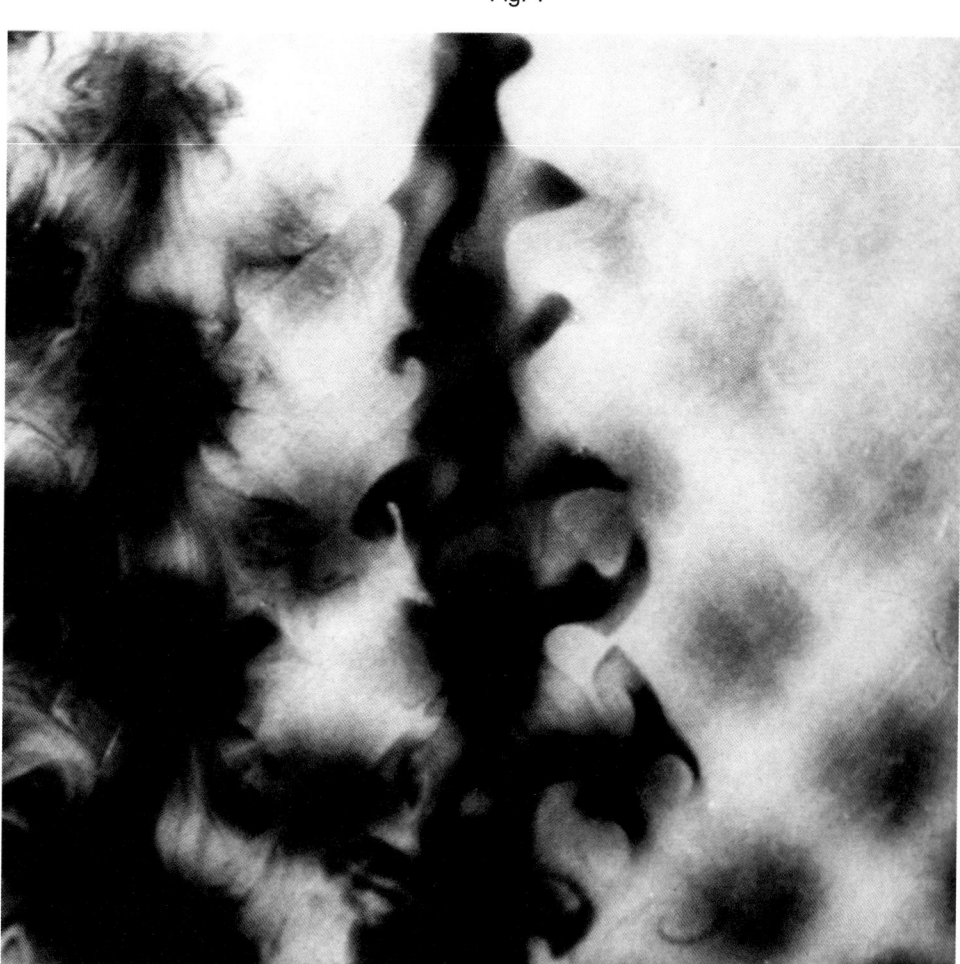

Fig. 2

3

Circulation in the Wave Train

If waves are generated in a viscous paste by vibrations, a strange phenomenon appears under suitable experimental conditions. At first the mass is seen to be rising and falling in waves. Then curious furrows appear which are particularly prominent in the wavecrest phase. The paste seems to be disappearing down these fissures. But where is it going? If a spot, say, on the side of the crest, is marked with a stain, it can be seen rising towards the furrow, and then actually disappearing down it, only to roll round and reappear again immediately at the side. By using the stroboscope (an instrument enabling rapid movements to be seen in slow motion), the process can be followed in detail. In Figs. 3 and 4 the paste is arranged in waves. The furrows we have described are visible on the crests. The waves must be imagined to be rising and falling, while *at the same time* the entire mass of paste is flowing symmetrically and bilaterally into the furrows, and then rolling up again only to disappear down the furrows once more. In other words circulation and undulation are combined. What is so striking about the spectacle, however, is the elegant way in which the whole process conforms to a uniform pattern. Each of the two processes, well defined in itself, follows an orderly and regular course without giving rise to turbulence. If one can imagine oneself, so to speak, transposed into the vibrating mass, one would experience the rolling and the up-and-down movement of the waves as *one* single self-contained process. At the same time there would be a symmetrical movement due to bilateral circulation. It goes without saying that the components can be separated, and the wave process and the circulation visualized apart by a suitable arrangement of the mass. And, indeed, this is necessary if the conditions are to be elucidated. What appeals to the imagination so strongly, however, is the simultaneous and orderly interplay of the two processes of undulation and circulation. At the same time it must be remembered that the phenomenon is generated by a *single* tone. This example brings the complex nature of vibrating media sharply into focus. It also encourages the observer to probe further because, besides undulating and revolving bilaterally, the mass is also pulsating. What our minds conceive as separate aspects is performed by Nature as a uniform process; the cymatic effects of pulsation, undulation and bilateral circulation in the viscous mass merge into a whole.

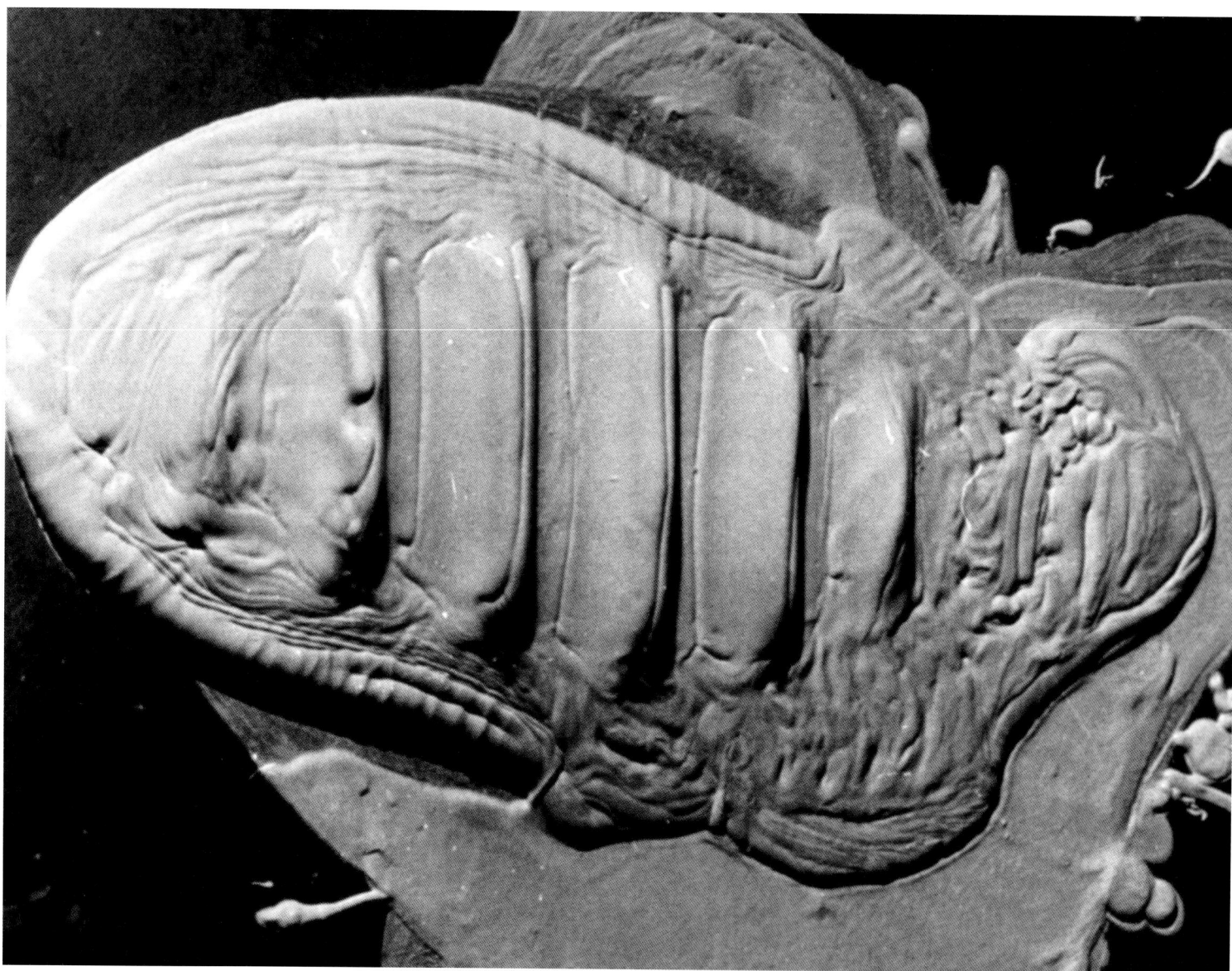

Fig. 3

Figs. 3-4
Waves generated by vibration in a viscous paste: on the crests of the waves there are furrows into which the substance pours in such a way as to start a rolling motion. This circulation is bilaterally symmetrical and can be made visible by using a marker dye. The processes of undulation and circulation are not merely simultaneous but also unified. The impression created is one of unity. (Approx. actual size.)

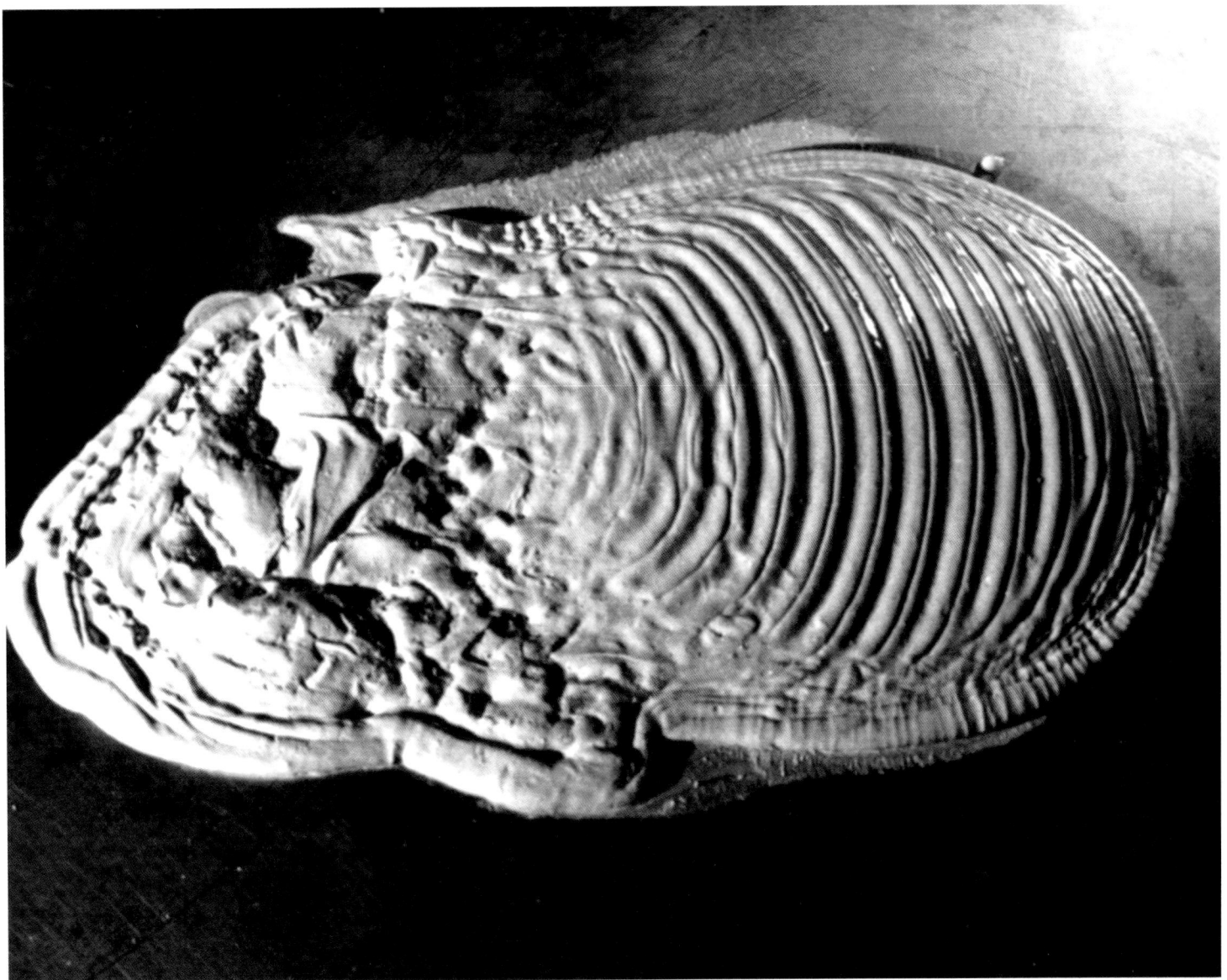

Fig. 4

4

Changes of Phase in Matter at the Same Frequency and the Same Amplitude

The next of the experiments introducing us to the richness of the world of cymatics shows how the properties of a medium change under the influence of vibration. For this purpose we use a mixture of salt and water which we excite by vibration. A typical and commonly seen effect is that the mass forms into a ball. In Fig. 5 the mass is shaped like the cap of a sphere. Around it can be seen the wave pattern impressed by vibration. The mixture is pushed together by the conglobing forces and, at the same time, the water is expressed from the brine. This brings about a change in the consistency of the medium. Fissures appear and the material breaks up (Figs. 6, 7, 8). The fragments are ejected and scattered. The scene is now one of total ruin (Fig. 9). The ejected fragments remain in the field of vibration, recover their flowability in the expressed water (Fig. 10), congregate again under the vibration to form the cap-like shape (Fig. 11), and once more build up into the regular round hill (Fig. 6) only to repeat the process of change, disintegration and, again, restoration. The cycle goes on repeating itself regularly. Thus we have the curious spectacle of the material undergoing wholesale changes while the exciting tone remains the same and does not even alter its volume. What we see is a cycle of contrasting phases at the same frequency and amplitude. This experiment shows, then, that one and the same uninterrupted vibratory impulse is capable of sustaining cyclical changes of phase.

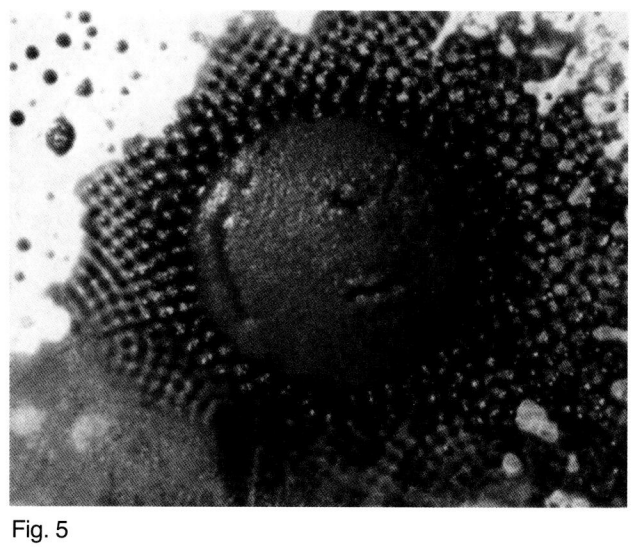
Fig. 5

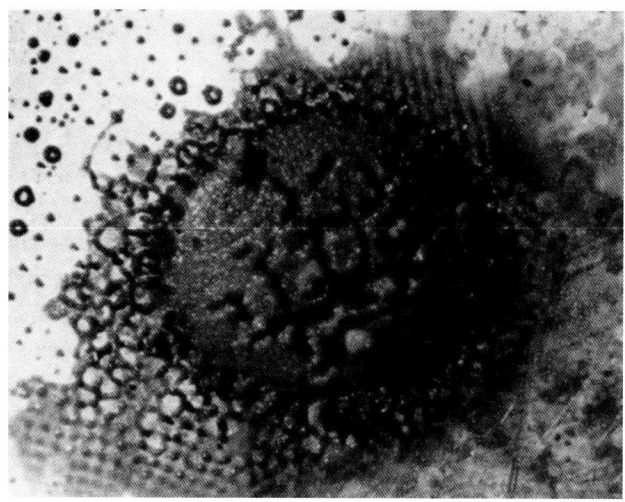
Fig. 6

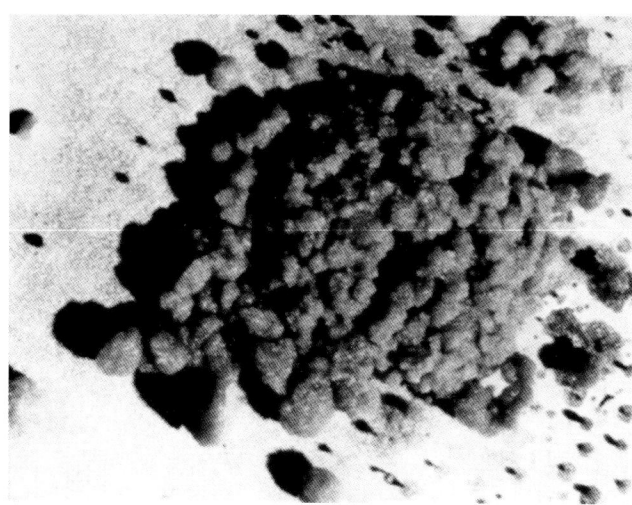
Fig. 8

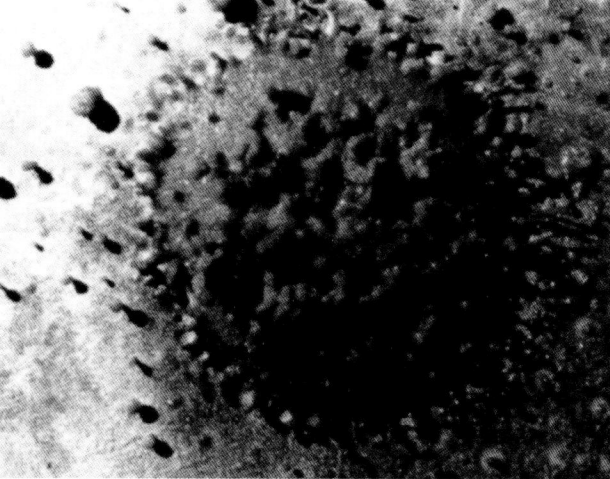
Fig. 7

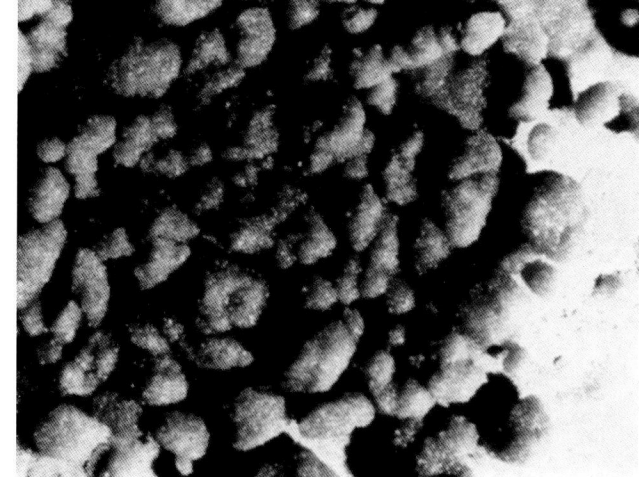
Fig. 9

Figs. 5-11
Vibration causes a mixture of salt and water to clump together into a circular shape (Fig. 5). More and more water is squeezed out by the clumping process and the brine changes in consistency. The mass begins to break up into bits (Figs. 6, 7) which are ejected (Figs. 8, 9) but then recover their flow properties in the vibrating water. (Fig. 10) The circular pile forms again (Fig. 11) and the pattern of the initial stage (Fig. 5) returns. What we have therefore is a cyclical process which is sustained while the frequency and amplitude remain unchanged. Although Figs. 6 and 11 are similar in appearance they actually show opposed phases of the process; in Fig. 6 the process leads to disintegration, in Fig.11 to the formation of a uniform ball. (Fig. 10 detail. Fig. 11 actual size.)

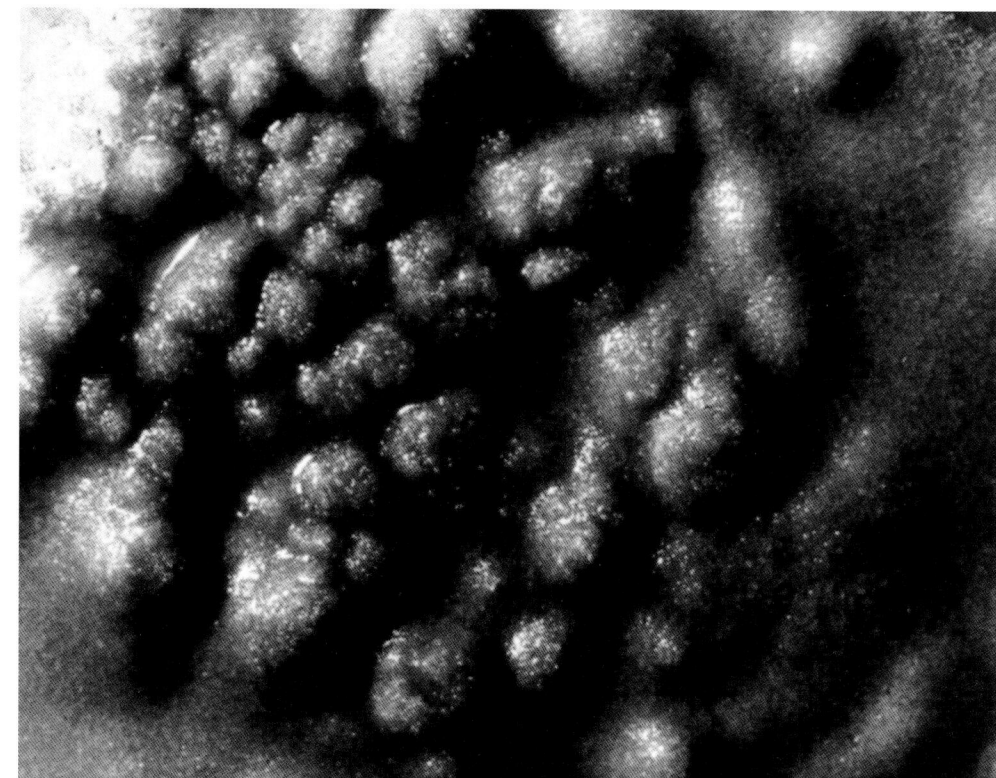

Fig. 10

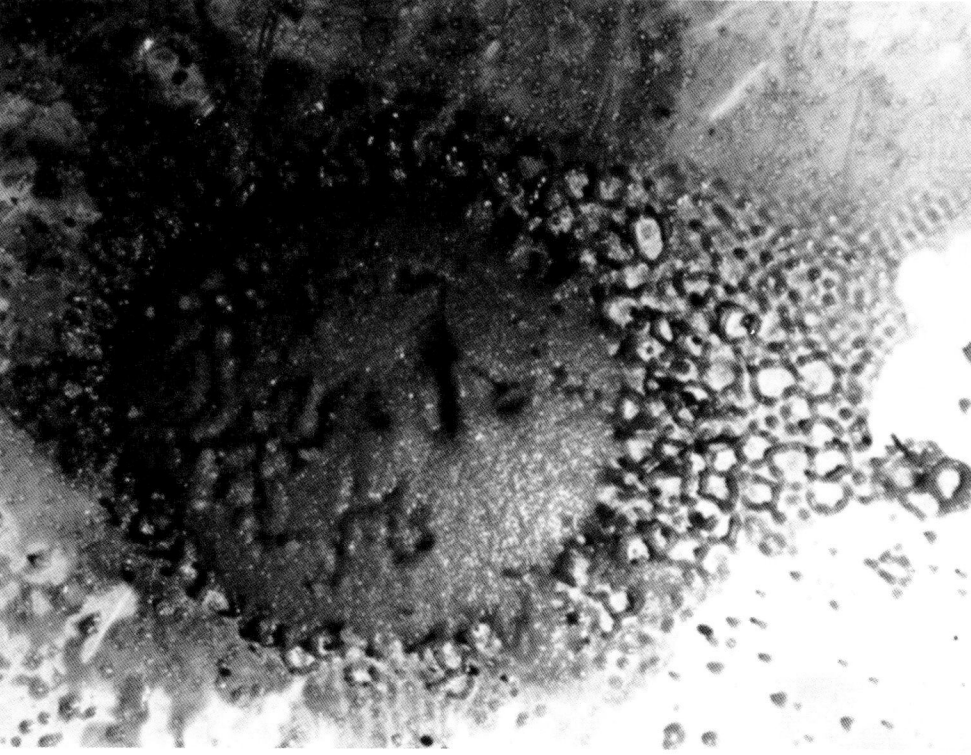

Fig. 11

5

The Influence of Vibration on Flowable and Solid Substances

In the following experiments flowable and solid substances are subjected to vibration. These experiments are comparatively simple, yet the phenomena are of extraordinary complexity. We shall try to describe a number of these processes with the aid of photographs. It is, of course, hardly possible to evoke the impression created by the original process. Like all the other phenomena illustrated in this book, these have also been filmed:* for the motion picture is a record of the *process* whereas still photography perpetuates the *structural pattern* as additional documentation.

First of all the substance is spread uniformly on the diaphragm. As soon as the tone starts, the whole mass comes to life. A relief pattern of hill-like waves appears (Fig. 12). But at the same time the substance is set in motion and conforms to the topographical features of the vibrational field and, all the time this is happening, the movements of both the substance and of the diaphragm can be followed in detail with the stroboscope. The layer of substance changes, growing thinner at certain points and piling up at others. At the same time these various regions begin to creep around. Fig. 13 shows the substance flowing along in this way, shaping itself into a cumulus-like configuration as it goes. In Figs. 14 and 15 marginal areas come into view. This whole "coast line" with its undulations, bays and promontories, etc., is caused by vibration. It frequently happens that the layers thin out as the process continues and then immediately clump together again. The processes have a pronounced tendency to repeat themselves. The way the individual

---

*Editor's note: Dr. Jenny did make several films of his experiments, highlights of which are now available on video. See promo pages at the end of this edition.

"floes" move around is determined by vibration. If at the same time the vibrating diaphragm is tilted, the masses slide along very easily because vibration reduces adhesion to the substrate. And then, apart from the effects on adhesion, the power of the substance to cling together is also diminished by vibration, and this reduced cohesion causes, as it were, the material to become more fluid. These two effects of vibration—reduced adhesion and reduced cohesion—play a crucial role throughout. It should also be noted that the vibrational field is so strongly impressed on the mass that in certain ones the material can be clumped and held against the inclined plane of the diaphragm. At such points it may even move contrary to the force of gravity. (This antigravitational effect of vibration is described in *Cymatics, Volume I* in greater detail). These movements, then, constitute a complex process: with the diaphragm flat they are caused solely by the vibration in conjunction with reduced adhesion and cohesion; with a tilted diaphragm there is interplay between gravitational forces

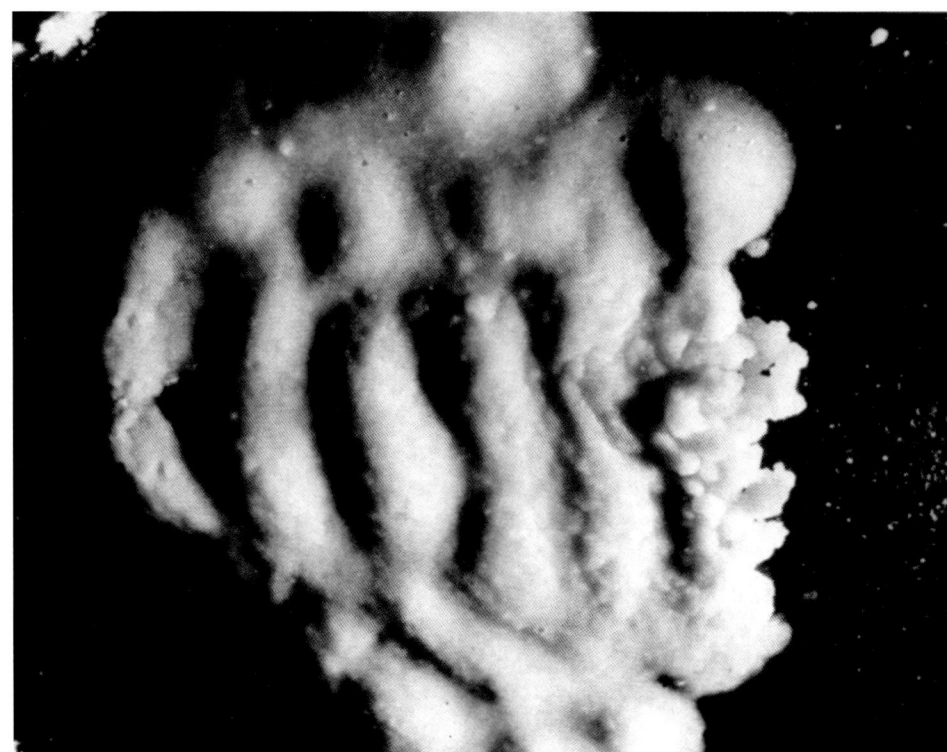
Fig. 12

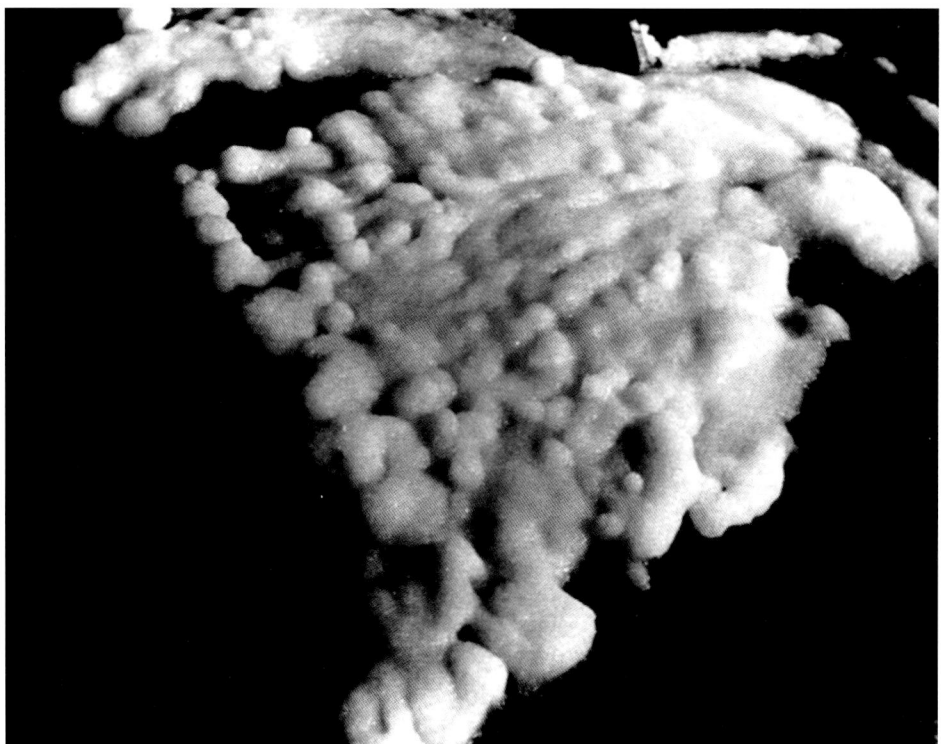
Fig. 13

Figs. 12-15
A flowable mass assumes a characteristic pattern under the influence of a high-frequency tone. Trains of waves take shape (Fig. 12). The substance forms into a lump which, because adhesion is reduced, glides around in one piece (Fig. 13). The substance (a mixture of salt and water) forms bays and promontories, spreads out thinly and is then thrown into folds which make a rugged relief. (Figs. 12, 13, 14 actual size. Fig. 15 enlarged detail.)

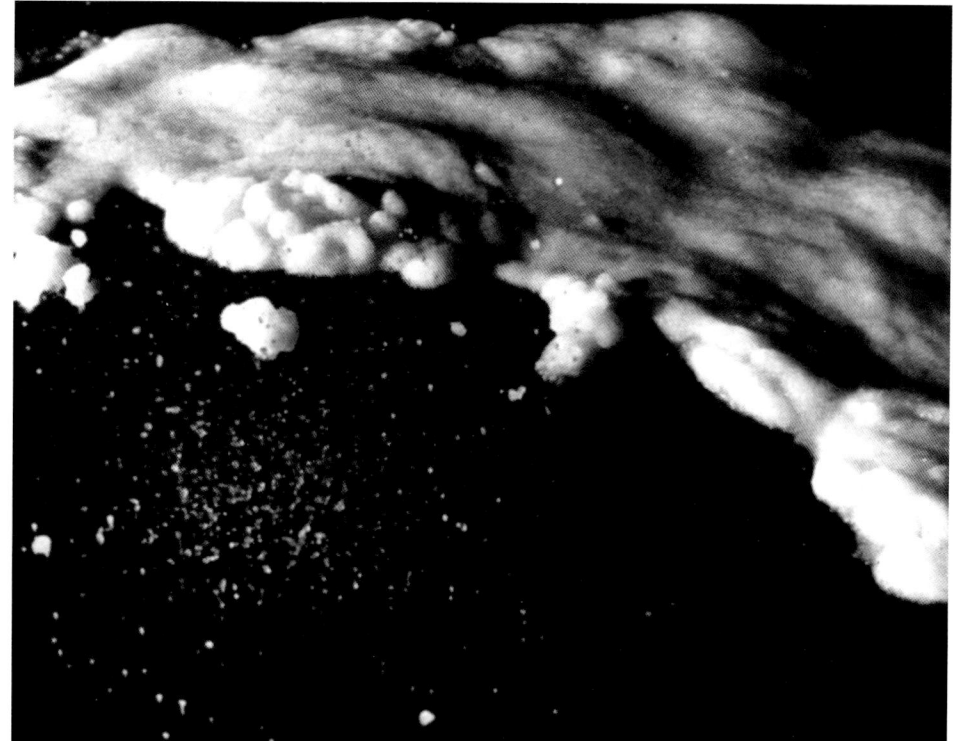

Fig. 14

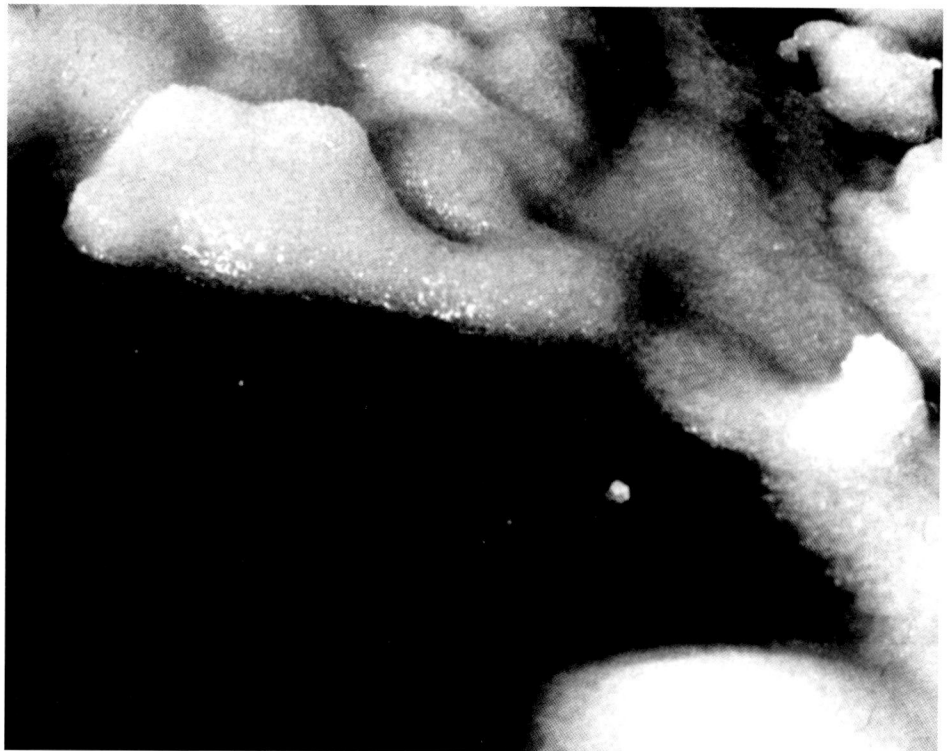

Fig. 15

and the effects of vibration. Thus the masses slide around, approach each other, and then withdraw again. However, there is no constancy in the processes involved. Indeed, phases of immobility alternate with phases of violent movement. A quiescent part is suddenly convulsed with activity. There are moments when, for a short time, "a tremendous amount happens." It is just the same with the motion of the material along the various paths. One of these complexes may be floating along quietly one moment and the next it is rushing madly forward. Directions also change. A simple forward movement (translational motion) may suddenly be followed by a swerve to the side. Again, in Figs. 12-15 small-scale versions of these configurations may be seen. Vibration is impressed on every part of the substance. Here again inactivity, acceleration, movement, counter-movement, direction, thinning and clumping, etc., alternate.

Another essential factor in the phenomenology is the nature of the material excited by vibration. Whereas a salt mixture was chosen for Figs.12-15, a viscous paste was used in the experiments illustrated in Figs. 16, 17 and 18, and the configuration obtained is entirely different. Here structural forms rise like abstract sculptures out of the paste, the pillar-shaped ones being in constant circulation. The processes we saw before are apparent here again. The masses glide around, they may flow together and then continue

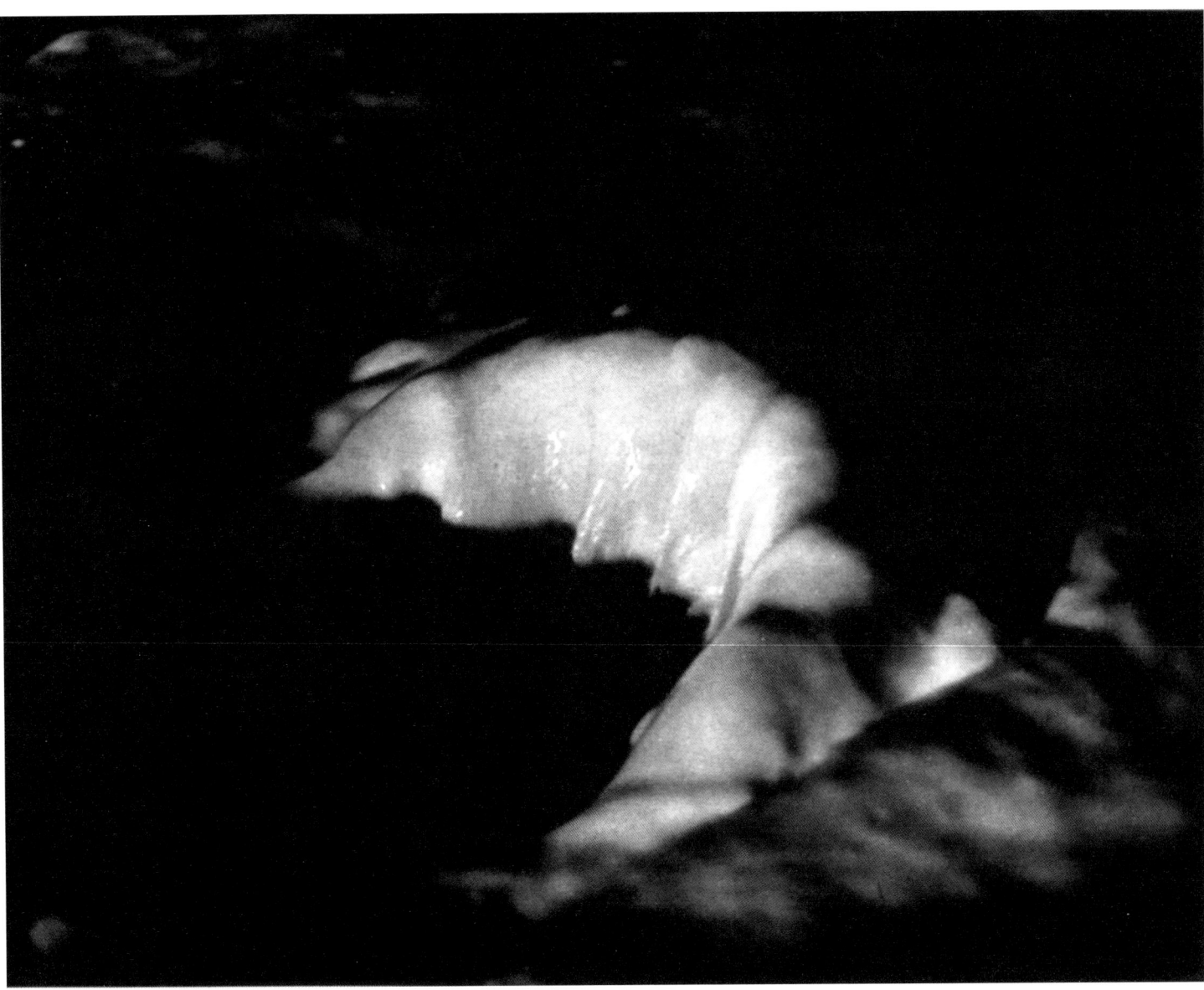

Fig. 16

Figs. 16-18
If a viscous paste is used instead of a salt mixture, it is also forced into characteristic shapes by the vibrations. The mass may be raised up like a plateau with walls where recesses and projecting ridges are carved out. At the same time the mass is circulating. These are not steady processes: there may be abrupt changes from stagnation to violent activity.
(Fig. 16 actual size. Figs. 17, 18 details.)

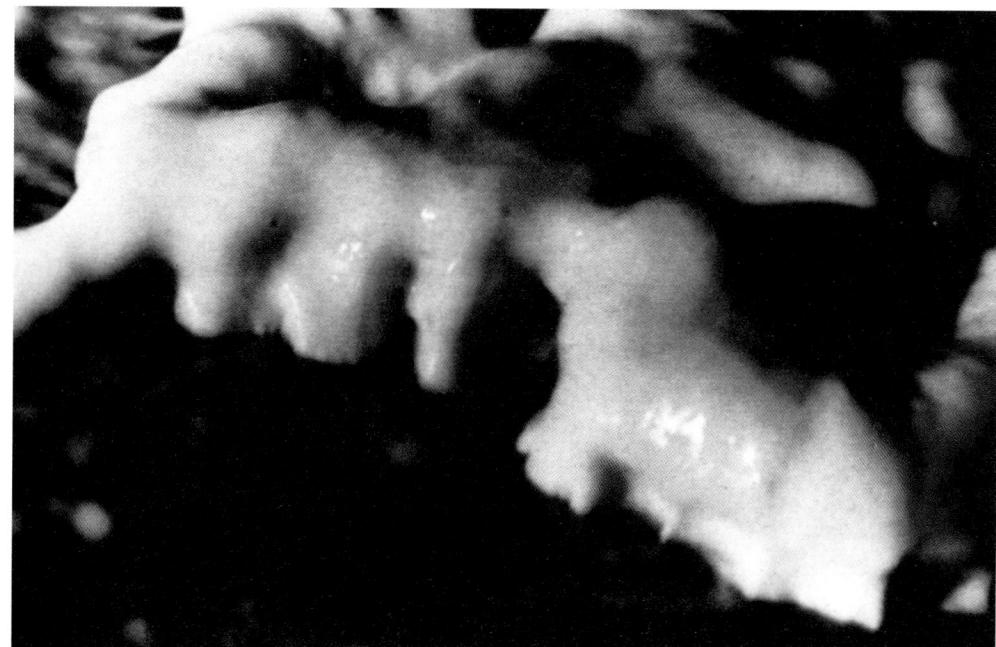

Fig. 17

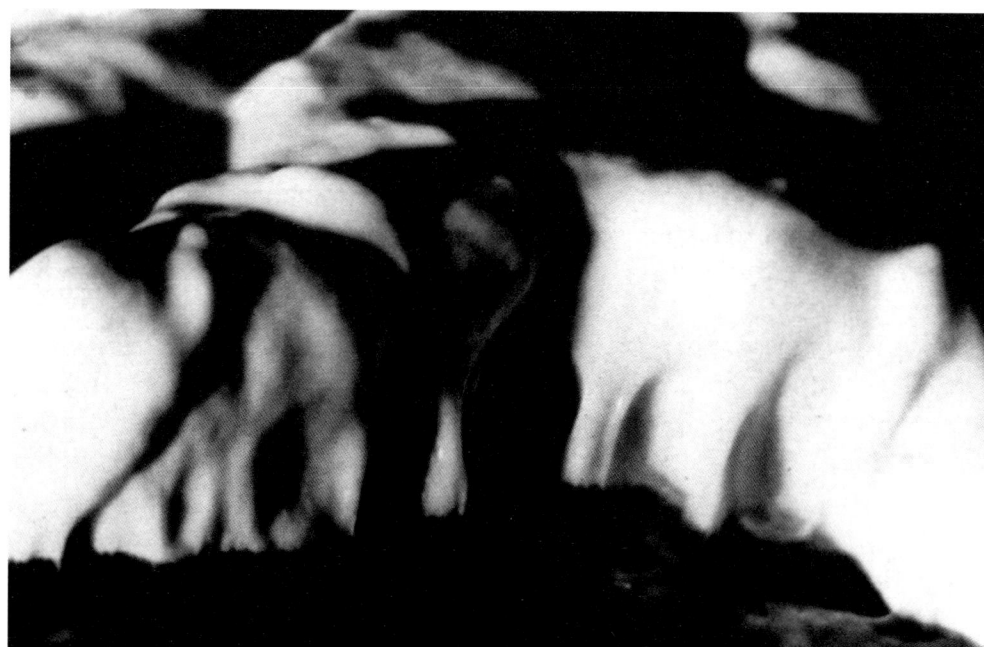

Fig. 18

as a single piece; then they spread out again and divide up. Each part once more reproduces the whole phenomenon in miniature. It displays uniform formation, circulates, and creeps around as a correlated whole.

If a more plastic substance is used, there is not only circulation but also overlapping and furrowing, because the tendency to confluence has been reduced (Figs. 19 and 20). The features of the process are more durably imprinted on the material, and over and over regions appear where circulation takes place with bilateral symmetry.

If the experiment is so arranged that a substance comes under vibration while it is solidifying, we may observe characteristic relief effects (Figs. 21 and 22). A complex pattern of folds appears and some of these move along in the form of parallel waves (Fig. 22). It is almost as if a landscape were shaped in relief and one could walk there in the mind's eye. On the left in Fig. 21 the front has swung sharply in a curve, marking a turn which took place when the substance was still in a more liquid state. The various forms appearing as the characteristic morphology of these processes, always reflect the specific properties of the substances. The relief has one form when the substance is more liquid, another when it becomes viscous, yet another when it is plastic, and another in the solidification phases. Also, as we have mentioned, the quality of the substance operates as a specific factor.

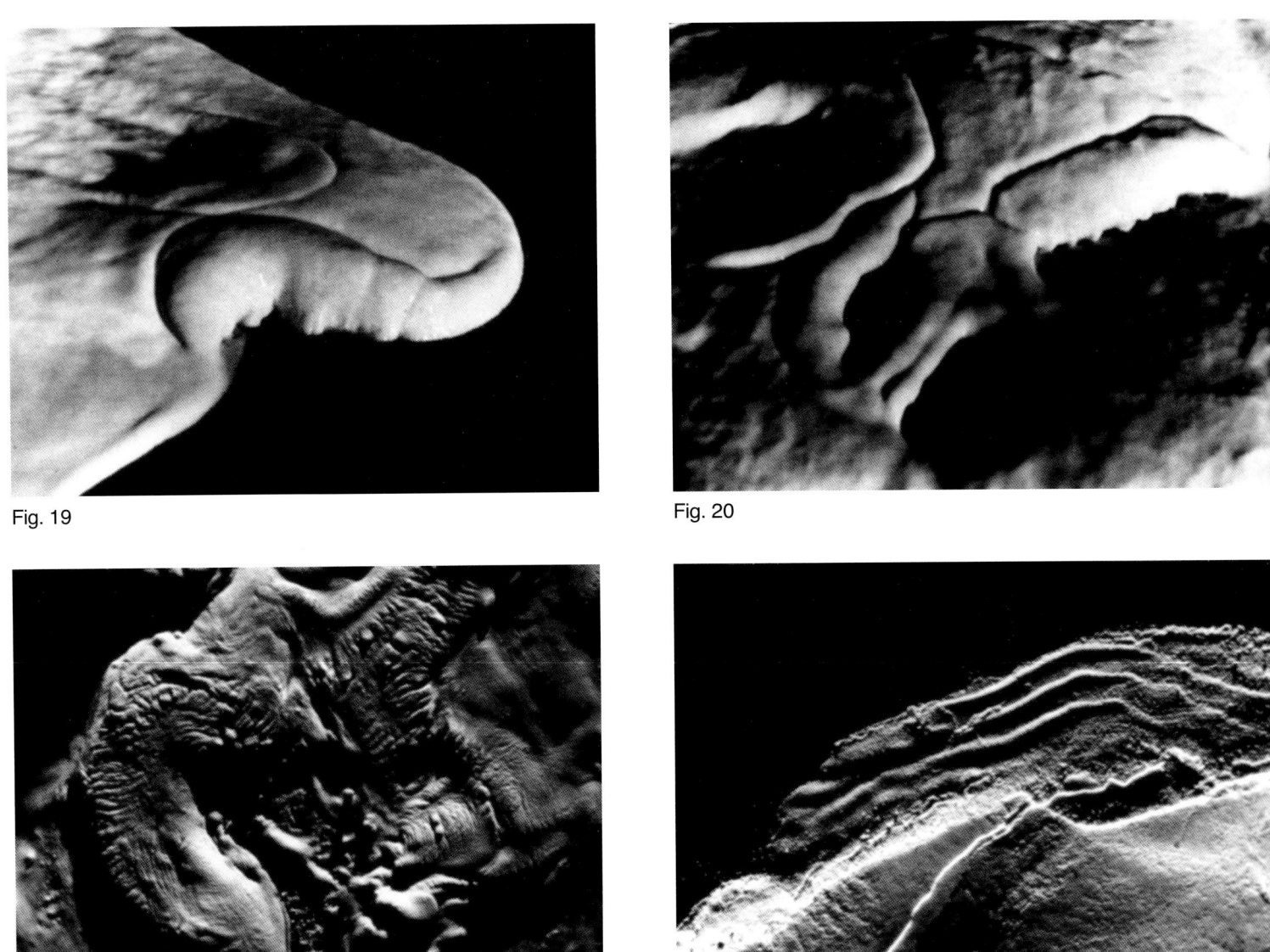

Fig. 19

Fig. 20

Fig. 21

Fig. 22

Figs. 19-20
The nature of the material is also an important factor in determining these vibrational effects. Here a plastic substance has been used. The creeping lobes may collide, one being thrust on top of the other, and produce furrowed valleys. Flowability is diminished and yet even here the substance continues to circulate. (Actual size.)

Figs. 21-22
These masses (plaster) have solidified under vibration. The relief is complicated in structure because each of the various stages of consistency the mass has passed through while solidifying, has left its imprint. There are large billow-like formations and tiny wrinkles, wave trains succeeding one another, and sudden changes in the direction of flow. It is as if the "history" of the process had been recorded in transverse and longitudinal folds. There is also a tendency for a latticework of folds to take shape. (Actual size.)

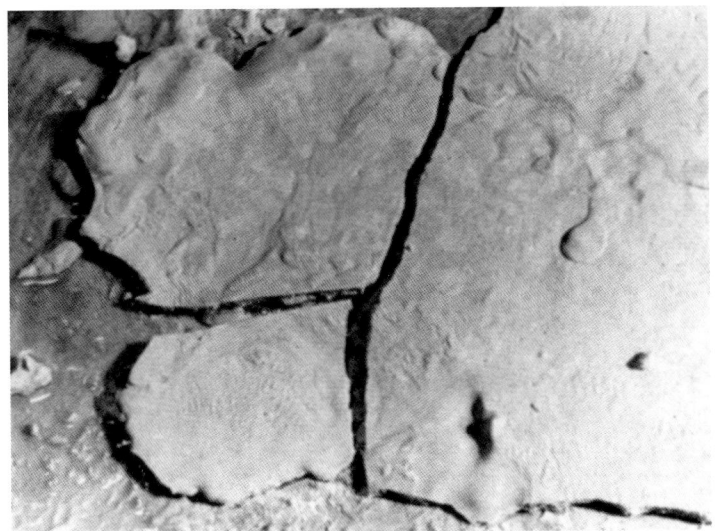
Fig. 23

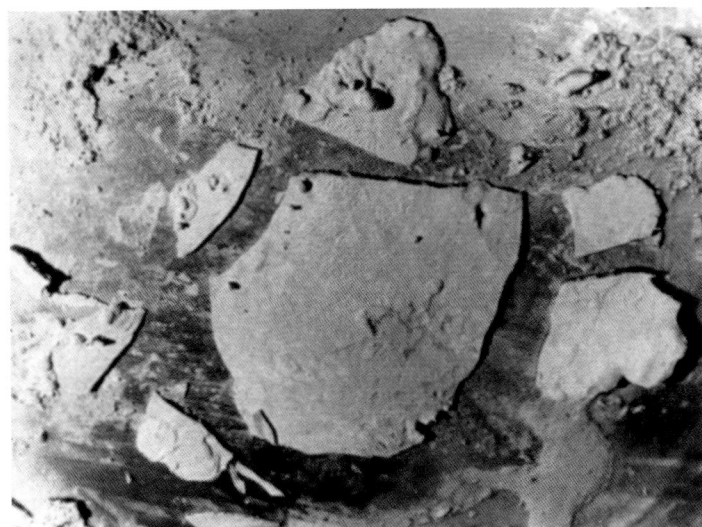
Fig. 24

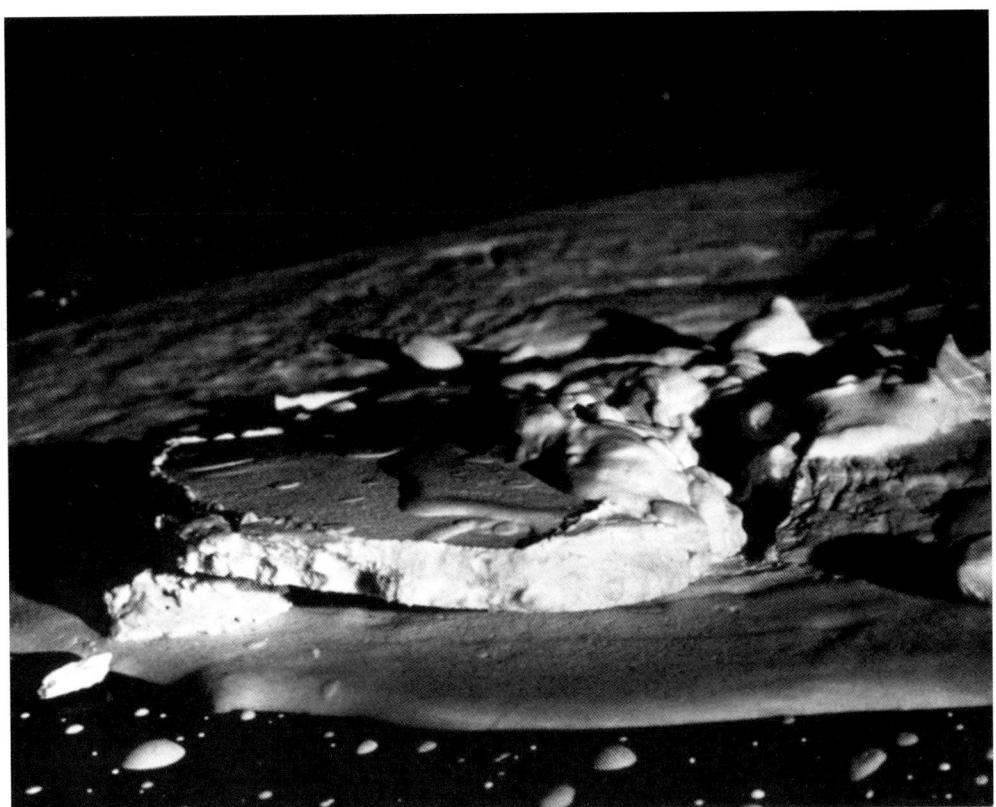
Fig. 25

Figs. 23-25
Subjecting solidified masses to vibration may cause them to break up. The fragments seen in Fig. 23 have been produced by vibration. These slabs or "floes" are also in motion. In Fig. 24 they have moved apart, but it is easy to see how they fitted together before. The slabs may also overrun one another. In Fig. 25 one slab has crept on top of another. The force driving the fragments apart and also causing them to be over-thrust is vibration and nothing else. These polar processes reflect the topography of the vibrational field and depend on the frequency and amplitude. (Approx. actual size.)

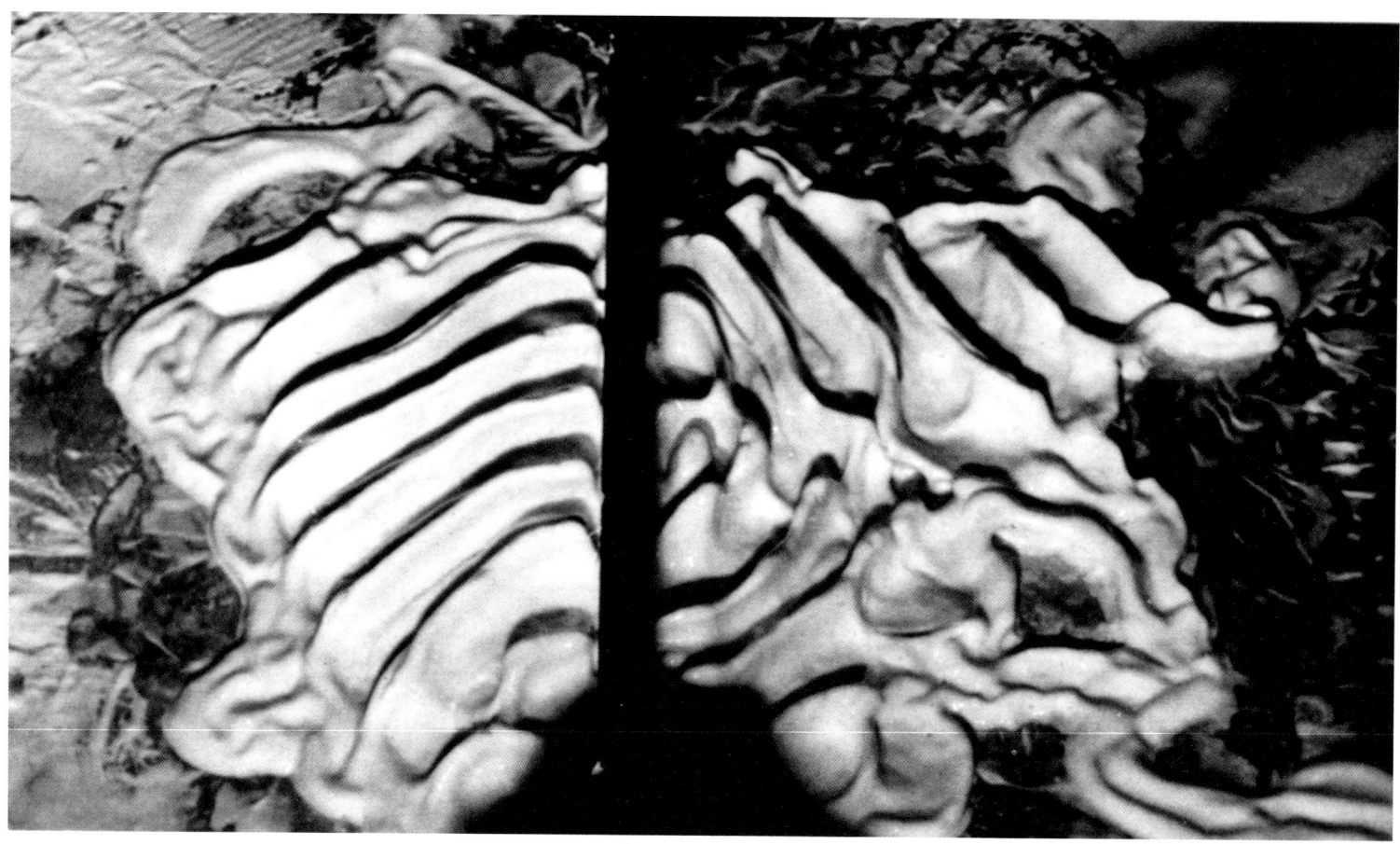

(wooden rod ↑)

Fig. 26
If the vibrating masses meet an obstacle in their path, series of overlapping waves appear and there is a tendency for the substance to start circulating. In an experiment of this kind there is no disintegration and the vibrating mass accommodates itself to the new situation by undulation. In Fig. 26 a wooden rod has been placed in the path of the undulating mass; on either side of the obstacle the waves fall into a regular pattern. However, if the substance is more liquid than that used here, characteristic turbulences are seen under these conditions. (Actual size.)

Once solidification begins, the situation changes radically. The masses break up under the action of vibration (Figs. 23, 24, 25), but the fragments still continue in motion. In Fig. 23 they are still close to each other, but in Fig. 24 they have somewhat "drifted apart." It is still apparent, however, that they originally fit together. Yet here again the pieces slide around in different directions. In Fig. 24 there is separation, in Fig. 25 approximation, which, when the lumps are solid, causes one "floe" to be thrust over the other. On the left in Fig. 25 the end of the slab has been raised. This process of overlapping and overthrusting is, of course, assisted by the reduced adhesion.

If an obstacle is placed in the way of an advancing front of viscous material, the latter is thrown into folds having a certain regularity, and the folds even overlap (Fig. 26).

Figs. 27 and 28 give the reader some idea of the enormous variety of cymatic effects. Here wave trains moving in different directions have woven a patterned tissue. In Fig. 28 an underlayer has first solidified and over it a pattern of folds has been subsequently formed and also solidified.

We can now, so to speak, draw up a list of the phenomena observed when flowable and solid substances are excited by vibration. Our spectrum of phenomena is a rich one. Whether the material is formed into clumps or flattened into layers, whether it is piled up or thinned out, whether it is thrown into violent commotion or remains inactive, whether it flows together or flows apart, whether its activities are repetitive or whether it assumes a steady shape— these phenomena all demand a flexible imagination in the observer. It is no use thinking, say, in terms of polarities; one must seek to entertain two opposites at the same time, for polarities proceed simultaneously; masses liquefy, separate, clump together and spread out, grow inactive and erupt, etc. The only efficient causes we can identify are the effects of standing and moving waves, of resonances and interferences, turbulences and circulations. What confronts us, then, is an unmistakable cymatic phenomenology. It is crucially important that, through our experience of these things, we should come to view them imaginatively and thus be able to formulate concepts concerning them.

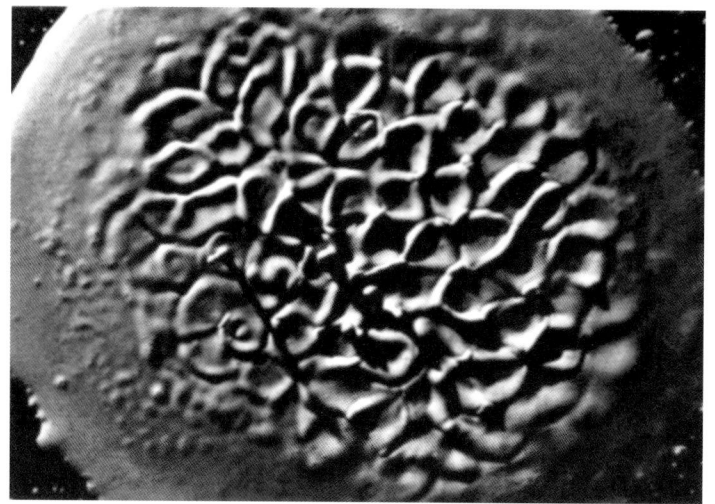
Fig. 27

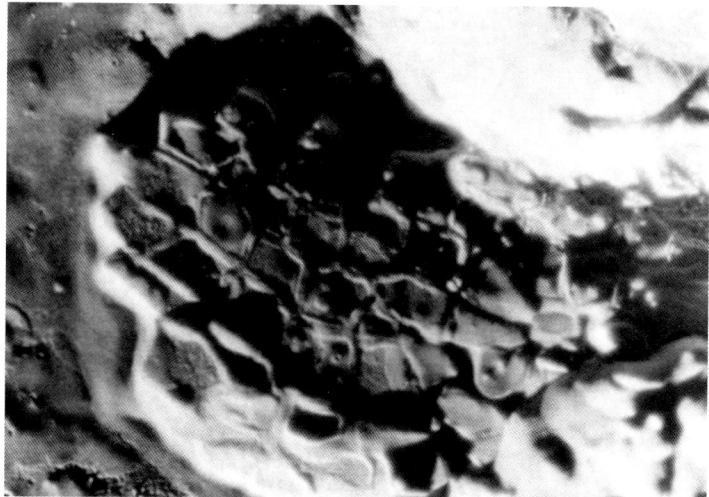
Fig. 28

Fig. 28a

Figs. 27-28a
These figures again exemplify the richness of the structured arrays occurring when a mass solidifies under vibration. The way the patterns are laid down depends on how great a change there is in their physical state. The festoon-like network in Fig. 28 and the sharply profiled ridges and valleys of the pattern in Fig. 27 are remarkable. Fig. 28a shows a curious pattern of areas bounded by lines and themselves sub-divided into similar patterns. This highly varied relief is again the result of vibration. (Actual size.)

6

The Oscillating Water Jet

The outflow pipe is excited by sound and this causes the emergent water jet to oscillate. As it transmits the periodic impulses it receives into space, the interplay of a number of factors can be seen. Surface tension causes the jet to retain a certain homogeneity. Where it is disrupted, the surface tension is evident in droplet formation. The main feature, however, is the enormous turbulence apparent in the periodic movement due to the high instability of the jet. This periodicity also appears wherever the jet is disrupted (Fig. 29, right). The instability of this interplay of forces can be seen very clearly with the stroboscope. For a brief moment individual drops have an incipient regularity of formation, but in an instant become turbulent again.

Curious processes involving attenuations and expansions can be seen. A spiral current appears in the jet time and again. In spite of the complex instabilities the pattern of oscillation still remains prominent. It involves a number of tendencies combined together: there is a tendency for the water to turn into drops, to form a homogeneous jet, and to flow in wave patterns, and at the same time a tendency to produce a regular vibrational pattern. None of these tendencies is ever fully realized; as soon as any single one becomes more precisely defined, it is caught up in the turbulence caused by the prevailing instability.

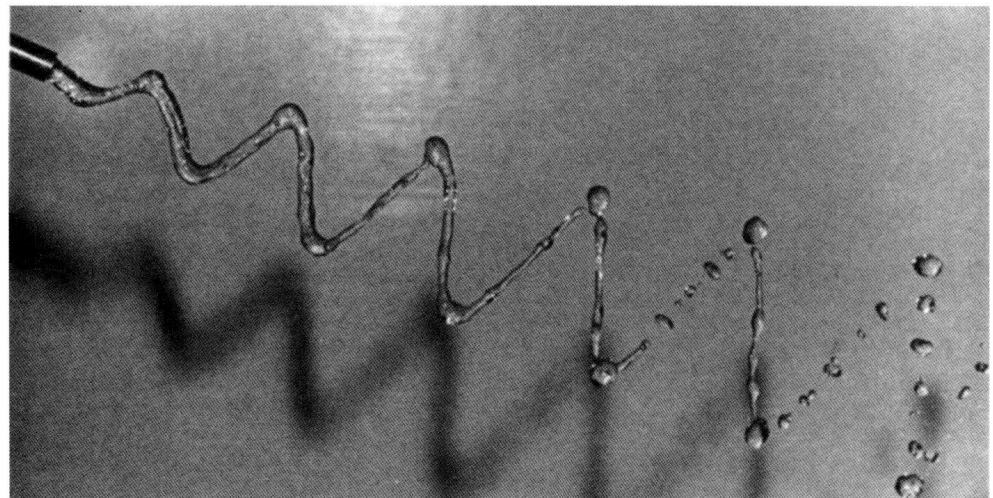
Fig. 29

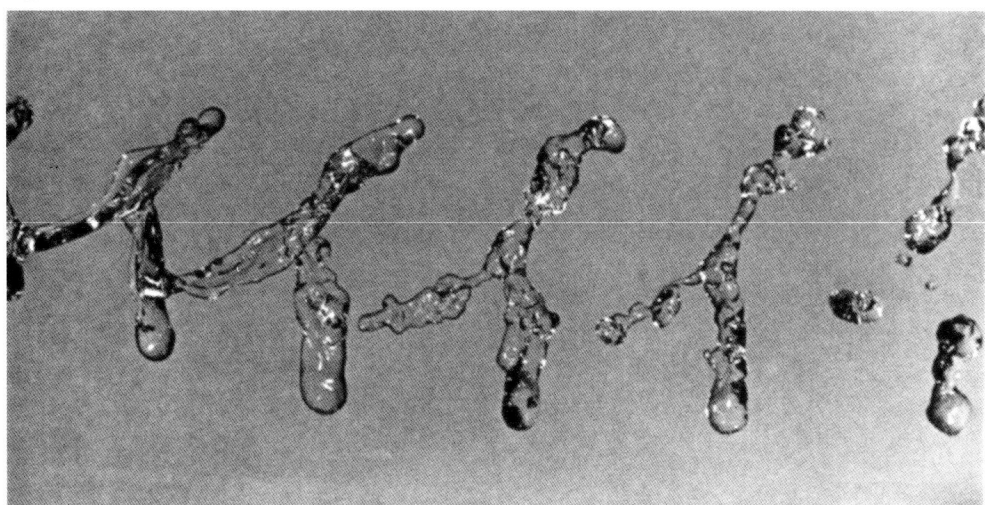
Fig. 30

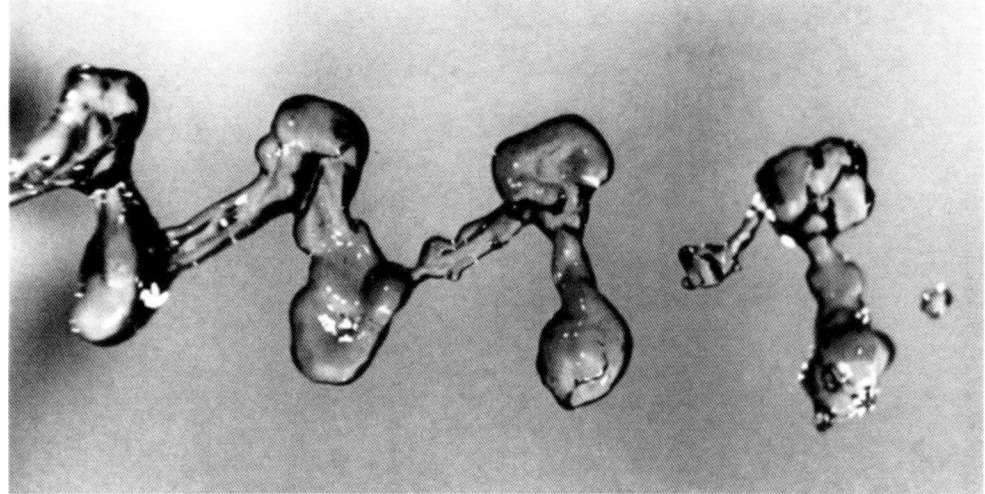
Fig. 31

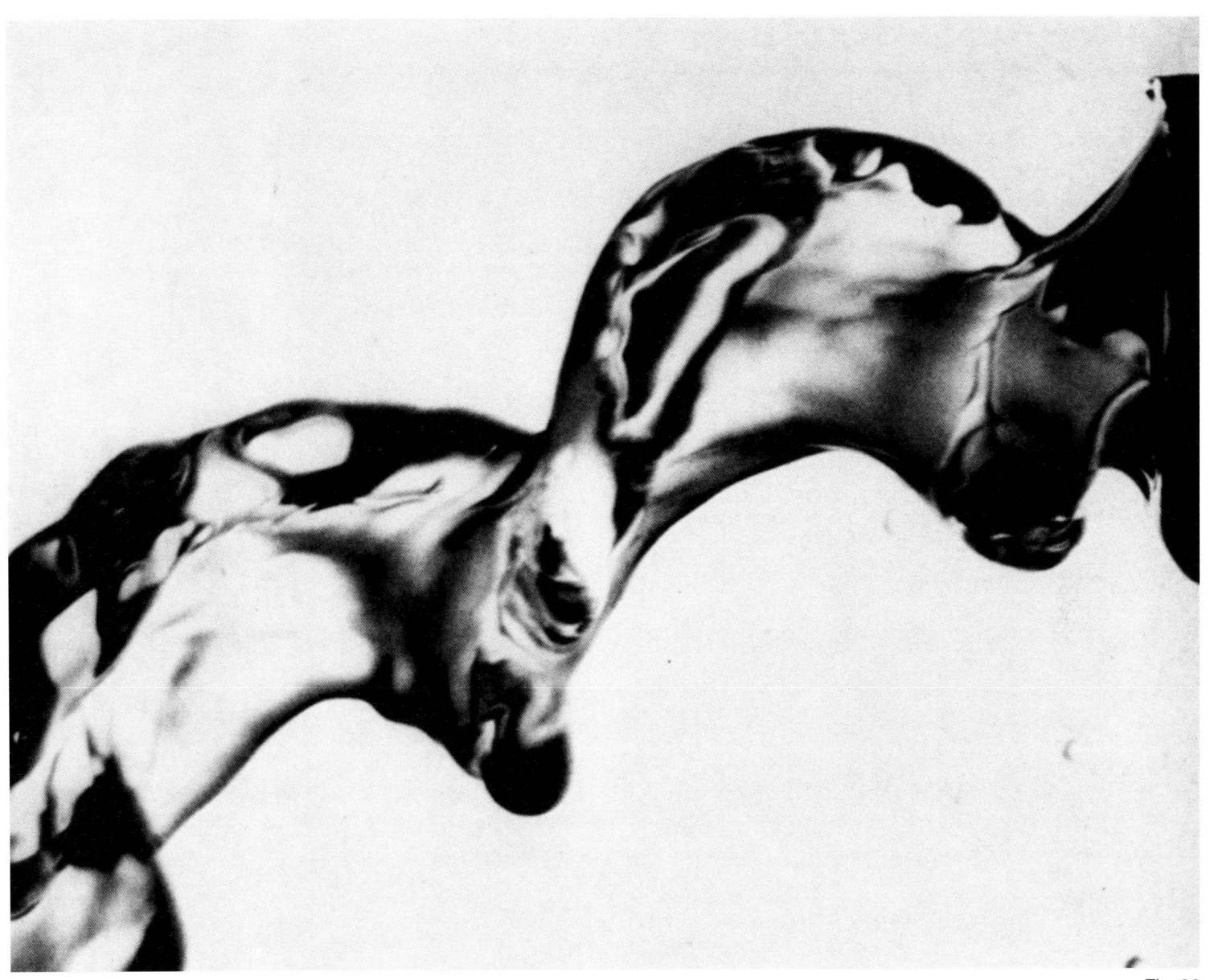

Fig. 32

Figs. 29-32
A vibrational impulse is communicated to the jet of water by the oscillating nozzle, resulting in a complicated interplay. First there is the undulatory path due to vibration, second the tendency for the integrity of the jet to be maintained by surface tension and for droplets to form when the jet breaks up. The stroboscope also reveals regular formations in the drops but, because of the extensive turbulence, they persist only for a fleeting moment. (Fig. 29 actual size. Figs. 30, 31, 32 enlarged.)

7

Vibrational Processes in a Capillary Space

To enable the complex conditions of vibration to be observed as a whole, we allowed liquid to flow into a capillary space bounded on one side by metal foil and on the other by a sheet of glass. On flowing into the space the liquid forms the familiar dendritic pattern. If the glass sheet is plucked away from the foil before vibration is applied, the pattern of dehiscence is seen.* As it is, "everything" remains in a state of inactivity. If the metal foil is now excited by vibrations the whole observed field comes to life. The glass sheet, of course, also vibrates in sympathy. Air enters the empty space at its periphery. Both this air and the liquid, which is stained, are affected by vibration. The inflowing air forms cavities in the film of liquid; the liquid itself puts out fingerlike processes. The vibration causes these formations to appear in series. They make, as it were, a woven texture. And at the same time everything must be visualized as being in lively motion. The processes float and divide. The liquid elements maintain their cohesion because of their surface tension. Fig. 34 shows this nascent process. The lower diagonal surface has not yet been involved. The air is entering from the top. Figs. 33 and 35 show individual parts of the field. The curious fingerlike processes, some of which are forked, resemble structures in a hollow organ of the body. They wave about in the air that's flowing around them. Cord-like and festoon-like processes also appear (Fig. 36) which look as if they were invested with folded epithelium. They display a fluctuating motion. In Fig. 37 the lowest part has not yet been affected. Above, an almost plant-like pattern is being woven. The impression of a woven texture is, of course, due to the uniform situation generated throughout the field by vibration. If the process affects a single liquid element spread out in the capillary space, say a flattened

---

*Details of the process of dehiscence will be found in *Cymatics, Vol. I* (Figs. 99-101, 104-105).

drop, the configurations seen in Figs. 38 and 39 appear. Once again we see fingerlike processes wafting about. There is interplay between the air and the surface tension of the drop. At the same time the air penetrates the body of the drop and turns it into foam. This foam circulates within the bounds of the liquid configuration and sometimes a pattern of bilateral symmetry can be seen. With its ceaseless movement the whole is reminiscent of amoeboid behavior. This impression is further enhanced as the formations move around. They proceed along their paths in correlation; i.e. they execute their movements while flowing as a whole. If they thrust out processes to one side, the rest of the body follows up behind. It slides and creeps as a *single whole* around the capillary space. If two such formations meet they may merge and form a new unit, or they may divide, with each component immediately functioning independently. The interaction of air and water can be observed here very well. The elements go through their activities as drops, as bubbles, as foam in a state of flux, in circulations and in fluctuations; while simultaneously the tendencies which dominate in the particular phase in which they exist at that moment, become evident. Vibration brings to light the direct interplay of these tendencies. The terms epithelium, hollow organs, and plant are used merely to convey some impression of the original experiment. No analogies or homologies are intended.

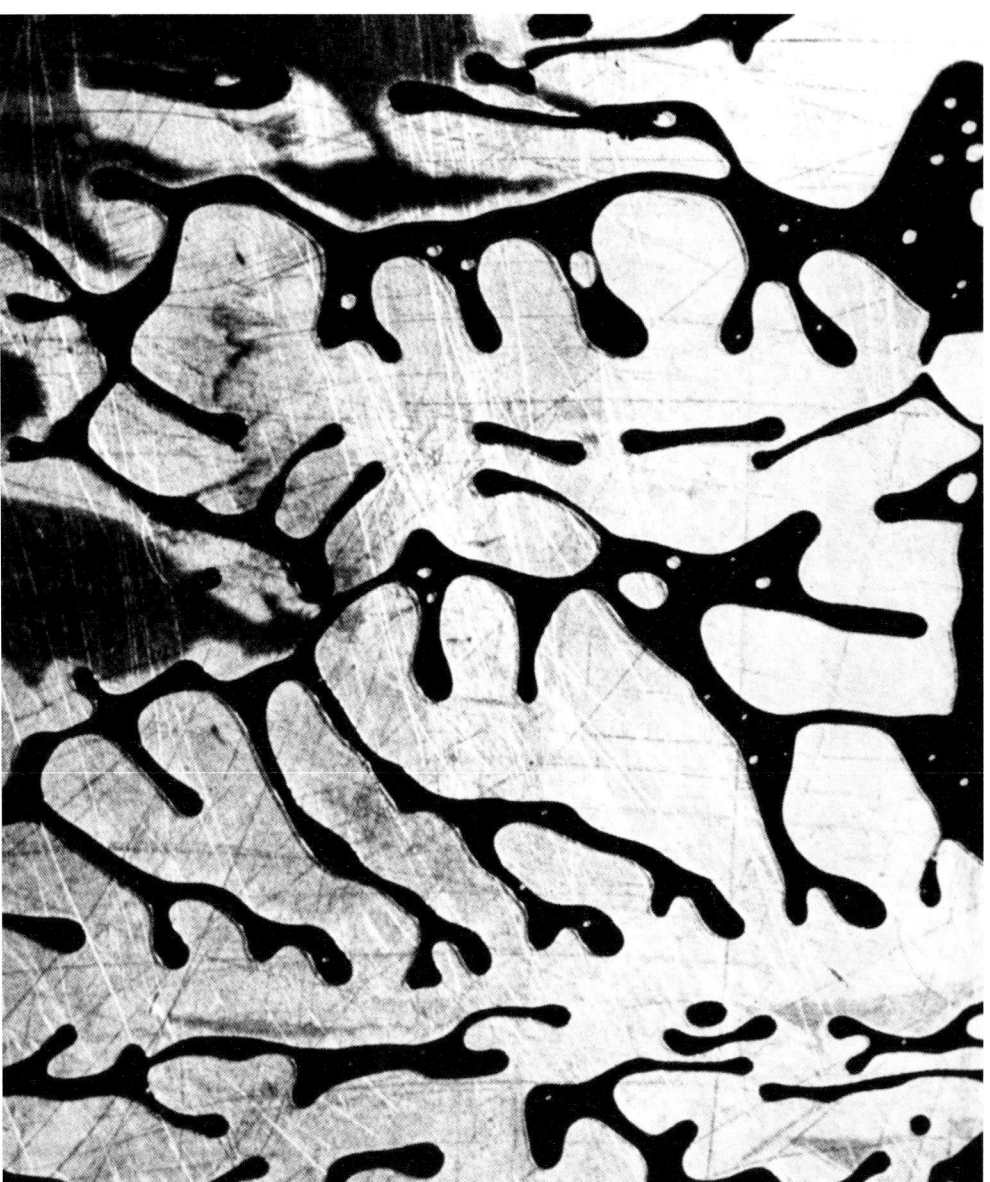

Fig. 33

Figs. 33-36
Air and liquid sandwiched between a sheet of glass and a metal foil, excited by vibration. Thus the walls, the liquid and the air are made to vibrate. In this complex system the air penetrates and forms cavities in the liquid, which in turn branches out into fingerlike processes that waft to and fro (Figs. 33, 34, 35). There are also garland-like cords which look as if they were invested with folded epithelium (Fig. 36). (Fig. 34 is reduced in size. Figs. 33, 35 and 36 are details.)

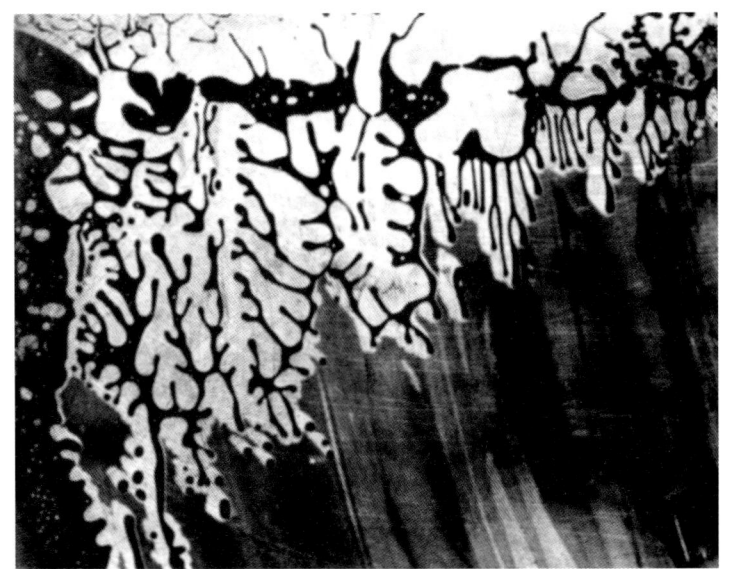

Fig. 34

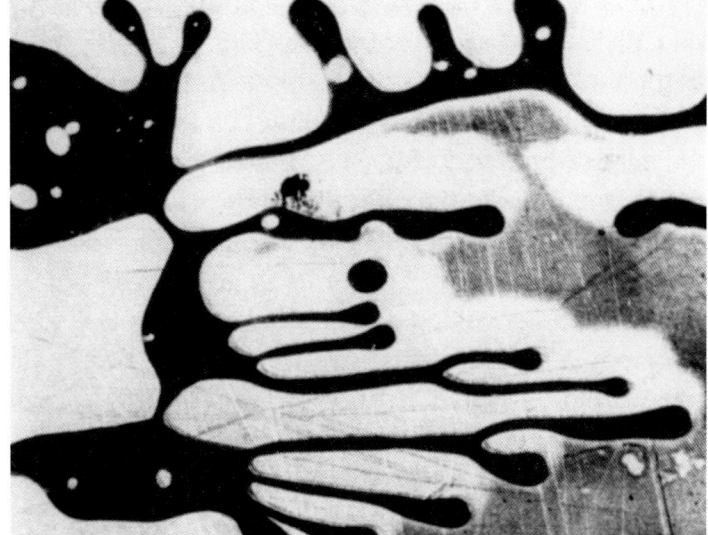

Fig. 35

Fig. 36

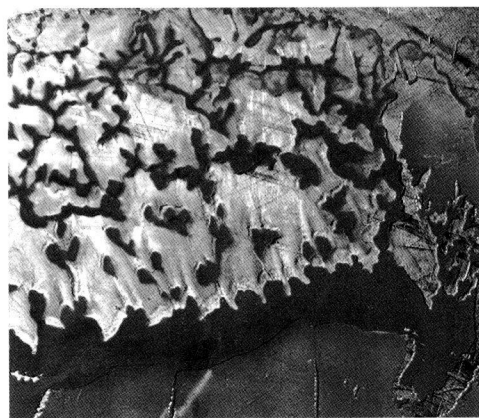

Fig. 37
What can only be described as a "vegetable" pattern has been evolved by the vibration. The forms sprout up but are continuously being shaped and reshaped. Some of them become detached from the substrate and produce structures such as can be seen in Figs. 38 and 39. The whole pattern most be imagined in motion, yet each of the processes is characterized by a typical morphology. "Everything" taking place, however, is the result of vibration.

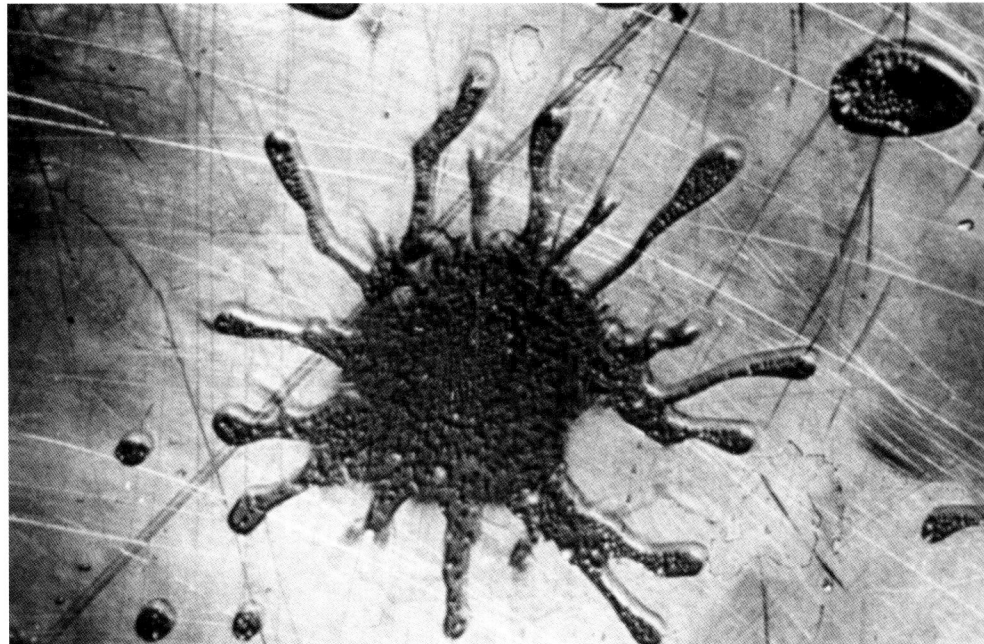

Fig. 38

Fig. 38-39
In these single flattened formations vibrating between the sheets of metal foil and glass there is again an intimate relationship between air and water. On the one hand the processes waft about in the air, on the other, the air, sometimes moving in circular currents, penetrates the liquid like a foam. There is also correlation between the liquid systems as they move around in the capillary space. (These figures are slightly enlarged.)

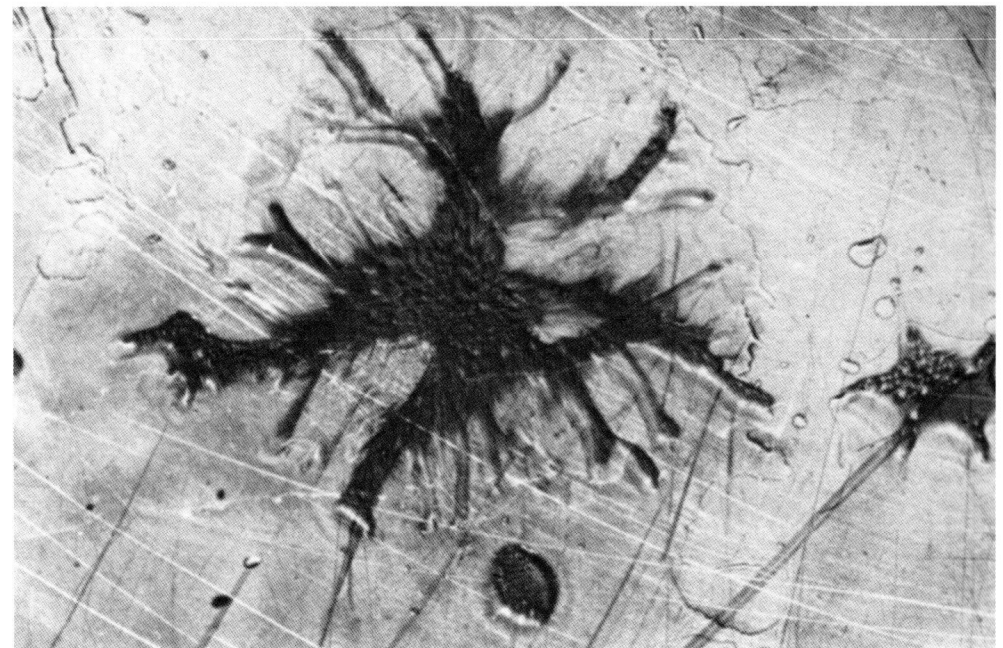

Fig. 39

8

Vibrational Figures Revealed by Schlieren Photography

Since it is difficult to visualize and comprehend the conditions generated by vibration in complex bodies (curved surfaces, bells, violins, etc.), the obvious thing to do is to use a method which manifests the processes directly in the object itself. To this end we chose the schlieren method in which incident polarized light is used to detect the stresses within in a transparent object by double refraction. We polarized the light of the stroboscope itself and allowed it to fall on a transparent sheet of plastic which was made to vibrate. With the aid of the analyser we could then observe what was happening in the plastic sheet under vibration, and found that characteristic vibrational patterns appeared (Figs. 40 and 41). It can easily be shown that these patterns are due to oscillation and not to other stresses, by following with the stroboscope the movements of the diaphragm generating the vibration and comparing them with the movements of the vibrating sheet, where the pattern of light and dark effects depends on the frequency employed. In this way the oscillatory process can be followed in the material itself. Many objects, of course, are not transparent, and consequently models must be made in the shape of these objects. They can be cast in artificial resin and must be free from internal stresses (well-tempered). In accordance with similitude relationships a true picture of the vibrational process can be obtained. Another technique is to study the reflected light of an object coated with a varnish giving a schlieren reaction. The process, with the models, however, yields very serviceable results.

Figs. 40-41
Vibration in a transparent sheet of plastic made visible by photoelastic methods. Polarized light and double refraction clearly disclose the tensions produced in the material by vibration. (Slightly enlarged.)

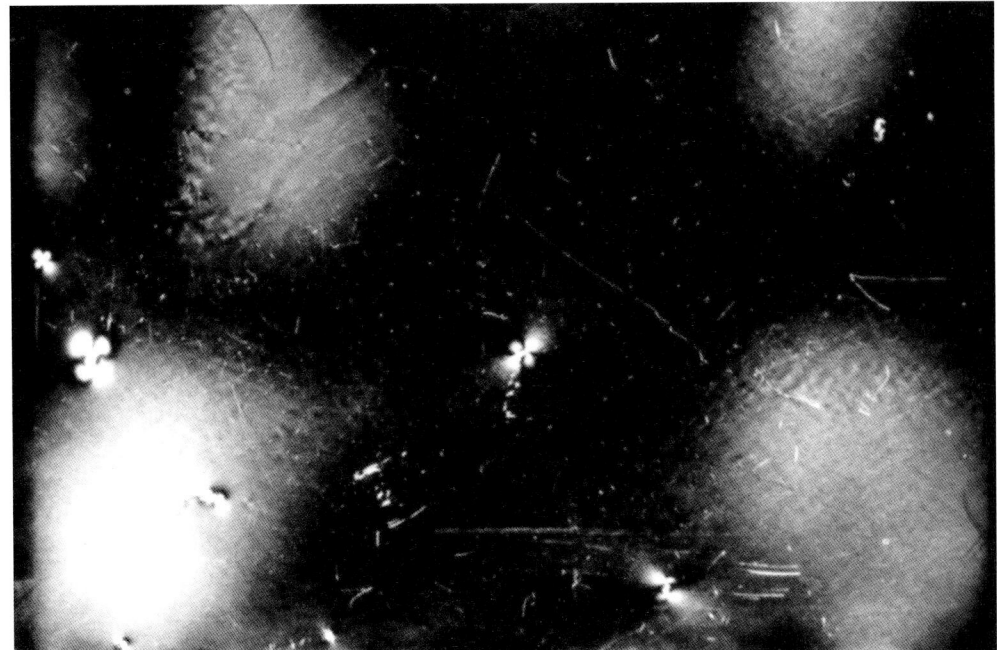

Fig. 40

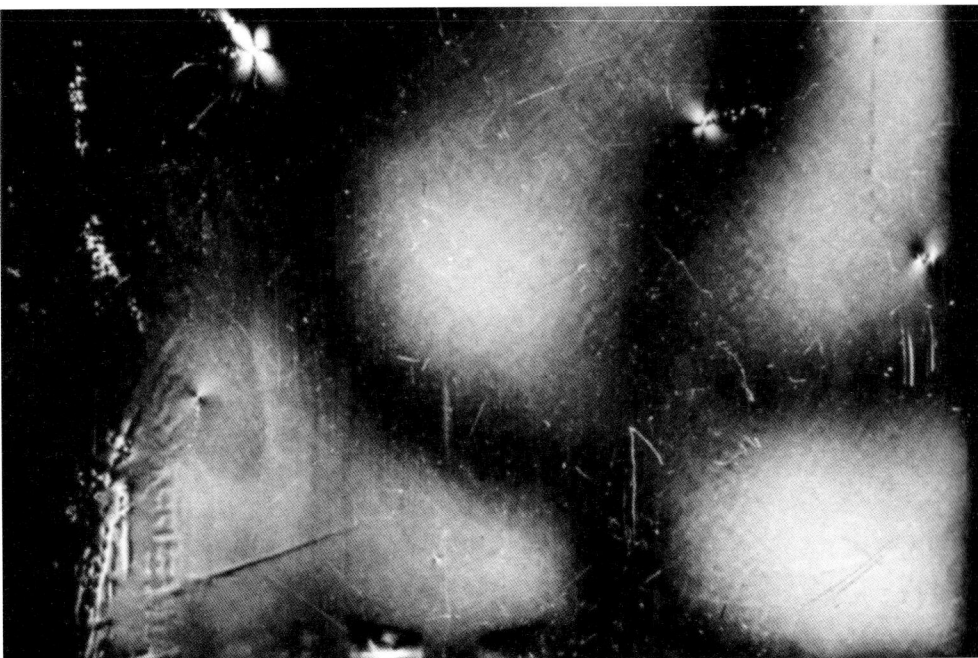

Fig. 41

9

Rotational and "To-and-Fro" Effects

Lycopodium powder (spores of the club moss) is an exceptionally homogeneous material, consisting of particles of uniform shape. "Cymatics, Vol. I" portrayed a series of experiments with this powder. The round heaps formed by vibration are in a process of systematic circulation: the particles at the bottom are transported from the periphery to the center and from the top back to the periphery again. Looking down on the pile we see the powder come welling up in the center and then streaming away again to the periphery. At the same time, depending on the intensity of the vibration, various specific regions and structural patterns are formed, and furrows may even appear, sometimes in a concentric arrangement. As the volume of the tone is increased, the material is flung out of the pile and small eruptions occur, particularly at its edge, but also within its boundaries. There is also an inflow of new particles which are likewise thrown up in small fountains. If we watch these processes carefully, we see that these powder fountains also trace out paths within the area of the heap, appearing in large numbers and scurrying over the curved surface. The tracks they follow conform to a mainly concentric pattern. Indeed, it becomes increasingly obvious that the rotary motion is a systematic process. While some of these eruptions rotate in a clockwise direction, the motion of others is counterclockwise. If two such eruptions collide, the process ceases. The experiment can even be arranged so that the particles circulate first to the left and then to the right as they spurt up in fountains. There is a real to-and-fro effect with the motion alternating direction in a highly regular way. Once the right adjustment has been made, the whole process is enacted at the same tone and the same amplitude, with the system moving to-and-fro in an orderly manner. There is no actual transportation of the particles; what

Fig. 42
A round heap of lycopodium powder (about 4 cm. in diameter) is made to circulate by vibration. At the same time two centers of eruption rotate at diametrically opposed points of the pile. Rotating first in a clockwise and then in a counterclockwise direction, this wave produces a kind of to-and-fro effect. The two points where the powder is flung up must be visualized as advancing through the quiet areas of the periphery, the very next instant. In their wakes activity immediately ceases again. The processes of rotation and circulation proceed uniformly. Frequencies can be set at which the rotation runs in one direction only, appearing as if a "diameter" were rotating.

happens is that the powder always spouts up at the same spot. Sometimes a connection can be seen between this pattern and the furrows formed by the circulating currents, but the rotary motion may appear without any such link being apparent. Conversely very lightly-marked furrows, which disappear almost immediately, can be seen behind the moving "fountains." A particularly striking fact is that these violent local reactions do not intrude destructively into the circulatory process. The phenomena proceed regularly side by side while a pattern of ordered unity is maintained. By very fine adjustment it is even possible to produce one circulatory phase without the other. The following phenomenon is then observed (Fig. 42): two small eruptions appear in the circular heap steadily circling at the opposite ends of a diameter. In the photograph they are circling in a clockwise direction. The spouting particles form a series of frontal wavecrests with a state of almost complete inactivity following in their wakes. One must imagine that the heap of particles welling up at the top of the picture is moving to the right, and the one at the bottom to the left.* The areas in between are meanwhile quietly circulating, but in a moment the disturbances will course through them again. The periods of revolution are, of course, much slower than the vibration. At a frequency of 66 cycles per second in one experiment, for example, there were about 30 revolutions per minute. It is particularly striking that the rotary waves move diametrically opposite to each other, and it is characteristic of the phenomenon that they move as *one* process along the same diameter.

What the phenomenon involves, then, is this: a number of eruptions move in a to-and-fro manner and obliterate each other, or else the eruptions move in a to-and-fro manner *one* pair at a time, each taking over from the other, or just one single pair rotates uniformly with no to-and-fro motion. Which form the process takes depends on frequency, amplitude, the quantity of material and the topography of the vibrating diaphragm.

*Editor's note: The dynamics of this fascinating phenomena are shown clearly in the video *Cymatic Soundscapes* and *Bringing Matter to Life with Sound*. See promo pages at the end of this edition.

Fig. 43

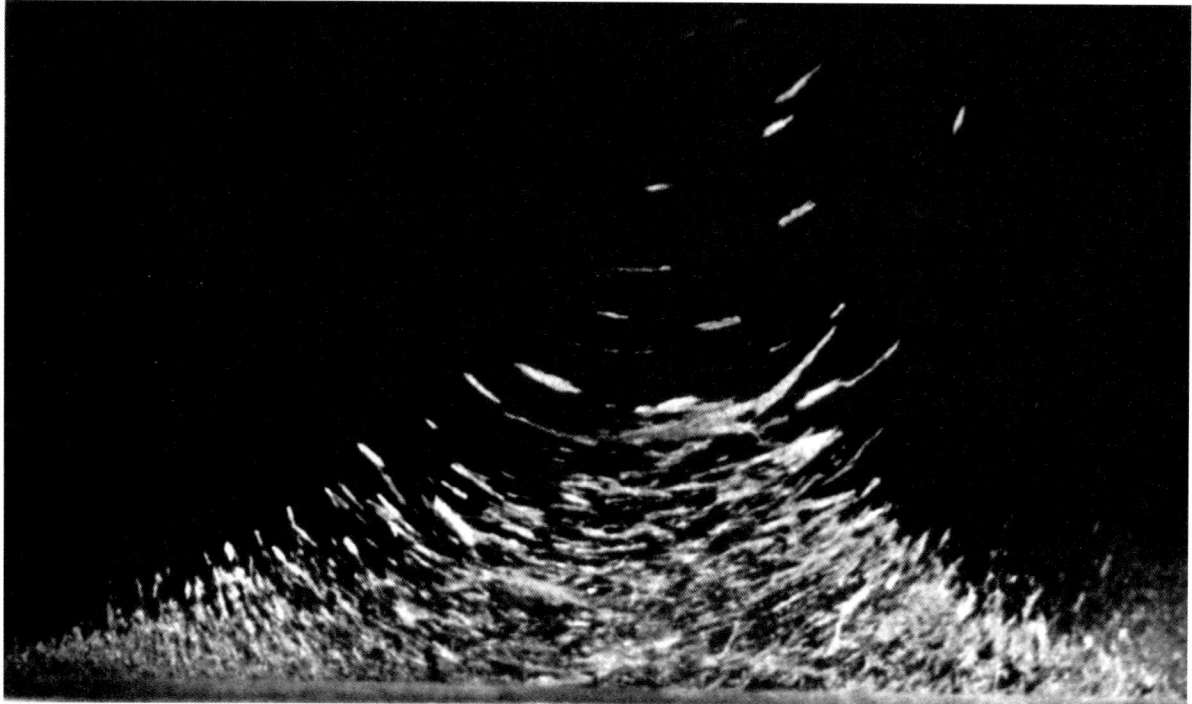

Fig. 44

Figs. 43-44
Oscillation reduces cohesion and adhesion. Accordingly these iron filings, acquire new degrees of freedom in their movement. The iron filings in these photographs must be imagined to be hovering and flying about, as it were, as a result of the vibration. This movement affords direct insight into the integrity of the magnetic space at all times. (Actual size.)

10

Ferromagnetic Masses in the Magnetic Field Under Vibration

As adhesion to a supporting surface is reduced when it is vibrating, and as a material, whether composed of particles or a viscous paste, becomes more mobile and liquid within itself under vibration, it is to be expected that forces will produce a different pattern of effects when adhesion and cohesion are reduced. This phenomenon was investigated by subjecting ferromagnetic material to vibration in a magnetic field.* First, fine iron filings are placed in a magnetic field and the supporting surface is made to vibrate. On the one hand the familiar phenomena of attraction make their appearance (Figs. 43, 44) and, on the other, the equally familiar polar figure (Fig. 45). Vibration which, of course, also operates in the air over the diaphragm, causes the particles, as it were, to be held in suspension. In Figs. 43 and 44 the iron particles in the magnetic field must be imagined to be moving with a certain degree of freedom. They hover and fly around in the magnetic space. A curious effect can also be observed in the polar figures. If the vibration alone were acting upon the iron filings, there would be conglobations in a process of circulation. But here there is the additional factor of the magnetic field impressing the polar pattern on the iron filings. But the action of the vibration does not cease here: the polar figures are also divided into tapering zones extending out to the periphery, all of which is in the process of circulation. In a radial direction there are masses of powder which are circulating at right angles to their long axis. Fig. 46 shows a single "roller" in a process of circulating several ways at once.

*Similar experiments were reported in *Cymatics, Vol. I* (Figs. 106, 107).

Further questions were: What happens if a coherent ferromagnetic mass in a magnetic field is subjected to periodic excitement? What happens to this mass if the magnetic field is even moved and its direction and strength are changed? How will this mass follow the magnetic pathways? How will it be affected by its own magnetizations? To find an empirical answer to these questions, a viscous paste was thoroughly mixed with iron filings, producing a homogeneous ferromagnetic mass. Using this mass, experiments were performed during which frequency and amplitude were varied and, more especially, the magnetic influence was also varied. The strength and direction of the magnetic field were changed, but most important of all, it was also shifted. The photographs in Figs. 47-60 capture moments from these experiments. The whole must be imagined in a state of continuous flow with the morphology caught by the camera clearly in evidence. Flowing sculptures might be an apt description.*

Omit the magnetic field and we are left with conglobations due to oscillation, pulsation and circulation. Now re-impose the magnetic field, and the mass pushes upwards, (Fig. 47) but does not burst asunder. The configurations appearing here are essentially spatial in character and it would be more appropriate to

*Editor's note: Many examples of this effect can be seen in motion in the video *Cymatic Soundscapes* and *Bringing Matter to Life with Sound*. See promo pages at the end of this edition.

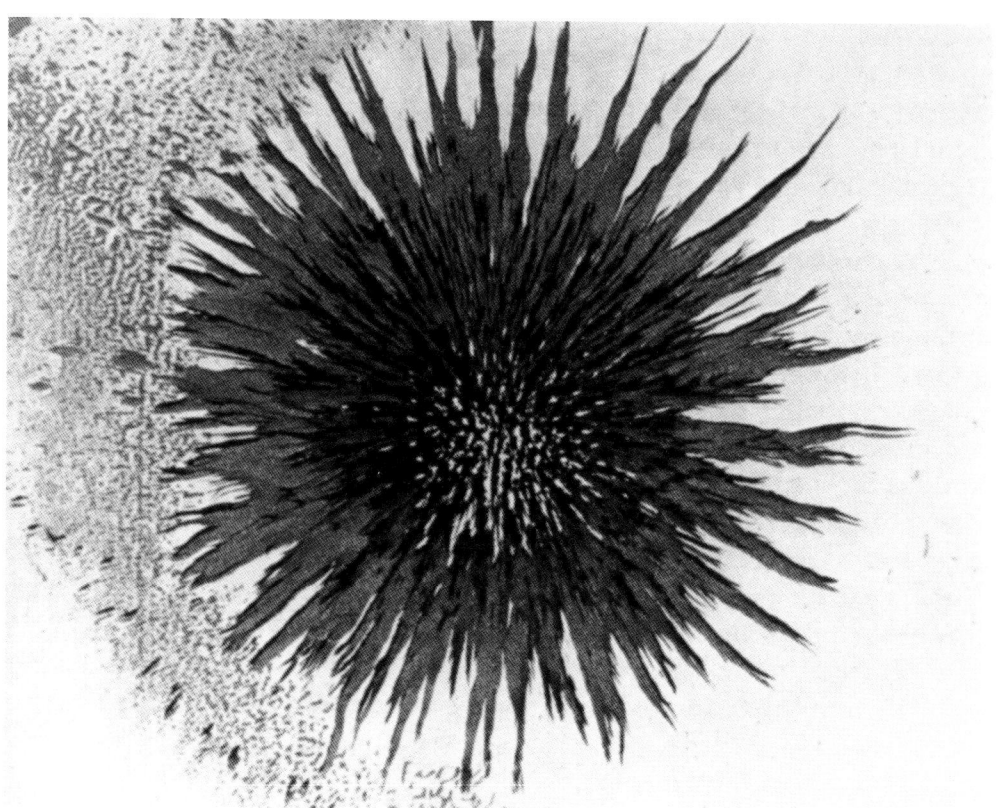

Fig. 45

Figs. 45-46
The figures around the magnetic poles are also "brought to life" by the vibration. Because adhesion and cohesion are reduced, these patterns are in continuous motion. Circulation is the most prominent feature. Fig. 46, which is slightly enlarged, shows such a "rolling" formation which has been created by circulation. The vibrational field confers a unified wholeness upon these phenomena which is apparent in the way they are continually being integrated.

Fig. 46

speak of *magnetic space* than lines or fields of force. Serial structures stand out from the general pattern (Fig. 48). Then the figures put out new branches (Fig. 49). They grow upwards like plants or leaves (Fig. 50). But the most prominent features of all are turning and helical formations in which a pronounced twist appears. The formation begins to rotate. In Fig. 51 a curved, S-shaped form can be seen. The photo has perpetuated a single moment. What actually is happening is that the formation is turning on its long axis and consequently the impression conveyed to the eye, for which the process is too fast to be resolved in detail, is of a dancing caduceus. This process is steadily continuous. When, however, the mass sprouts "leaves" (center, Fig. 50), and "tips" appear with like magnetism, these repel each other and the whole figure collapses. At the same time the outgrowths flutter and rotate. A spiral motion can often be clearly seen in the rising columns.

Under other conditions there are upfoldings which rise up to form arches (Fig. 52). These structures tower up and tend to flow along a path (Figs. 53, 54, 55). Then they thrust out again into space (Fig. 56), and in the interplay between the cohesion of the mass and the magnetic force they spread, grow thin, and peter out. Forms tower up displaying the configuration wrought by magnetic force and oscillation (Figs. 57, 58, 59). Large leaf-like walls take shape and sway to-and-fro (Fig. 60) in the magnetic field.

Figs. 47-52
To make visible magnetic and cymatic effects in space, iron filings were mixed with a viscous liquid. This "emulsion" was then subjected to vibration and a magnetic field at one and the same time. The masses rear up (Fig. 47), or grow into serial, branch-like, or leaf-like forms (Figs. 48, 49, 50). Sudden repulsions can be observed in these budding structures when two of the processes are both magnetized alike. Often a twisting movement appears. In Fig. 51 the small "S-shaped" figure is rotating and in the experiment looks like a tiny dancing caduceus. (Approx. actual size.)

Fig. 47

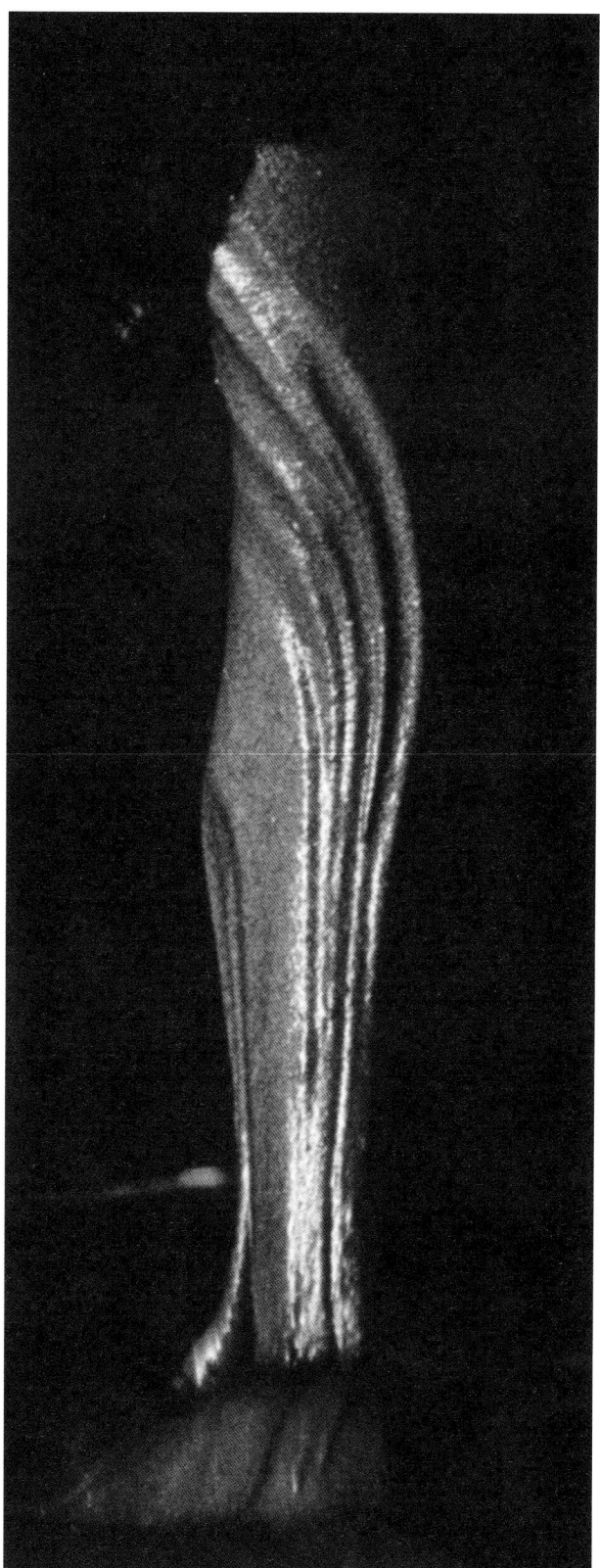

Fig. 48

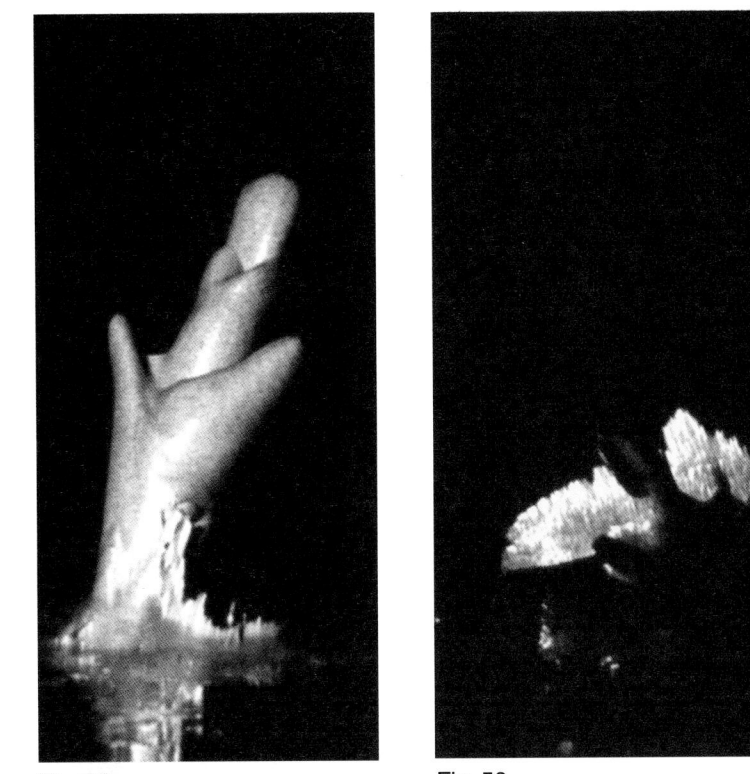
Fig. 49

Fig. 50
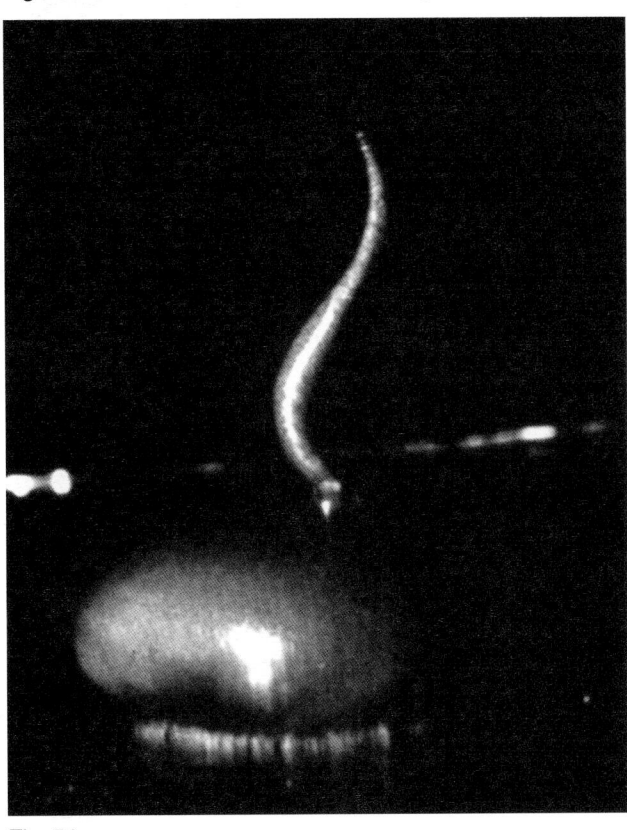
Fig. 51
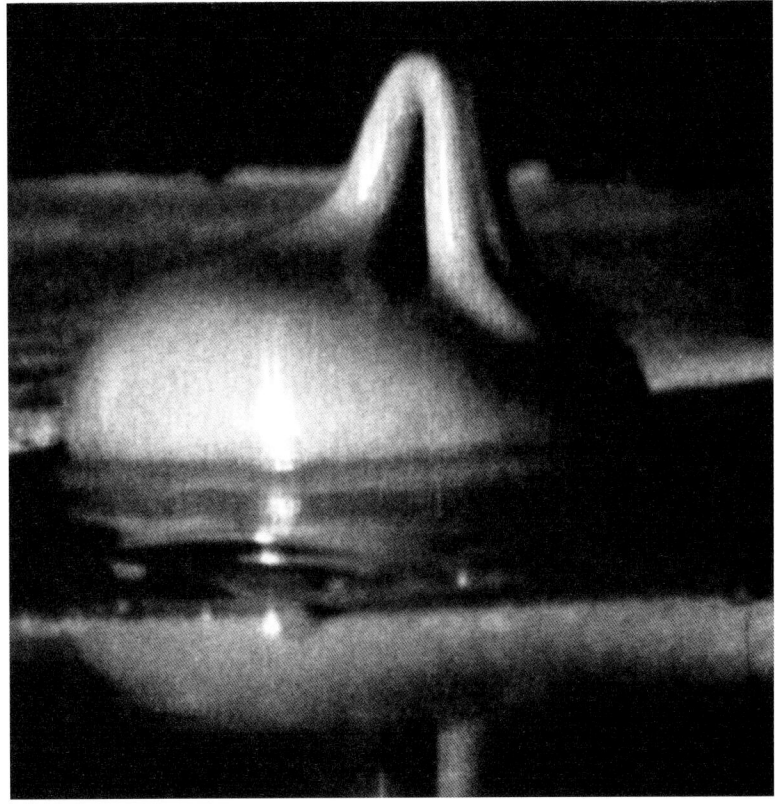
Fig. 52

Here Nature reveals an abundance of sculptured forms, and all of them, it must be remembered, are the result of vibration. If the tone ceases, the mass "freezes." Looking at these vibrational effects, it would be no exaggeration to speak of a true magnetocymatics with its own dynamokinetic morphology. Experiments like this based on pure empiricism, stimulate the imagination and develop one's ability to sense oneself within such a space permeated by unseen forces.

It might be added that we have started experiments with physical plasma. A small element reacting magnetically could be made to vibrate by the impingement of sound. We have not yet succeeded in producing configurations because of the small dimensions, but appropriate structural patterns may be confidently expected. Experiments of this kind are being continued.

Fig. 53

Fig. 53
A ferromagnetic mass placed in the magnetic field can conform to the pattern of its lines of force when excited by vibration. Adhesion and cohesion are reduced. Under these conditions the characteristics of magnetic space are made visible. What we see in Fig. 53, is not a rigid arch but, so to speak, a flowing piece of sculpture.

Fig. 54

Figs. 54-56
These figures show the plastic pattern of movement displayed by a ferromagnetic mass in a magnetic field under the influence of vibration. The mass flows in the magnetic space and reflects its configurations. It writhes, rears up, and stretches out but always in a way that reflects the situation in the magnetic field at that particular time. (Actual size.)

Fig. 55

Fig. 56

Fig. 57

Figs. 57-60
Depicted here are additional processes taking place in magnetic space. The vibrational field confers certain degrees of freedom on the material by reducing its cohesion and adhesion. The mass becomes more "fluid," more liquid. These flowing sculptures differ from each other in an extraordinary variety of ways. They clump together and are then drawn out in serpentine threads which squirm and writhe, but there is no real continuity about the performance. The flow often stops abruptly and then begins again with a rush. The flat surfaces may begin to flap and flutter. If the fingerlike processes have the same magnetic polarity, they are mutually repelled: the whole structure collapses and then begins to take shape again. In this way the relationships prevailing in the magnetic space are revealed in their gradients and varying strengths. (Figs. 57, 58, 59 actual size. Fig. 60 somewhat enlarged.)

Fig. 58

Fig. 59

Fig. 60

11
The Morphology of Lichtenberg Figures; A Carrier System

Constant concern with periodic and serial phenomena gives the eye a certain expertise. Everywhere it discerns repetitions and regularity or tendencies to these qualities, and at every turn one encounters quasi-periodic and quasi-serial formations and sequences of events. However, such features of regularity need not involve any real vibrational field. In *Cymatics, Vol. I* (p. 95 ff.) we drew attention to periodic processes in which vibration in the true sense plays no part. Rhythmic precipitation (Liesegang rings), dehiscence of plastic masses, and diffusion in fluids were described. Phenomena coming under this heading occur in large numbers. Electric leakage discharges and discharge paths play a similar role. It must always be remembered in this connection that wave processes may also occur at the same time as, for example, in the rupture of solids, in the penetration of bodies by shot, and the explosive deformation of bodies, in all of which typical fracture forms and rupture contours appear. But quite apart from such simultaneous wave phenomena, cases like these also show a typical tendency for discontinuities to recur and for a pattern of regularity to appear even in the absence of a proper vibrational field. The familiar Lichtenberg figures (Figs. 61 to 63) are a further example of this kind. Normally they are formed by leakage discharges imprinting themselves directly in the photographic layer. The question of interest here is: How does such an electric discharge take place? What determines the actual pathways it follows? It is at once apparent from the figures that the electricity does not travel like an ordinary flash; instead we are struck by the way in which the elements are regularly repeated in the figures. One shoot comes after another, branch follows branch. What we have is a true dendritic pattern. There is a certain tendency toward repetition at intervals, although the figures do not display any periodic or

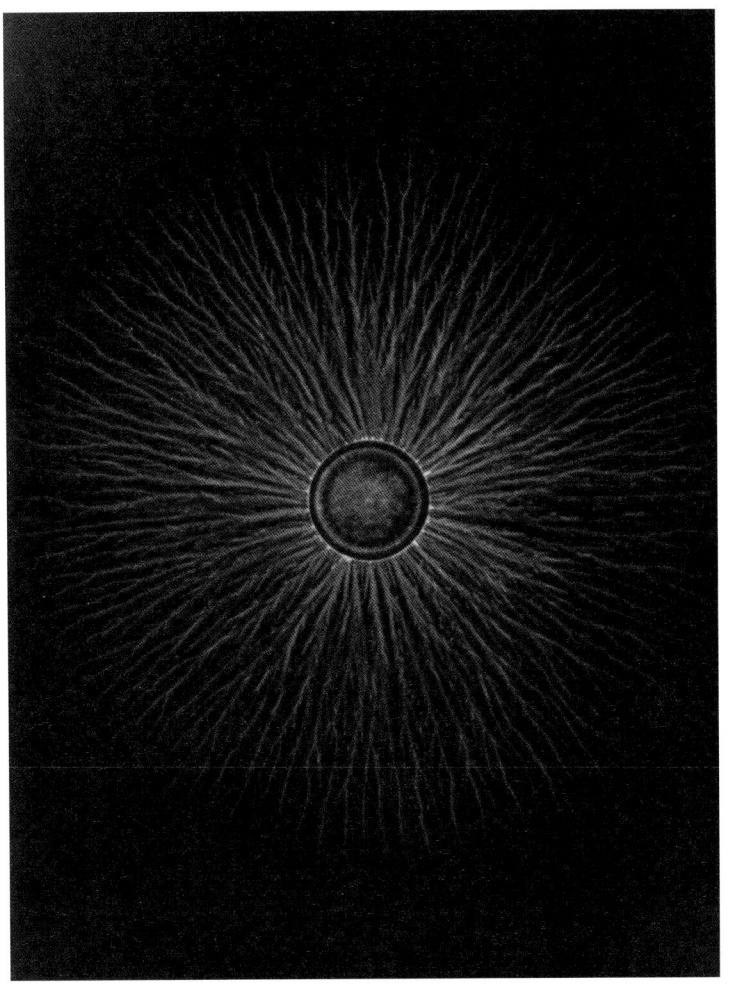

Fig. 61

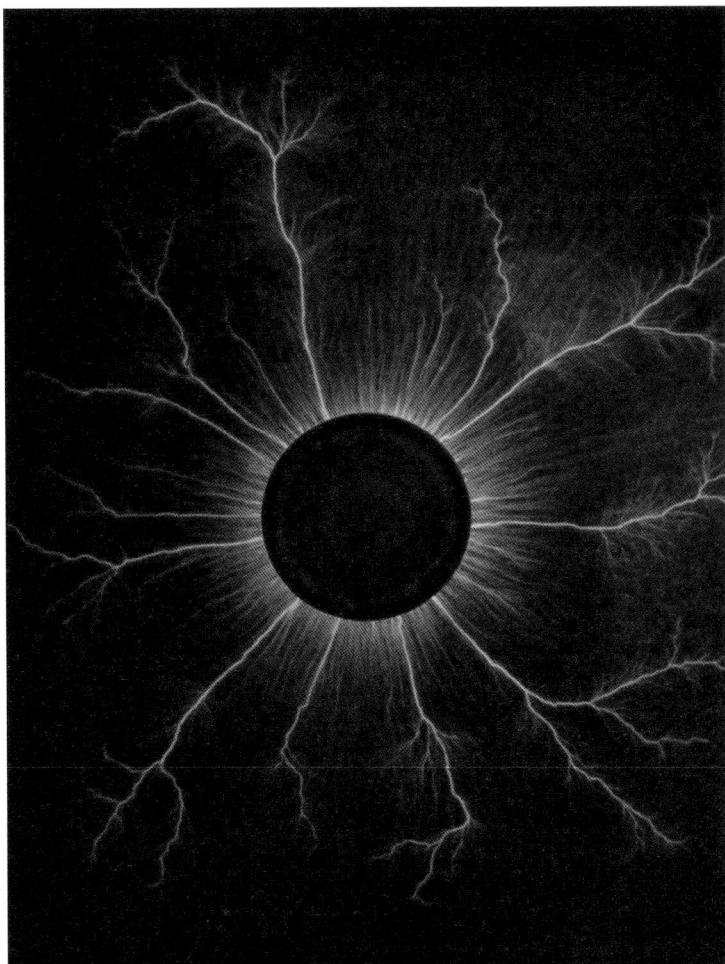

Fig. 62

Figs. 61-63
Here we see Lichtenberg figures (electric leakage discharges). These patterns are produced by the familiar method of allowing the electric discharge to imprint itself directly onto the photographic film. There is a clearly marked repetition in the morphology of these "current paths" where ray is followed by ray, and branch by branch, etc. A kind of serial arrangement appears in the discontinuities in which no actual vibration is involved. (The discharge took place at about 15 kV. Actual size.)

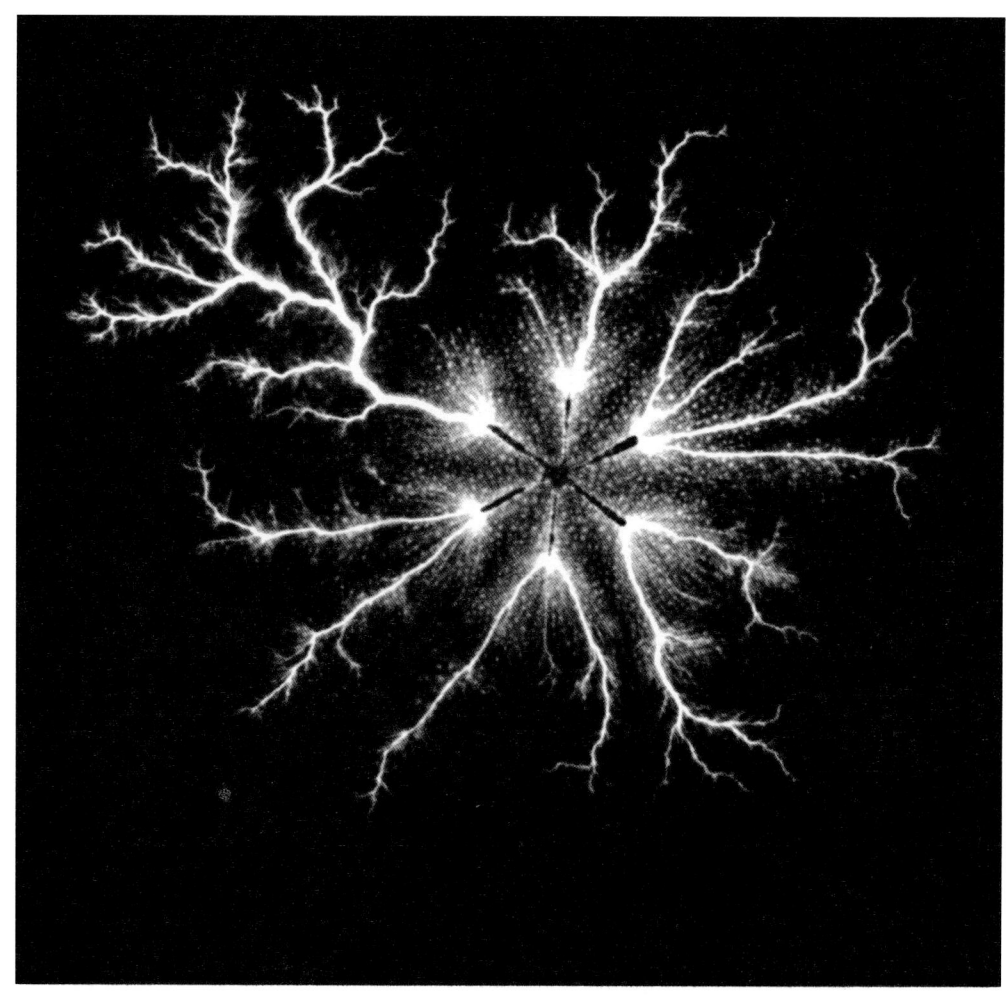

Fig. 63

harmonic pattern. Apart from this, the discharge involves electrical wave processes and serial current impulses. Although neither periodicity nor harmonics are in evidence, the figured elements are shaped in such a way that polar situations are repeated. Similar patterns are found in the bifurcations in the structure of plants, shrubs and trees, and the same key pattern can be traced down to the finely marked detail in these figures. The granulations and spots which have imprinted themselves on the film (mainly in Fig. 63) are very regularly arranged. The point to note about these well-known Lichtenberg figures is that their morphology is characterized by the regularity of the discontinuities and thus by the tendency of their elements to be repetitive. They serve to draw attention to one of Nature's most pervasive phenomena which might be formulated as follows: even without vibration in the narrow sense of the word periodicity and seriality are disclosed in Nature. The picture that offers itself to our eyes is one of repeated discontinuities. The elements in the areas where the phenomena transpire reveal a tendency to recurrence and the media in which they are embodied display alternating phases. In this context we might refer to the quasi-crystalline properties which have been experimentally demonstrated in liquids. What is involved here is not a chaotic confusion of atoms and molecules but a turbulent interplay of elements of a different order (quasi-crystalline "structure" of liquids). Indeed, even in gases variations of density have been disclosed by the diffraction effect (M. v. Smoluchowski, 1916). Similarly, in amorphous masses there are certain regularly recurrent configurations of "nearest neighbors" with regularly occurring distances between neighboring elements. Observations of this kind alert us not only to the fact that Nature appears to be discontinuous throughout, but also that there are tendencies to repetition in the discontinuity. And even if wave phenomena in the strict sense are not apparent in these regularities, there is nevertheless a broad spectrum running from the strict repetitions of polar phases to seesawing, turbulence, alternating fluctuations (quasi-crystalline groups and clusters, density fluctuations, spatially regular configurations, etc.). Of particular interest in this context are Brownian molecular movements with their fields of oscillation.

In Figs. 64 and 65 we combine vibrations and electric leak discharges. The figures are visualized by iron filings. Needless to say, the discharge passes through the conducting masses of particles and thus follows the pattern of the sonorous figures. Even in an apparently commonplace experiment like this there are some interesting discoveries to be made. A close examination of the photographs will reveal the curiously segmented discharge paths with their more or less regular articulations. Moreover, the granulations present in the Lichtenberg figures are also seen here. Near the main pathway there are suggestions of wave groups (particularly in Fig. 64).

Our experiments are certainly not at an end when there are phenomena of a periodic kind displaying such a multiplicity of features without involving any vibration as such. Experimental work is in full swing. Even though there are already intimations of an overall set pattern of regularity— repetitive discontinuities and regular tendencies to repetition as structural patterns and processes— it is of the very greatest importance to scrutinize every corner of the experimental field. How else in complex biological systems can we discern chains of respiratory enzymes and their structures, cyclical reaction mechanisms in cytochemistry and molecular sequences and transport systems, unless our minds are tutored to interpret periodic, quasi-periodic, or approximately periodic processes correctly. Figs. 66-72 are photographs of experiments which represent a proper "carrier system." The dance performed by elderberry balls between electrodes is a well-known phenomenon. We utilize the usual experimental set-up but dip the electrodes in oily liquids. Then to the liquids we add crystal particles representing a wide variety of organic substances. Switching on the current starts the familiar and yet astonishing performance. The 'particles of different substances begin to migrate from one electrode to the other. If two particles meet, both lose their charge and retrace their path to the electrode from which they set out.

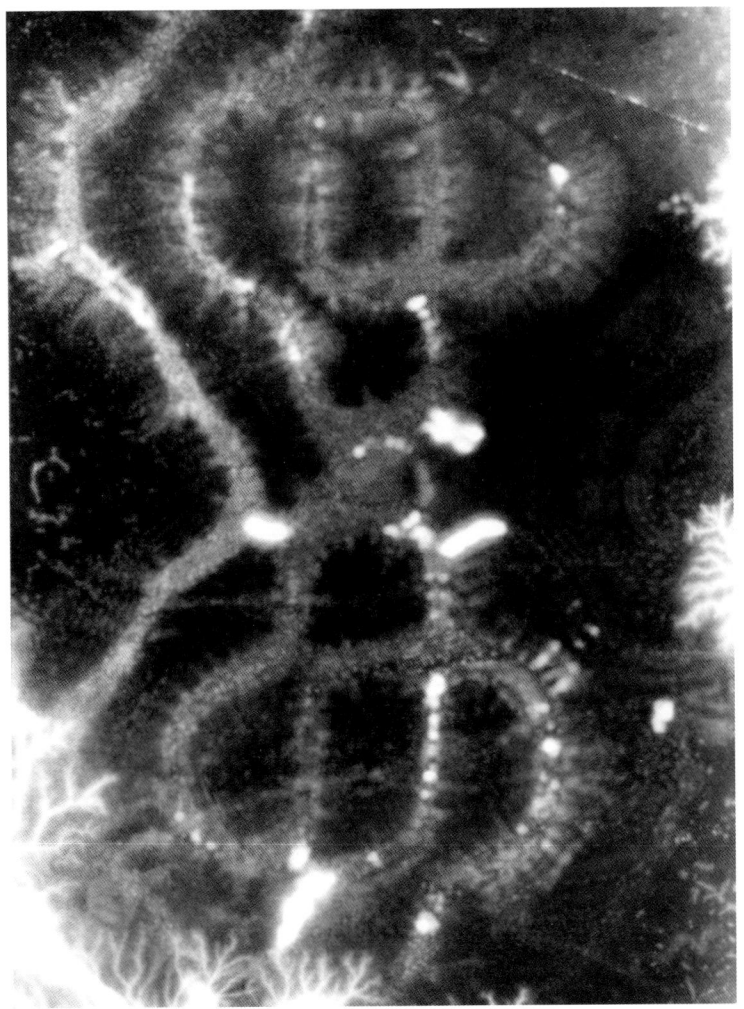

Fig. 64

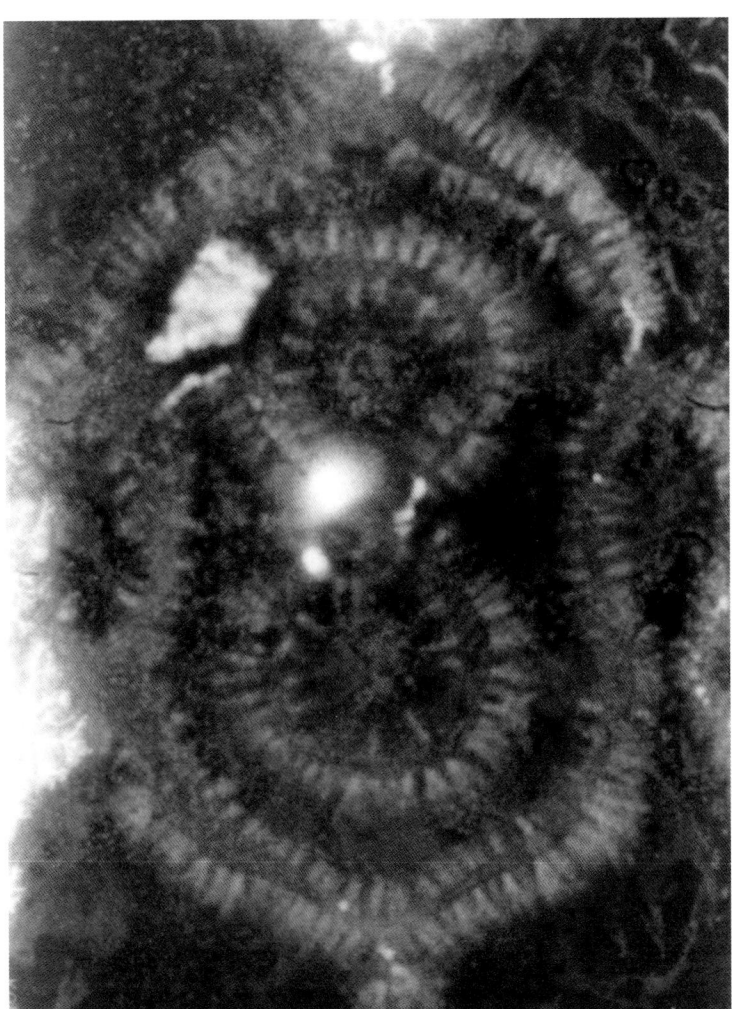

Fig. 65

Figs. 64-65
Here the discharge took place through sound patterns created with iron filings and the current naturally followed the structural layout. It is interesting to examine the morphology— the granulations, segmentations, wave families, etc.— in detail. In these experiments systems are revealed in which there is a complex interplay of the factors involved, viz. electric discharge and vibrations in this case. Each detail deserves close scrutiny. (Actual size.)

Fig. 66

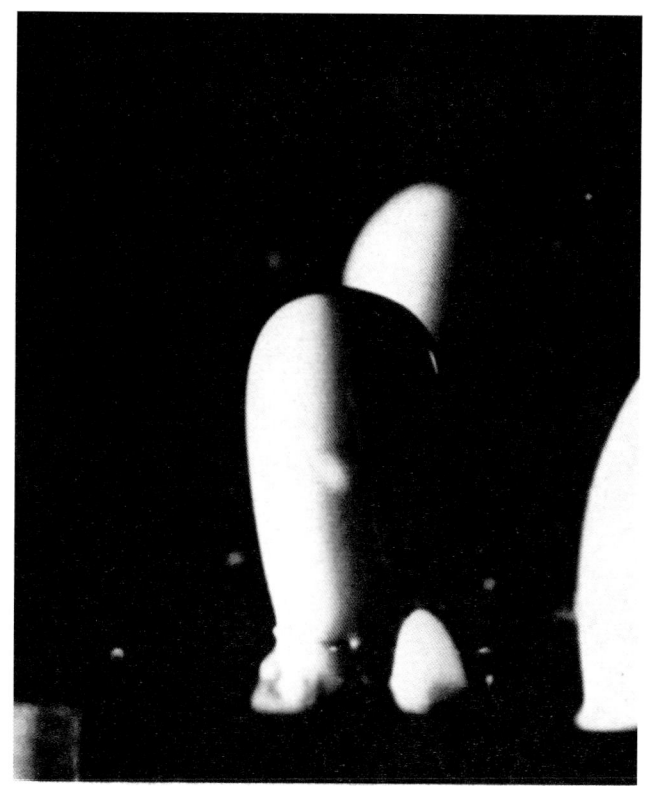

Fig. 67

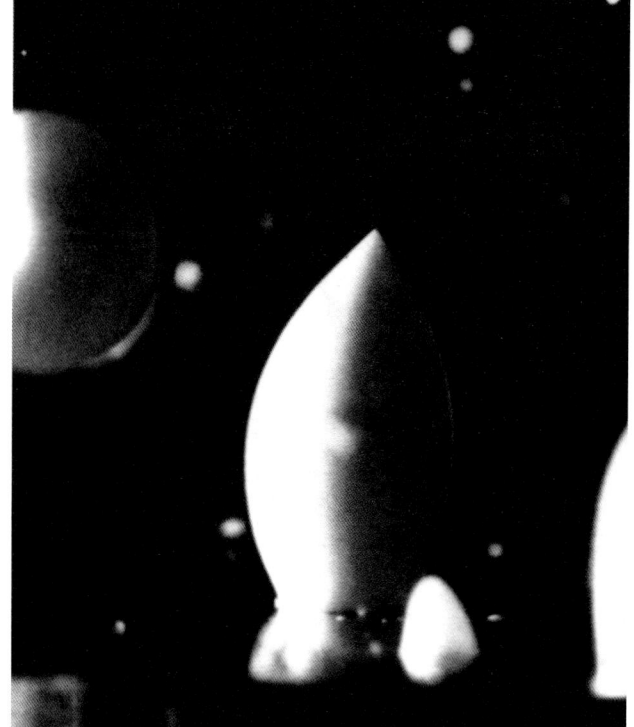

Fig. 68

Fig. 69

Figs. 66-69

A proper carrier system. Particles of many different kinds migrate to-and-fro between the electrodes. If they meet, they transfer charges and go back to where they started. Cream was the material used in the experiment photographed here (Figs. 66-69). Fig. 66 shows conditions at rest. If the current is switched on (about 10 kV), the material becomes distorted (Figs. 67-68) and then actual transport begins (Fig. 69). The electrodes have been dipped in oily substances. (Enlarged.)

Figs. 70-72

Migrant particles in the carrier system (salt crystals, various organic particles) are often strung together and form thread-like or pillar-like structures (Figs. 70-72). The kinetic carrier system changes into a structural one. Phases occur in which both migration paths and transport structures are present at the same time. What we have in this carrier system (Figs. 66-72) is periodic activity involving the repetition of polar phases, although there is no vibration in the strict sense of the word.

In Figs. 66-69 ordinary cream is used. In Fig. 66 everything is at rest. In Fig. 67 the current is switched on. The drop of cream changes shape (Figs. 67, 68). Then the regular migration between the electrodes begins (Fig. 69). There is no violent storm of particles, but instead a "regulated" current transport. This phenomenon inevitably evokes the idea of a "carrier system." Certain variations in the conditions cause the particles to join together in rows. They are strung in threads which grow and bridge the gap (about 3 cm. in the experiment photographed) between the electrodes (Figs. 70, 71, 72). Then structures take shape in the same experimental field, showing that a kinetic "carrier system" can be converted into a structural "carrier system." The threads may take the shape of a string of pearls (Fig. 71). If a pillar collapses or a thread snaps, the migration starts once more from the beginning and the thread is spun again. Future observations of such a carrier system will not be restricted to physics alone. How can we set about observing biological carrier systems (muscle, kidney) and what kind of conditions must be watched for? These are the next questions to answer. Transformations of migratory paths and transport structures are instances of the phenomena for which we must watch.

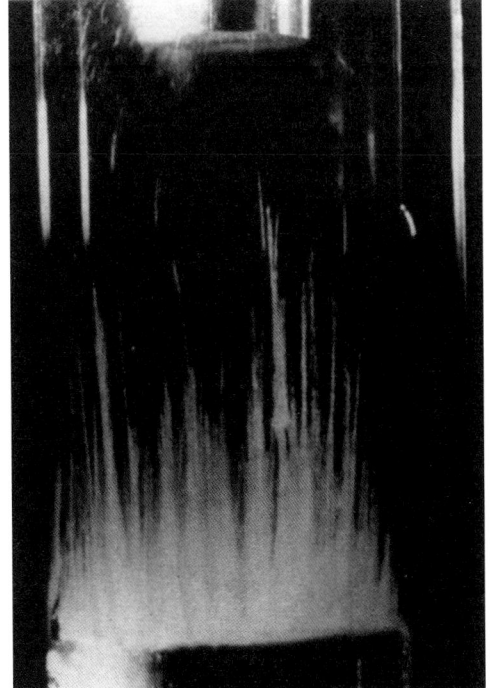

Fig. 70

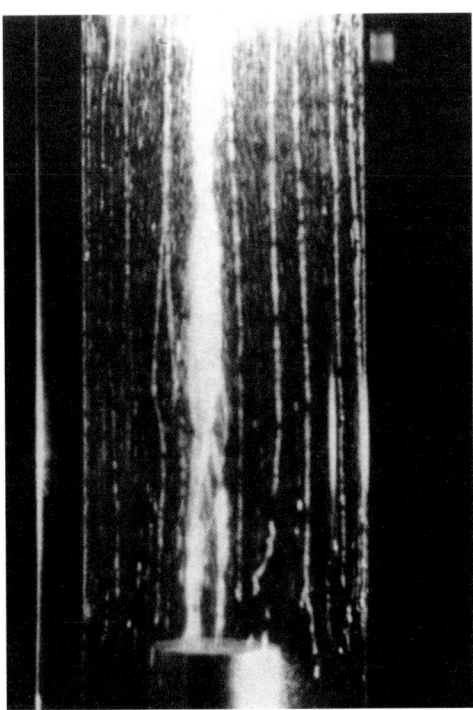

Fig. 71

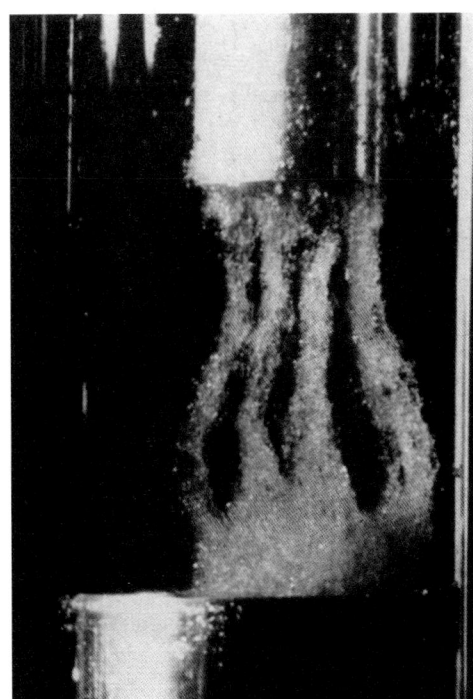

Fig. 72

12

Polygonally Pulsating Drops / Three-Dimensional Structures Under Vibration

Drops excited by vibration oscillate in a polygonal pattern*. Tetra-, penta-, hexagonal and other formations appear (Figs. 73-79). To see what is happening in detail, slow-motion pictures have been taken at 1500, 2000, 2500 frames per second. In this way the process can be visualized as a whole. We find that the *whole* drop pulsates. The liquid moves as a whole as it passes through characteristic phases. It rushes to the periphery, then back to the center, then to the periphery again, and then back to the center and so forth. Figs. 74, 75, 76 show a quadratoid pulsating drop of water on a vibrating metal foil. In Fig. 75 the mass of the liquid is at the periphery; in Fig. 74 it is beginning to flow back to the middle. Fig. 76 shows an intermediate phase. In Fig. 79 the water is at the periphery; Fig. 77 is a photograph of an intermediate phase. Slow motion, however, reveals other details. On its return from the periphery to the center, the water changes in shape, so that, for example, in the pentagonal arrangement, the "pentagon" in the central phase appears to have been rotated through 180°. A kind of inversion or transformation takes place in which one portion of the pulsating mass is pushed into and through another. Apart from the extreme phases, there is also a process of repulsion when the water undulates back into itself. Figs. 73 and 78 show a trigonal and hexagonal form of pulsation.

To continue our studies of these vibrational phenomena in space, we impressed vibrations on soap bubbles, which again pulsated in a regular manner (Figs. 80 to 86). The spherical form imposed by surface tension undergoes changes. The photographs record different phases and show how the various zones fluctuate to-and-fro between curved and flattened forms; Figs. 80, 82, 84 and 86 show an incipient curvature at the top; Figs. 81, 83 and 85 a flattening;

*In *Cymatics, Vol. I* an account was given of polygonally pulsating drops of mercury (e.g. Figs. 114 and 118-124).

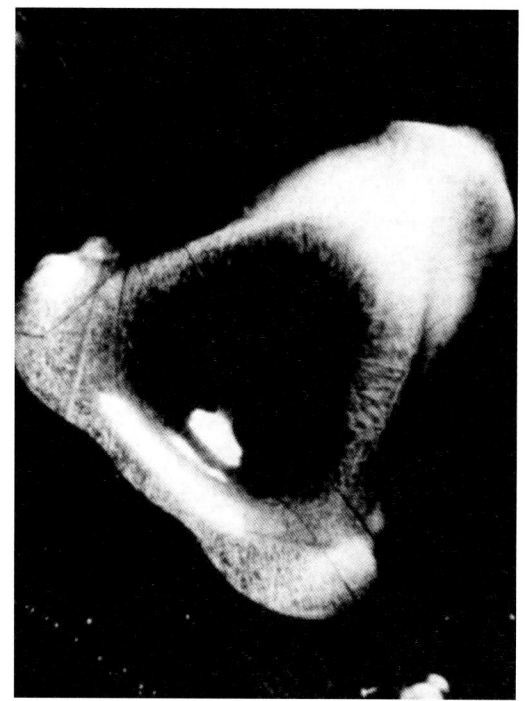

Fig. 73

Fig. 74

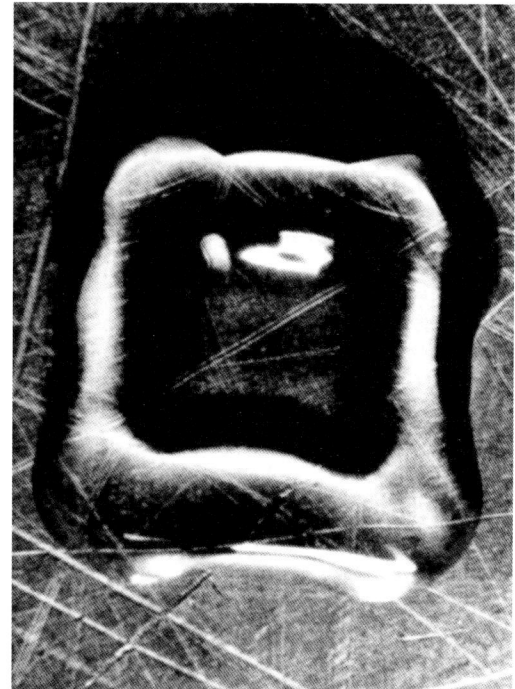

Fig. 75

Fig. 76

198

Figs. 73-77
These photographs show drops of water which pulsate as polygonal figures when excited by vibration. The figures may be trigonal (Fig.73), tetragonal (Figs. 74-76), or pentagonal (Fig. 77). When the drop pulsates, all the liquid flows to the periphery (e.g. in Fig. 75) and then back to the center (in Fig. 74 the mass of water is piled up in the middle; Figs. 76 and 77 show an intermediate stage). Slow-motion films show how the liquid is, as it were, pushed into and through itself. In between there is a phase of repulsion. The structures visible under and beside the drops belong to the vibrating metal foil. (Photos enlarged several times.)

the next moment the position is reversed in conformity with the rhythm of the exciting frequency. When the tone is lower, there are fewer zones of oscillation and vice versa. Thus Figs. 80-81 are produced by a lower tone and Figs. 83 and 84 by a higher tone. In the photographs the soap bubbles are viewed from the side— a so-called "meridional" view; when viewed from the top— an "equatorial" view— regular forms of pulsation can be seen. Hence these pulsating soap bubbles are systems which vibrate in space in mathematically regular arrangements. If *whole* soap bubbles (Figs. 85 and 86) are produced, this fact becomes particularly clear. It would be absolutely accurate to call this a hexagonally pulsating spatial configuration.* It might be added that there are, at the same time, flow patterns in the lamellae— which are again the result of vibration. We shall encounter such flow processes again in harmonic vibrations and take a closer look at them. Drops and soap bubbles tend to be spherical because of surface tension, and the regular vibrational forms are able to develop in interaction with this tendency. But we also find tendencies to regular formations in masses of powder and paste-like substances. There is a certain orderly arrangement about the pulsating heap of lycopodium

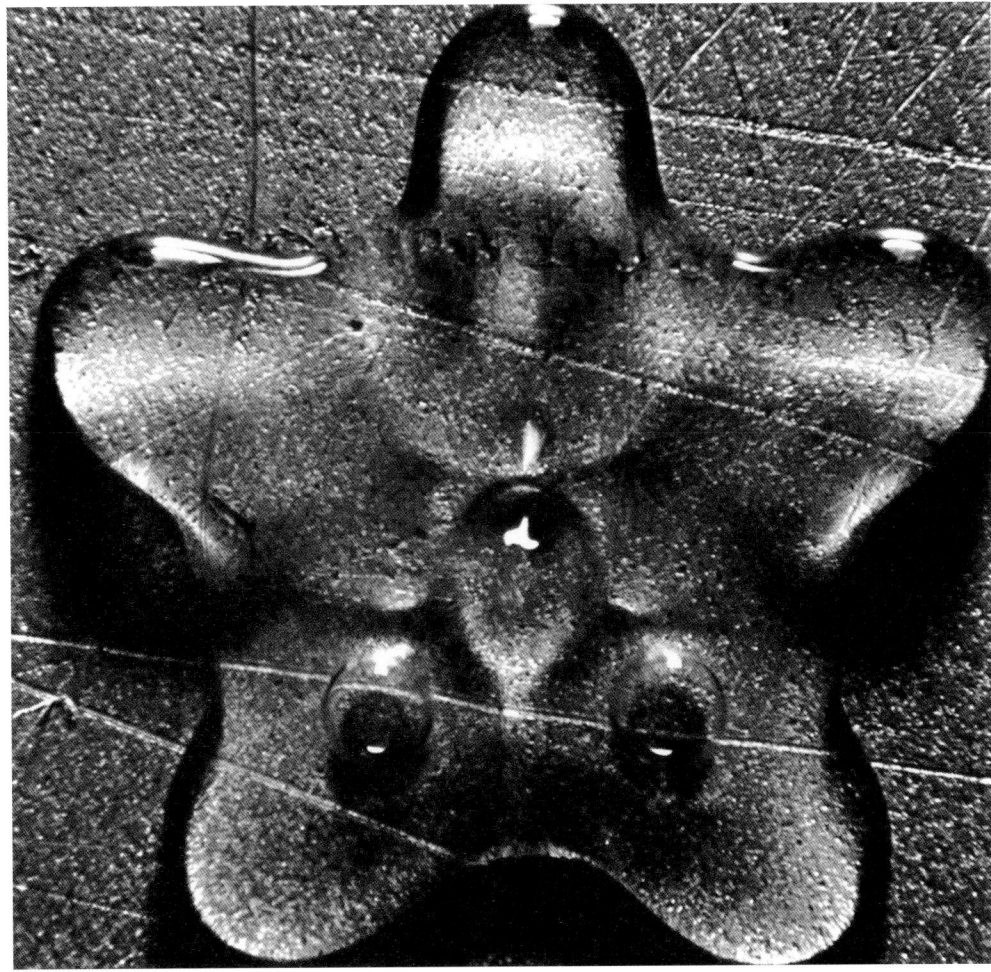

Fig. 77

*In *Cymatics, Vol. I* an account was given of polygonally pulsating drops of mercury (e.g. Figs. 114 and 118-124).

Fig. 78
Hexagonal pulsating drop. The camera has caught the liquid in an intermediate phase. The next moment it will be at the periphery, whence it will return to the center in regular pulsations. It must be remembered that the phenomenon is invoked by sound vibration and thus also has an auditory aspect. The formation depends on frequency and amplitude, and also on the properties of the material. (Photo enlarged several times.)

Fig. 79
A pentagonal drop pulsating. The liquid is at the periphery. In a moment it will pulse back to the center, thereby passing through the phase illustrated in Fig. 77. But the liquid does not collect uniformly in the center, for (as can be seen in slow-motion films) it again assumes a polygonal, pentagonal pattern, only this time the pentagon is rotated 180°. (Photo enlarged.)

Fig. 80

Figs. 80-86
Soap bubbles were used to show how three-dimensional shapes are structured by vibration. The zones of pulsation bulge and flatten alternately (Figs. 80-84). The higher the tone, the greater the number of pulsating zones to be seen (Fig. 80 is a lower tone than Fig. 83). The pulsations of the bubbles can be seen from the side and also from the top, and so it is proper to speak of regular polygonal vibrations in three-dimensional shapes. Figs. 85 and 86 show whole soap bubbles and again the impression created is one of pulsating "polygons." (Figs. 81-84 actual size; Figs. 80 and 85 enlarged.

powder in Fig. 87, likewise in Fig. 88. Fig. 89 shows the round heap of particles in a stellate formation. Fig. 90 shows a pentagonal arrangement of segments in a viscous paste. Naturally surface tension does not operate in the heaps of particles or this viscous mass, but there are conglobations in these phenomena which are maintained by the vibration. The conglobing force of the vibrations imparts a uniform system to the heap of lycopodium powder and the blob of viscous paste; and it is in this homogeneous system that the regular vibrations are manifested as a mathematical arrangement. What underlies the symmetry and mathematical order of these vibrational effects will be examined in our further studies in cymatics. The essential point is that the uniformity of the system is a condition precedent to such symmetrical and mathematically ordered formations whether it is due to surface tension or a unifying force, as for example, the conglobing effect of oscillation.

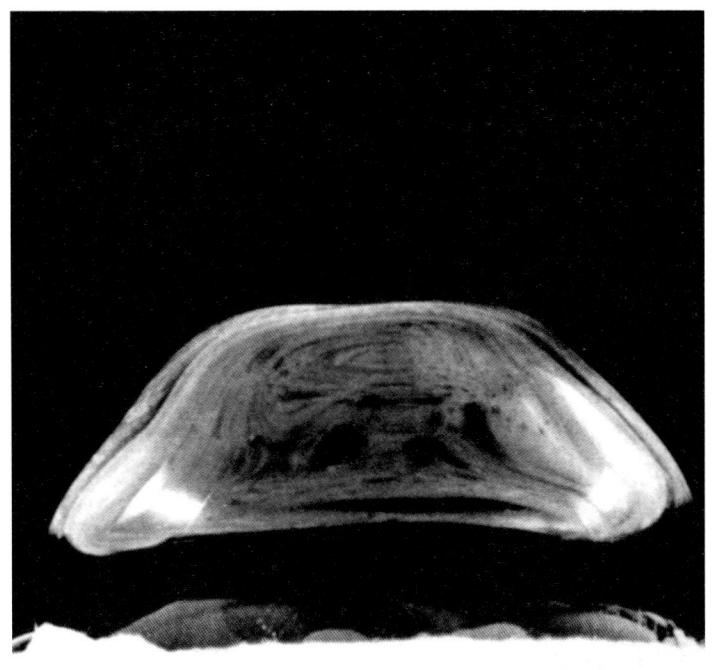

Fig. 81

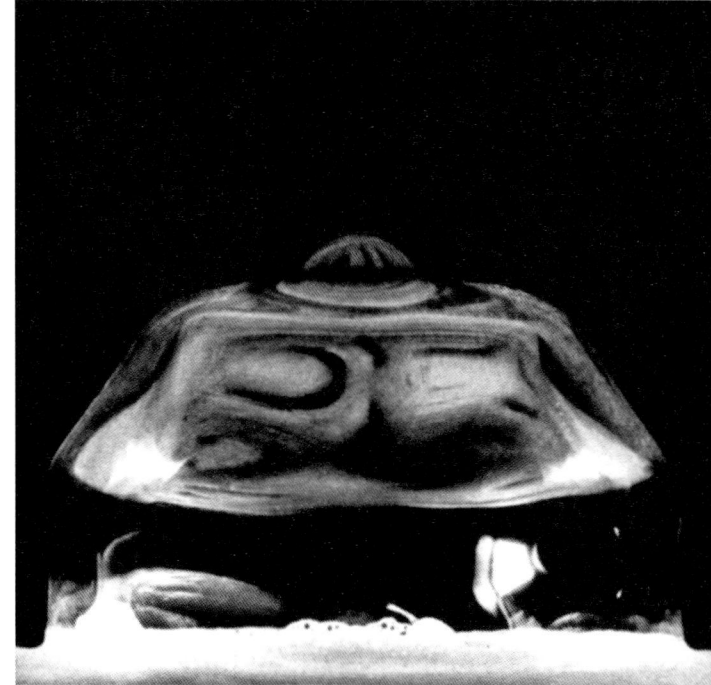

Fig. 82

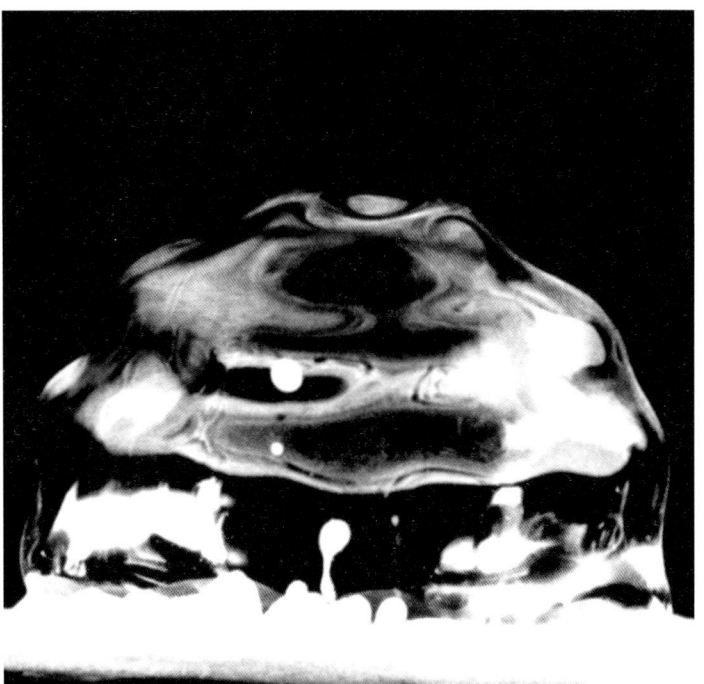

Fig. 83

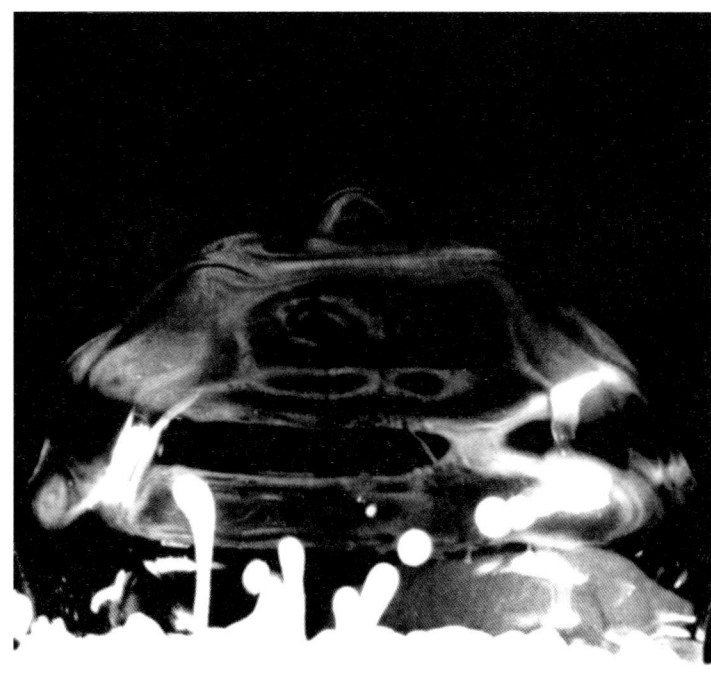

Fig. 84

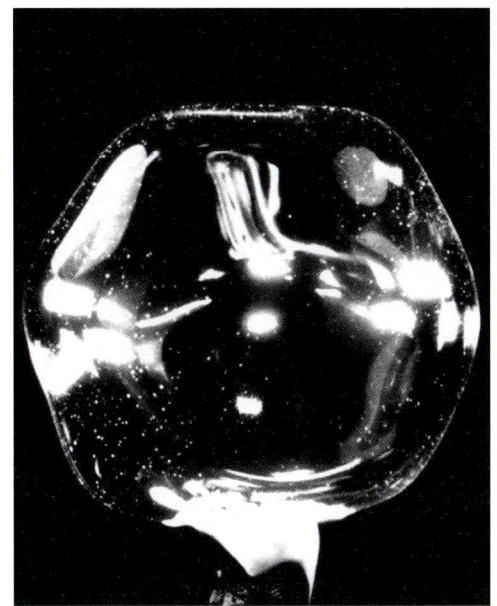

Fig. 85

Fig. 86

203

Fig. 87

Fig. 88

Fig. 89

Figs. 87-90
Geometrical arrangements can also be seen in vibrating heaps of particles and viscous masses. Although there is no surface tension here of the kind that holds the soap bubble together, the oscillation exerts a conglobing force and this bestows a certain unity upon the system. Vibrations therefore give rise to regular patterns with a tendency towards symmetry. In Figs. 87, 88 and 89 heaps of lycopodium powder have been jiggled into geometrical arrangements by vibration; in Fig. 90 a viscous paste has been divided into sectors. Here again it is the conglobing force that confers uniformity upon the structure.

Fig. 90

13

Deflection of the Electron Beam Due to the Interaction of Oscillations at Two Different Frequencies / Pathways of the Mechanical Pendulum

To obtain a general picture of oscillatory forms in their symmetry and mathematical regularity we utilized the cathode ray oscillograph, which allows us to study the ways in which the electron beam is deflected when oscillations at two different frequencies are applied. At the same time the phase of the horizontal beam was displaced by 90° in relation to the vertical beam in order to trace on the screen closed oscillograms, produced by mutually perpendicular harmonic motions. First the familiar Lissajous figures are generated for various ratios of horizontal to vertical frequency. A wide variety of commensurable frequencies can, however, be combined in turn and thus produce what is virtually a complete series of periodic figures. Moreover, the strength (amplitude) of the two frequencies can be mutually adjusted so that the visualized curves go through corresponding transformations. Although these experiments are familiar enough, one example may be described in greater detail. We begin with a frequency of 400 cps. alone for excitation; then we introduce a frequency of 160 cps. at a steadily increasing amplitude while that of the 400 cps. frequency is gradually reduced. While this is being done, a heptagonal figure then runs through a series of systematic transformations (Figs. 91-104), with the pattern of curves changing according to the way the amplitudes are modified. By these means, figures can be generated which display the greatest diversity in the number and proportion of their phases. We have, among other things, been able to generate the formal vocabulary of Gothic tracery. It would therefore be correct to say that these architectural forms actually embody intervals as figures, thus verifying Goethe's dictum that "architecture is frozen music."

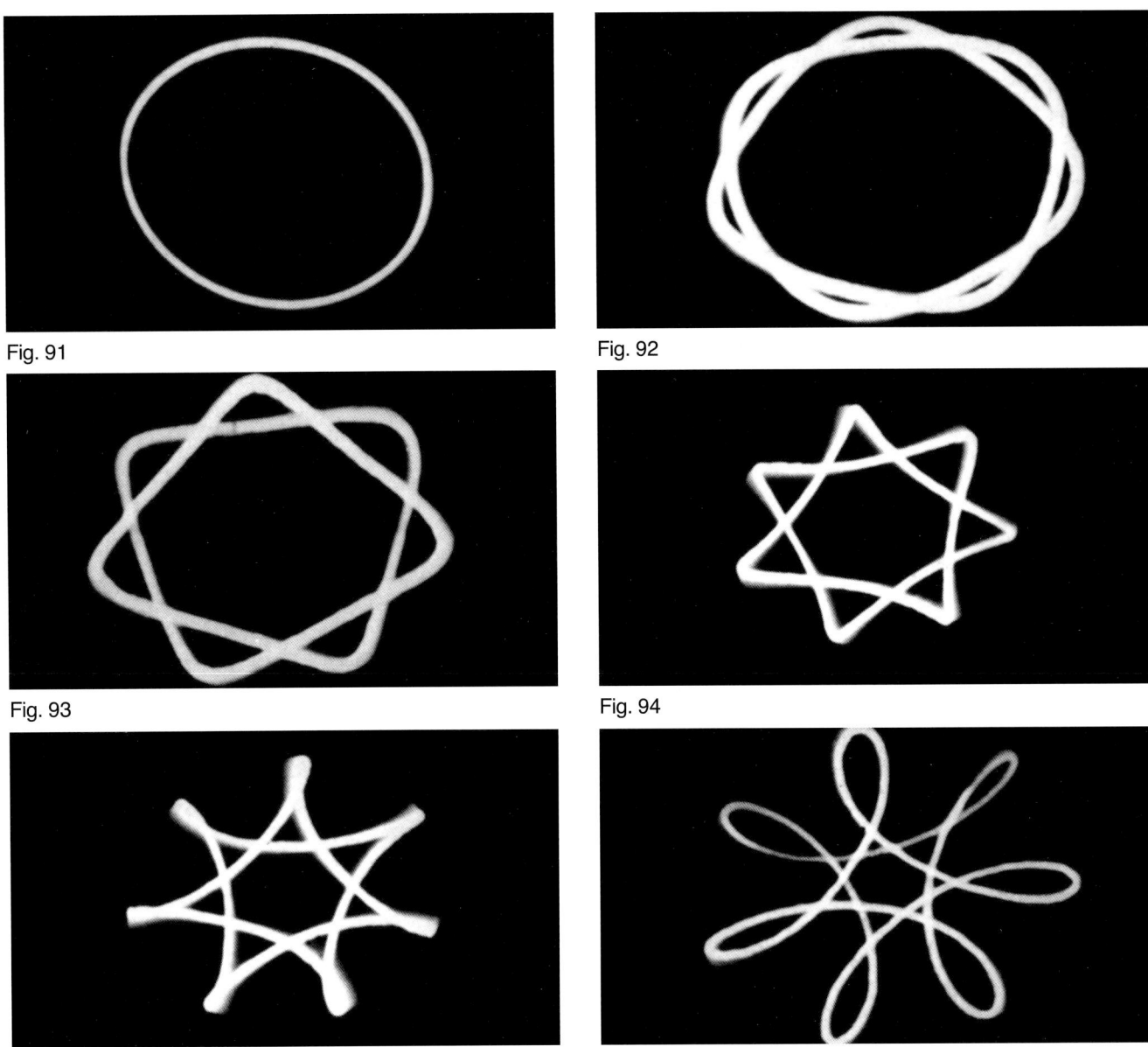

Fig. 91

Fig. 92

Fig. 93

Fig. 94

Fig. 95

Fig. 96

Figs. 91-104
The interval between two frequencies (400 and 160 cps.) is visualized with a cathode ray oscillograph. The frequency of the electron beam is first set at 400 cps. (Fig. 91); the second frequency of 160 cps. is then introduced gradually. While the amplitude of this frequency is increased, that of the other is reduced until only the frequency of 160 cps. is operative (Fig. 104). The interaction of the two frequencies produces a beam pathway exhibiting symmetry, number and proportion. As the amplitudes of the two frequencies are varied, the heptagonal figure goes through the transformations seen in this series of pictures.

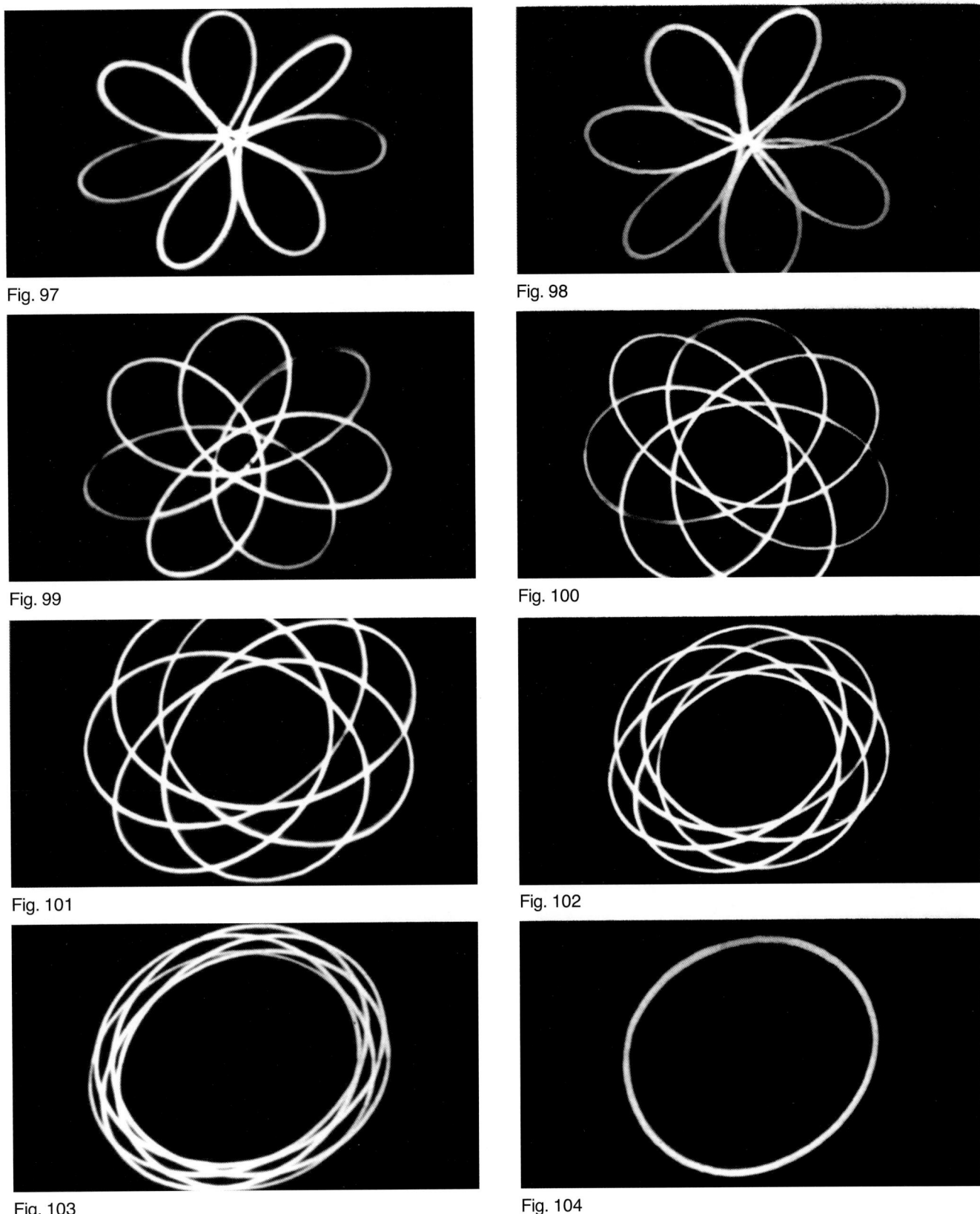

Fig. 97

Fig. 98

Fig. 99

Fig. 100

Fig. 101

Fig. 102

Fig. 103

Fig. 104

The series of observations is particularly enriched when the bent paths of the electron beam are deflected by magnetic fields. The paths are not, as it were, destroyed, but their courses become enormously complex, so complex indeed that the eye can no longer make sense out of them. Fig. 105 reproduces the Lissajous figure of the octave in one phase (ratio of frequencies 300:600, 1:2). We arrange for two magnets to impinge on this figure, and the path is altered accordingly. One aspect of this oscillogram, which is constantly shifting as the magnetic field is moved, has been caught by the camera in Fig. 106. We could, of course, only photograph the screen, and since, under the magnetic influence the curve does not lie wholly in one plane, it seems to break off. In actual fact, however, it joins up with itself in space according to the basic impulse of the octave (Fig. 105). A further example: the minor seventh with its frequency ratio of 540:300, 9:5 (Fig. 107). Figs. 108 and 109 represent the deflections of the electron beam caused by two magnets. The Lissajous figure of the minor seventh is now hardly recognizable. Manipulation of these pathways is so instructive because in complicated vibrational processes of a structural or kinetic kind, there may lie basic processes and basic impulses which are, as it were, hidden under the interplay of the impinging factors.

We also included in our studies experiments with the mechanical pendulum so that we could observe the symmetrical, mathematically regular forms generated by its oscillatory ratios. Although the observations described here are familiar enough, a figure or two which were inscribed by the pendulum itself with a light source attached, are reproduced to remind the reader that such ratios between vibrational frequencies (intervals) exhibit regularities in their number, proportion and symmetry (Figs. 110, 111, 112). It should be noted that the numerical ratio runs through the entire figure as in a diagram. Figs. 110, 111 and 112 display 5-, 7-, and 8-fold patterns, respectively.

To digress for a moment, these pendulum pathways also reveal instructive changes if motion is introduced into the point of view of the observer. For example: first we allow the pendulum to swing with a light source attached to it; second, the camera is made to rotate with the shutter open; and third, the rotating camera is made to travel on a vehicle in a straight line (translation). The photographs seen in Figs. 113, 114 and 115 are the result. They show the pattern of curves which an observer would see if he were inside the camera. We feel that such processes are instructive because they reveal the *real* relationship to the eye and afford us practice in following them. Of course, we can resolve such movements into their individual components and abstract them from their context; but in actual fact there are no such abstracted standpoints either in physical or, still less, in organic systems. If we wish to take mental possession of these various factors in their true reality, we must— when all our analyzing, dismembering and dissecting are done— always make the imaginative effort

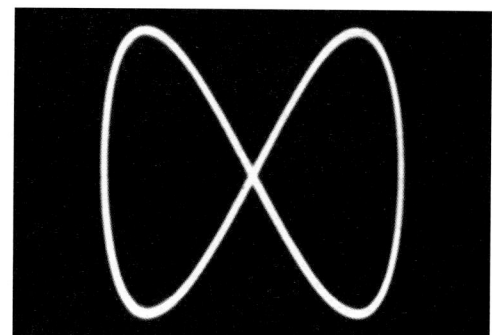

Fig. 105

Figs. 105-106
Using the cathode ray oscillograph we can produce a phase of the Lissajous figure of an octave (Fig. 105). The frequencies are 300 and 600 cps. We then submit the electron beam to the influence of two magnetic fields, resulting in complicated deflections of its pathway (Fig. 106). But the curve is still a continuous pathway; the discontinuity is only apparent because the three-dimensional pathway cannot be reproduced on the screen of the oscillograph. So we must remember that the figure in Fig. 105 underlies that in Fig. 106.

Fig. 106

to see everything as a whole. Such an aim can be realized through the experiments documented here.

The series of experiments outlined in this section afford an assured synopsis of the way in which vibrations can appear in patterns displaying symmetry, number and proportion. These basic facts provide us with a clue for understanding the harmonic vibrations to which we shall turn in the next chapter.

Fig. 108

Figs. 107-109
First we see a phase of the Lissajous figure of the minor seventh, frequencies 540 and 300 cps. We exposed this electron beam curve to two magnetic fields in different orientations (Figs. 108, 109). Complicated pathways result but the curve of Fig. 107 is basic to both of them. Figs. 108 and 109 are patterns made by the continuous oscillatory curves of the electron beam. These curves are, of course, oscillating in three dimensions and so their image apparently disappears from the screen of the oscillograph in some places.

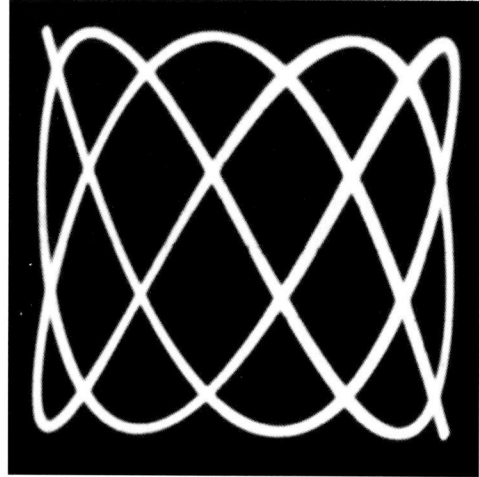

Fig. 107

Fig. 109

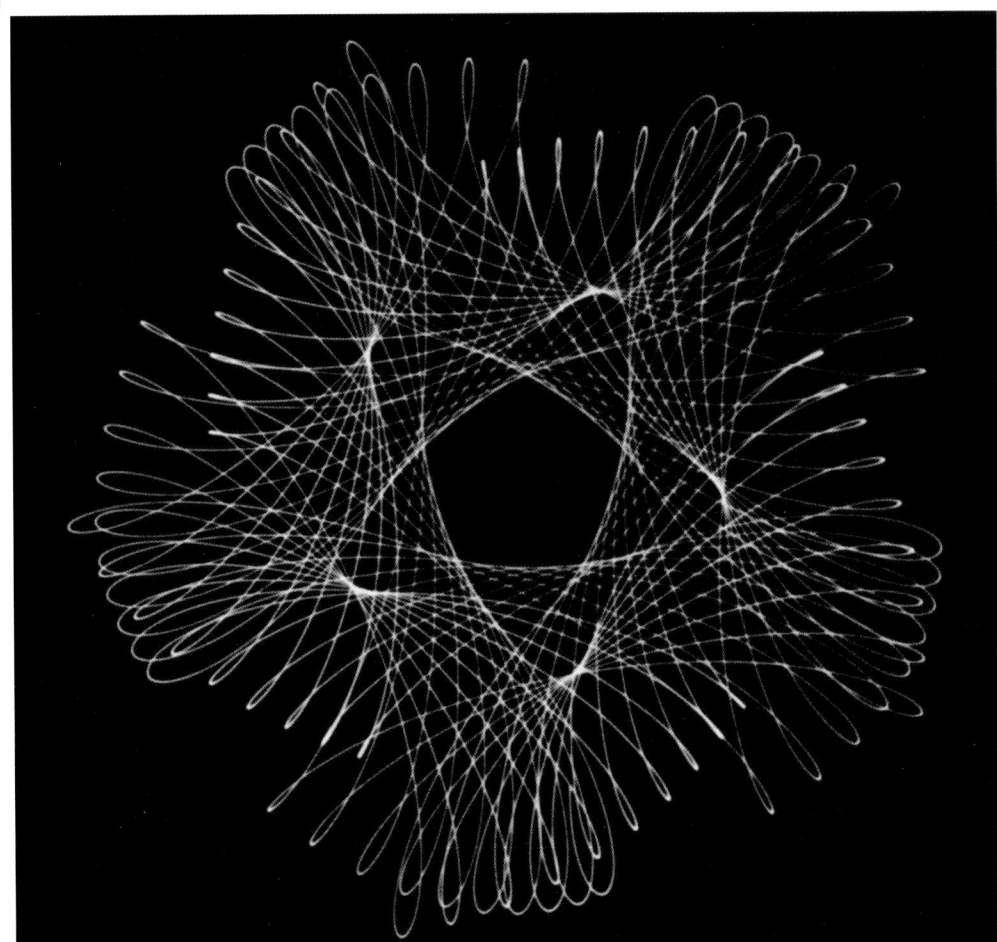

Fig. 110

Figs. 110-112
Here again a familiar method has been employed in which a mechanical pendulum, to which a light source is attached, inscribes its pathway directly into the camera. The point of this is to show that oscillatory processes involve patterns exhibiting number, proportion and symmetry. The pendulum pathways seen here describe patterns with features of regularity in all their parts. (Fig. 110 pentagonal, Fig. 111 heptagonal, Fig. 112 octagonal.) Harmonic elements are visible in every phase of the vibrational process— in the central area of the figured pathway as well as at the periphery.

Fig. 111

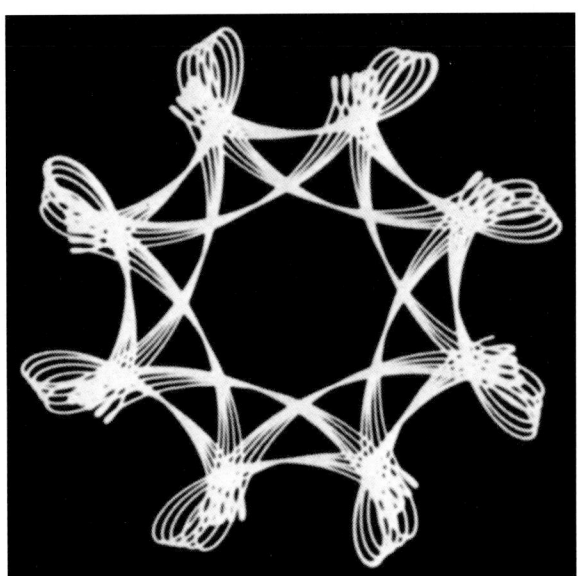

Fig. 112

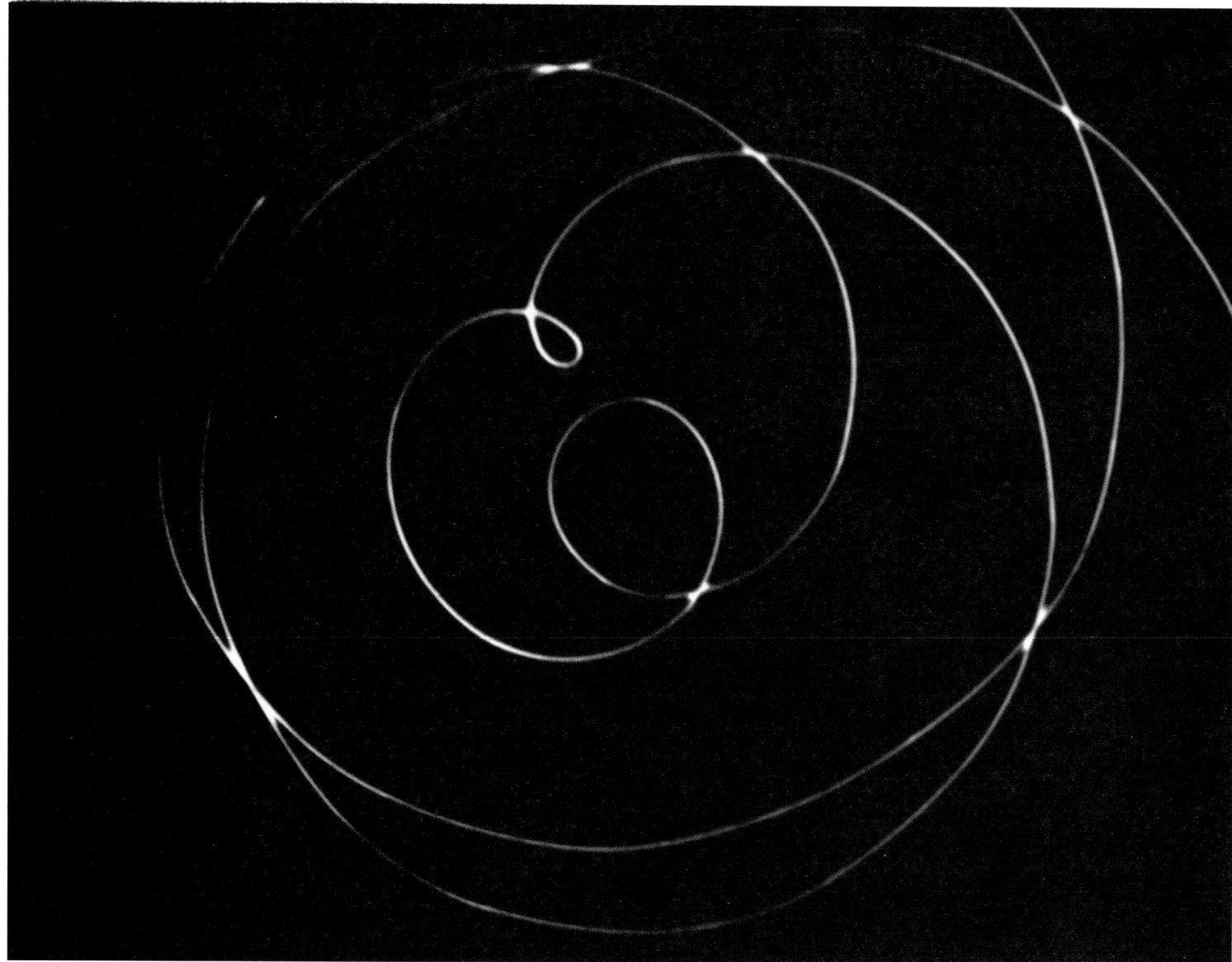

Fig. 113

Figs. 113-115
To get a better idea of the reality of vibrational behavior and to be able to visualize the complex relationships between oscillatory systems, a swinging pendulum to which a light source was attached was photographed by a rotating camera, which itself moved along a straight line. Figs. 113-115 show vibrational patterns as they would appear to an observer inside the camera under these conditions. Observations of this kind focus attention on the dynamic relationships existing between regularly revolving and rotating systems.

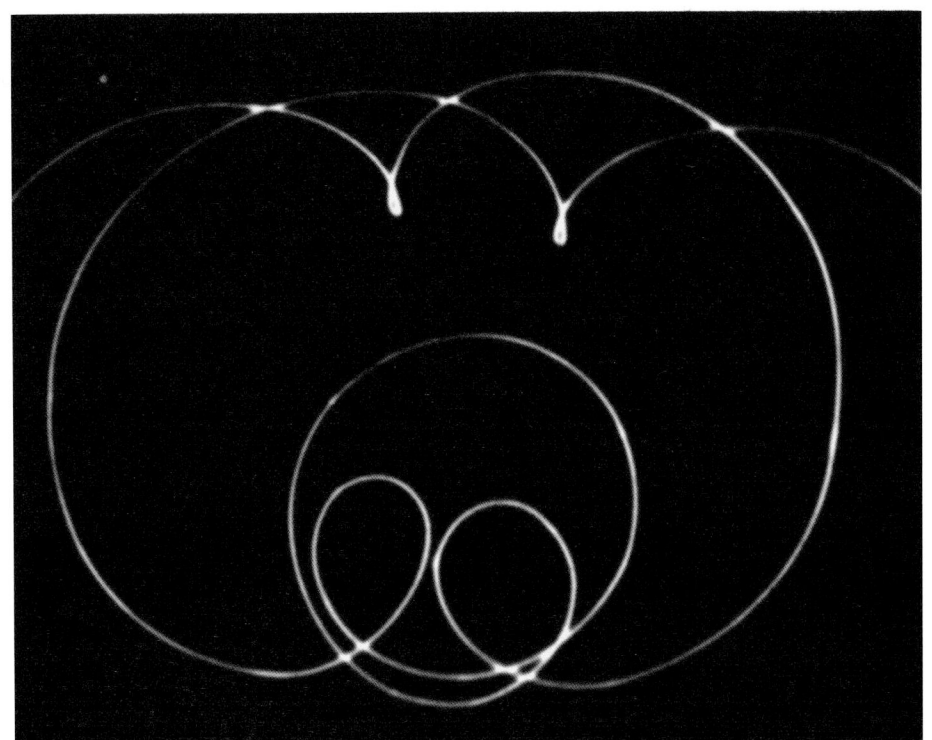

Fig. 114

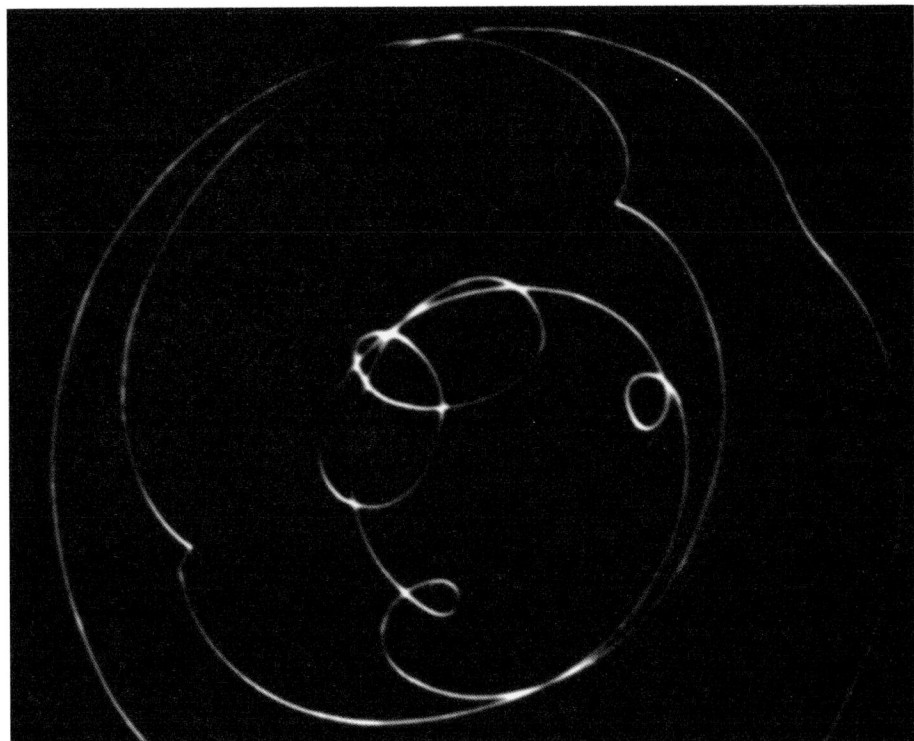

Fig. 115

215

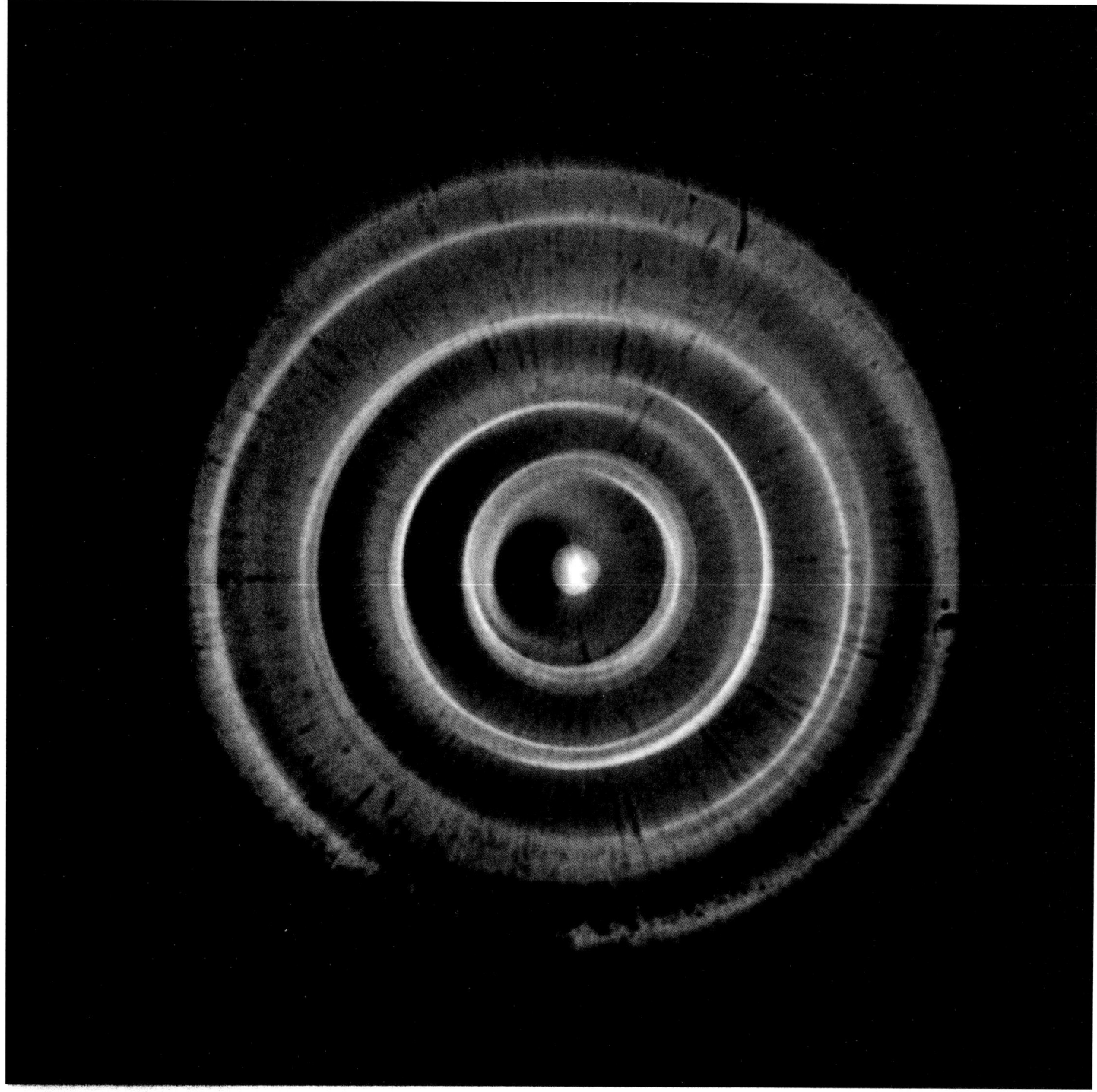

Fig. 116
The concentric circles in this small liquid system (about 1.5 cm. in diameter) are generated by the action of sound. The circles are pulsating and at the same time the liquid is circulating with bilateral symmetry. Pattern, dynamics and pulsation create the impression of a unified whole.

14

Harmonic Vibrations in a Concrete Medium

Once again we must ask: How do vibrations proceed in a concrete medium? What kinds of effects do wave phenomena produce in a specific material? We shall attempt to answer these questions by experimenting with liquids. As a method we shall use the schlieren process, which consists essentially of observing transparent media via transmitted light. If there are inhomogeneities in a liquid, the light is deflected accordingly and any configurations present are made visible. Almost all the following photographs reproduced here have been taken by the schlieren method. The light is projected from underneath and passes through the liquid, and we look down on the processes taking place therein. By adjusting the length of the lenses we can observe the phenomena at various focal planes within the vibrating liquid. Many variations in the design of the experiment are possible: drops as such can be excited by a tone, or liquid can be held in shaped containers, or whole films of liquid can be excited by vibration. It will be found that basically the same fundamental relationships and effects are brought to light.

How does such an experiment proceed? First we have a drop of water at rest. As soon as we excite the drop with sound, concentric rings are formed (Fig. 116). If we increase the same tone, there is an *abrupt* change and a quite different pattern emerges. The configuration is entirely different. If we make no changes in the experimental conditions, the phenomenon persists. If we increase the volume of the sound again, then another configuration appears. The transitions are always *abrupt*. In this way it is possible to obtain a whole series of structural patterns by keeping the exciting tone (the frequency) unchanged and varying the volume (amplitude). (Figs. 117-142 are records of such structures obtained in various experiments.)

Fig. 117

Figs. 117-120
The figures produced by irradiation with high-frequency tones display bilateral symmetry. The elements on the left and right are counterparts of one another (See also caption to Figs. 117-136).

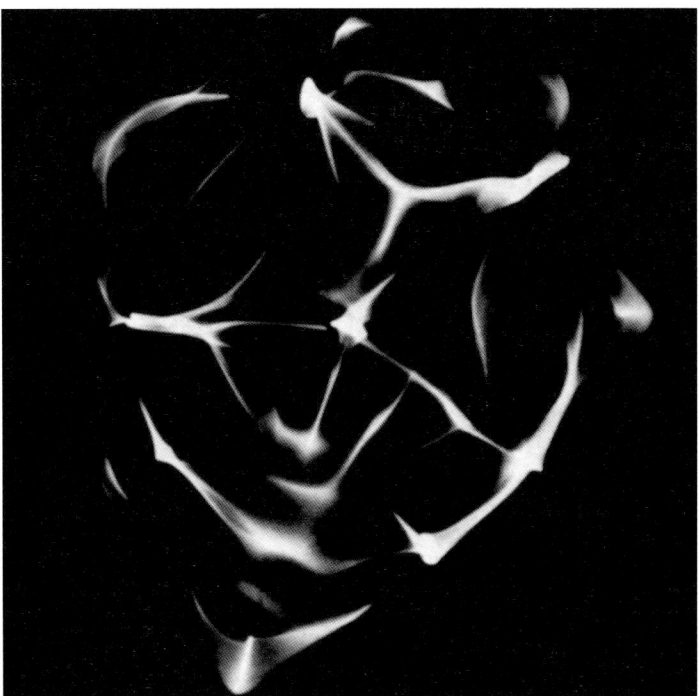

Fig. 118

Fig. 119

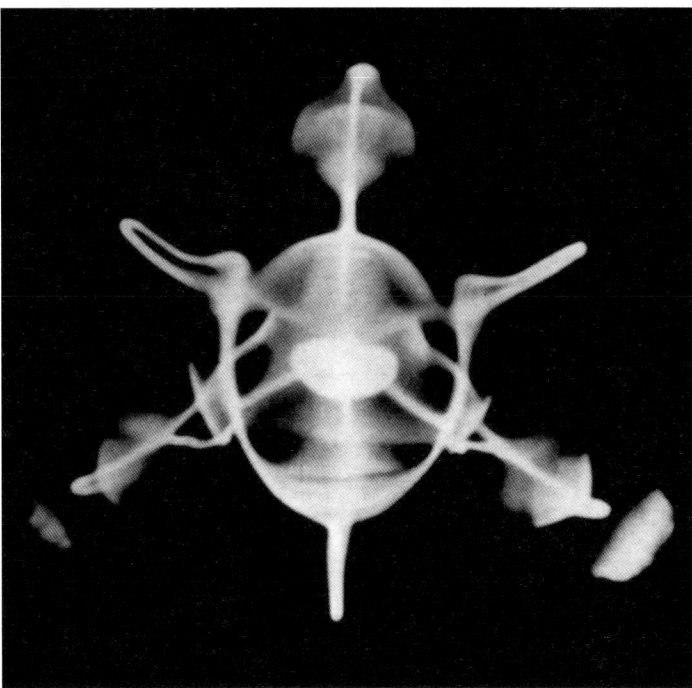

Fig. 120

Figs. 117-136
(General) When the amplitude producing the pattern in Fig. 116 is increased, figures of the kind seen in Figs. 117 to 136 appear abruptly. They are extremely regular with regard to number, proportion and symmetry. These vibrational conditions arise in the liquid itself: they are the creation of its own natural vibrations. The whole array is flowing and pulsating, and yet these kinetodynamic processes are also fitted into the pattern of symmetry. These phenomena, displaying regularity in every category, we call harmonic oscillations. If the amplitude is again increased, (producing a crescendo for the ear) another harmonic figure of the same type appears with no intervening stage between. We can now look in detail at these formations which are, one and all, created by sound. To observe them we use the schlieren method, which involves transmitting light through the liquid from below, thus rendering its structures visible. In these pictures, then, we are looking down from above. By moving the focal plane we can study the characteristics of these figures. The experiments were performed with various liquids (ether, benzine, alcohol, water, and various oils), and in each case the same basic relationships were found, thus showing that we are dealing with the operation of a precise set of laws. These experiments can be replicated at any time, with each particular pattern resulting from the particular frequency used, the properties of the liquid, the quantity of it involved, and of course, the amplitude. As far as the emergence of harmonic oscillations is concerned, it makes no difference whether we vibrate the liquid as free drops, in shaped vessels, or as a film.

The question arises: Where do these figures come from so abruptly? Do they originate in the liquid itself or do we inject a tone medley, a frequency spectrum, into the medium through the agency of the vibrating element (the diaphragm)? One of the ways of throwing light on this question may be mentioned here. First we identify the kind of tone we communicate to the diaphragm. It is a "simple" sinusoidal tone [sine wave] which we visualize in the cathode ray oscillograph. Utilizing this tone we can produce this entire series of vibrational patterns. If we now take the vibrational motion of the diaphragm alone and determine its oscillation in a second cathode ray oscillograph, we shall obtain the same oscillogram as with the tone communicated to the diaphragm. In other words, the diaphragm vibrates in a simple manner: the structural patterns appearing abruptly in the liquid are actually formed there. What is involved is their natural vibrations. Thus we see in front of us the result of complex periodic vibration, a musical tone becoming a "visible" figure in which one or more intervals are featured. One must always bear in mind that these phenomena are generated by sound. If the sound is removed, the whole picture along with its dynamics will disappear and return again immediately when the sound is restored. These phenomena are subject to definite laws and are repeatable at any time.

There is another simple way of showing that the liquid builds up its own oscillatory pattern, and that is by observing it with a stroboscope. If the stroboscope is adjusted in such a way that the oscillating diaphragm is always still during flashes (because the same phase of the oscillation is always illuminated by the flash) it can be ascertained that the liquid still goes on oscillating in conformity with its ordered systems while the surface of the medium and the membrane remain still (stroboscopic effect). The vibrational process going on there, using the same frequency but at an increased amplitude, is clearly built up on the basic oscillation.

The experiments were performed with the most varied materials (ether, alcohol, water, benzine, glycerin, turpentine, paraffin, egg white, etc.). In principle the same basic processes take place, whatever the material involved, although, of course, individual factors vary the phenomena. The first and most essential of these is the starting frequency, but the quantity of the material and the thickness of the film also enter as factors, and it is also important whether the drops are allowed to vibrate freely, in mountings, or in films. Most crucial of all, however, are changes of amplitude (modulation). In the experiments presented here only audible frequencies are used. Yet, vary the conditions of the experiment as we may, the phenomena always display uniform characteristics, and it is to these that we shall now turn our attention. In this connection we would stress that we are deliberately refraining from quantifying individual parameters or analyzing the phenomenon quantitatively. Justified and necessary though such processes may be in their place, here they would merely distract us from the reality of the matter. We shall find that in actual fact it is the phenomena themselves that require us to leave and observe them in their completeness. Only through observation of this kind does the whole phenomenon become revealed little by little, and it is only by looking at it like this that we come to discern more and more, new things which form an essential part of it. If the phenomenon is dissected, total realities are missed, and we are left with disconnected and unreal partial aspects. Not that we simply have to gape at the phenomenon— far from it; on the contrary, our senses must be continually alert. Along these lines, we will now attempt to approach the wholeness of the phenomena step by step.

Let us start with any of the pictures, say Fig. 121, and try to take a closer look at its characteristics. What immediately strikes us is its extraordinary regularity. We begin to count. The countable elements follow in a systematic order, and this order dominates the whole picture. First of all we describe the structures as they are when they are static, without taking account of the movements and pulsations we shall meet later. It must be realized that the photographs have been taken in such a way that a whole phase sequence has been recorded. And yet the emergent order exists in all phases. Even if our photographs are selective for individual moments, hexagonal regularity, as in Fig. 121, is a constant feature. In this photograph "radii" can be seen around the central point. Then follows a zone in which two "hexagons" interpenetrate. Further towards the periphery the situation is repeated; here the "hexagons" have rounded curves. The radial principle prevails throughout the area. At the periphery we again find two circles with a petal-like form, the outer one, once again, having 12 countable elements. To do justice to the phenomenon we should have to describe each phenomenological category in turn; each of them is fitted into the prevailing order. Thus we find regularity in form, in spatial configuration, in chronological sequence, in centration, in symmetry, in the relation between exciting and excited elements, in the integration of whatever is continuously taking shape, etc. The effect of sound and the patterns which arise, involve the whole essence of the phenomenon. We can, for example, quote all ten categories of Aristotelian metaphysics: they all appear embodied in the vibrational order which penetrates the phenomenon uniformly through and through. Let us call vibrational effects which are so comprehensively ordered *harmonic*. We find this harmonic character in the various regular figures: hexagonal (Figs. 121, 127, 135), pentagonal (Figs. 122-125), tetragonal (Figs. 129-134), trigonal (Figs. 126, 128), heptagonal (Fig. 136). Well-defined bilaterally symmetrical figures are also generated (Figs. 117-120). On looking at these formations one can properly speak of symmetrical diagrams, displaying mathematical order; but these are not merely diagrams, they are concrete reality.

Fig. 121

Figs. 121-125
Hexagonal, pentagonal and trigonal figures. Note how number, proportion and symmetry are found throughout the figures. They have all the qualities of a diagram but are in fact "alive" in a fluid medium. The photographs have been taken in such a way as to record a whole sequence of phases. If individual phases have been singled out by the camera, they are nevertheless integral parts of the patterned process as a whole. We call a figure hexagonal, say, although there are 12 elements in the surrounding circles, for closer examination reveals that there are in fact 2x6 elements, likewise there are 2x5 elements in the pentagonal figure. These phenomena show clearly that they are harmonic in all their categories: number, proportion, form, centering, symmetry, and also in their dynamics, pulsations, transformations, polarities, etc. (See caption to Figs. 117-136.)

Fig. 122

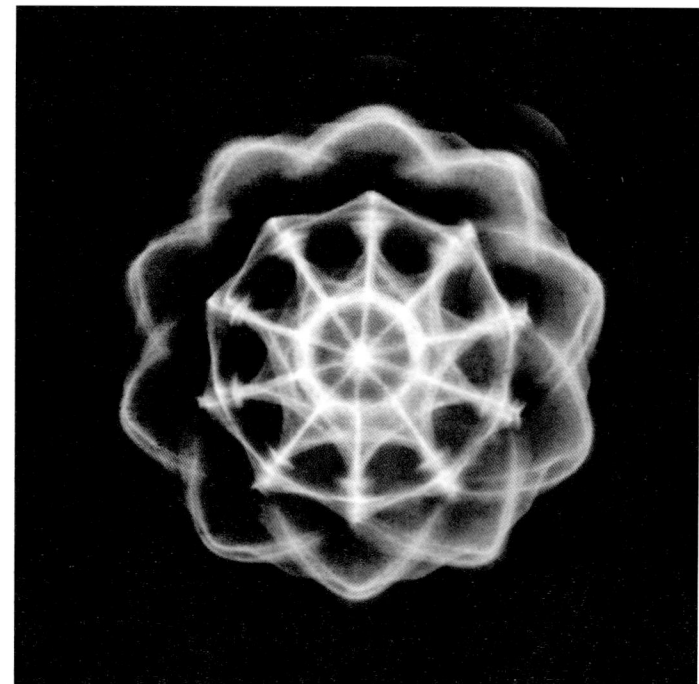

Fig. 123

Fig. 124

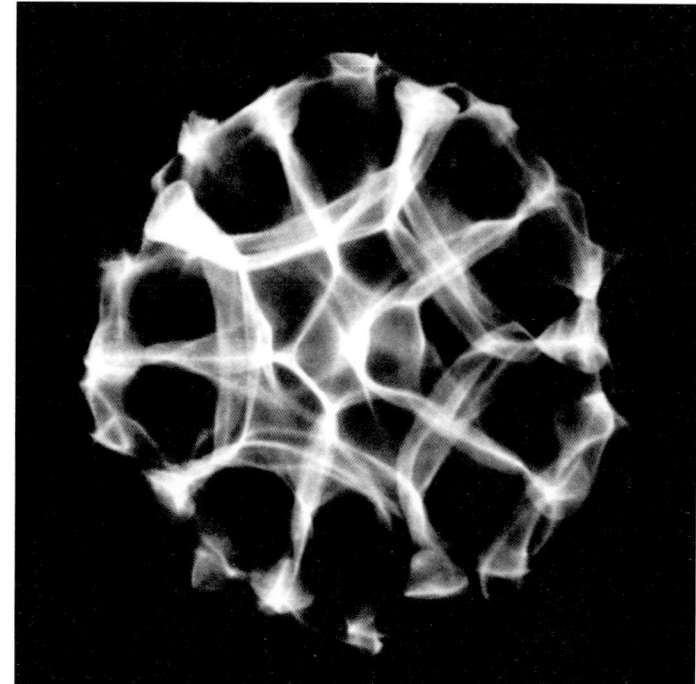

Fig. 125

Fig. 122 shows diverse elements arranged in a quintuple group; they are located on various circles. At the same time they are seen to alternate. The radial fivefold figure alternates with its fivefold quintuple counterpart: the fivefold external circles also alternate (rule of alternation). It is apparent at once that the rule of equidistance also holds true. Incidentally, Fig. 122 shows a freely vibrating drop. The outer circular boundary, which can just be discerned, is the limit of the drop and not a shaped container. There is a plenitude of structures and figures. Note the formation of squares in the tetragonal forms (Figs. 129-133). Bounding forms alternate with radiating forms. Always a particular principle of configuration is characteristic within one circle, while in another circle a different principle prevails, although conforming to the same general dynamic. Fig. 128 shows two configurations of this kind, each arranged like a triangular diagram. What we are really doing, then, is studying the morphology of vibration, or, in other words, drawing up an inventory of all the variety of forms in which it appears. We are not concerned with the play of the subjective mind but with the "objective play of Nature", or with physics. Although it is always "the same", there is an enormous range of mutability in which each individual pattern nevertheless has its precise morphological characteristics. The resultants of harmonic vibrations are at all times so strictly law-ordered that it is possible to draw up a systematology of morphogenesis. What one must bear in mind is, that given a specific set of conditions, Nature produces this form only and no other. Nothing here is diffuse and indeterminate; everything presents itself in a precisely defined form.

The more one studies these things, the more one realizes that sound is the creative principle. It must be regarded as primordial. No single phenomenal category can be claimed as the original principle. We cannot say, in the beginning was number, or in the beginning was symmetry, etc. These are categorical properties which are implicit in what brings forth and what is brought forth. By using them in description we approach the heart of the matter. They are not themselves the creative power. This power is inherent in tone and in sound.

Tone and sound are, so to speak, the entelechies which are active here. The figures represented are, of course, in no way rigid: they pulsate. Let us take the simple example of Fig. 116. The circles of waves pass to-and-fro between center and periphery. But the same pattern of activity also occurs in the phenomena seen in Figs. 117-142. The stroboscope is used here to visualize details of the sequence through which the phases pass. In this way it is possible to comprehend the continuous to-and-fro motion of the liquid. The structural patterns are also affected by this pulsation but not in the sense that they grow turbulent and interfere destructively with one another. They go through regular transformations, they pass into each other, and change periodically in polar phases. The polygonal pattern disappears,

Fig. 126
A trigonal configuration is predominant in this figure. A number of variations on this basic principle appear in the diagram. Again it must be remembered that currents and pulsations also play a part and conform strictly with the trigonal structure. The processes involving pattern, hydrodynamics and pulsation form a unified whole. While the tone persists, the phenomenon continues in motion as a total situation perpetually seeking wholeness.

Fig. 127
A hexagonal arrangement. The detailed diagram is completely subordinated to this ordering principle. The hydrodynamic process is also reflected mathematically in this family of curves.

Fig. 128
In this diagram, a trigonal oscillation is observed. Different formal patterns are created with different frequencies, but in all of them a strict harmonicity prevails. (See Fig. 143 for clarification.) (Photo slightly enlarged.)

Figs. 129-134
Here we see tetragonal figures. Although the boundary is round (whether it be the boundary of the drop or the enclosing vessel), the structures appearing are square. Even if 8 elements can be counted in the surrounding circle, the diagram itself always has 2x4 elements. Whatever form these elements take — edges, diagonals, perpendiculars, radiations, fans, leaves, etc.— the system of the pattern in terms of proportion, number and symmetry is preserved in its entirety. Nor must one lose sight of the fact that there is the same orderliness in the way everything pulsates and flows. Rotary waves often emerge and set the whole pattern turning. The elements also stand equidistant from each other and in an alternating pattern, like those forming the flowers of the phanerogams (See also caption to Figs. 117-136; the small photos have been slightly enlarged.)

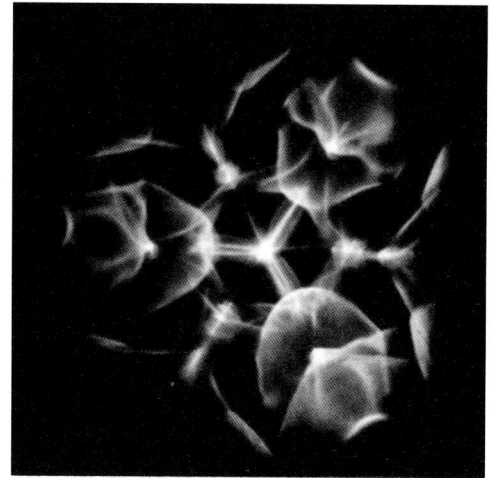

Fig. 128

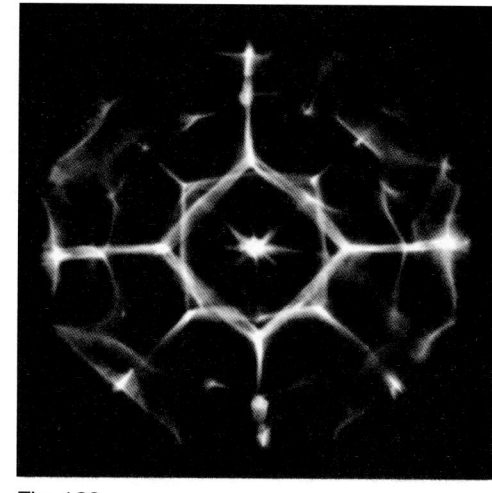

Fig. 129

Fig. 130

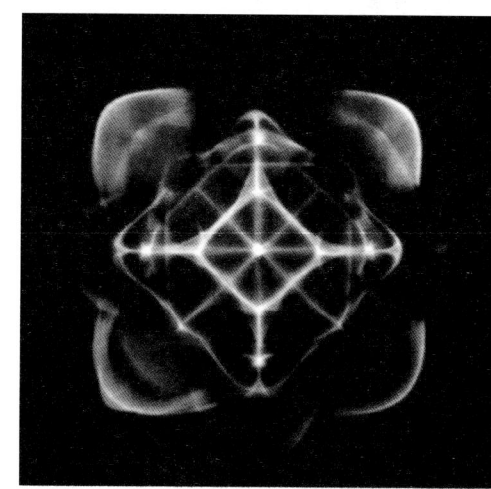

Fig. 131

Fig. 132

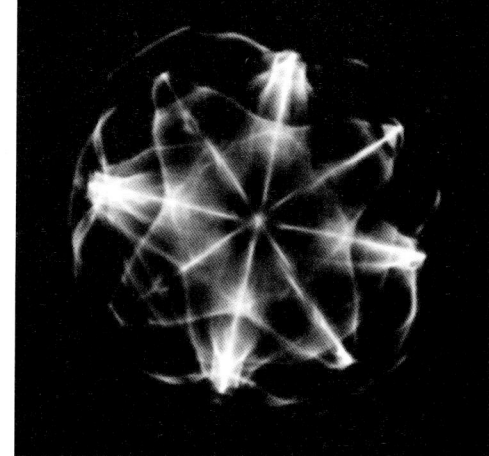

Fig. 133

Fig. 134

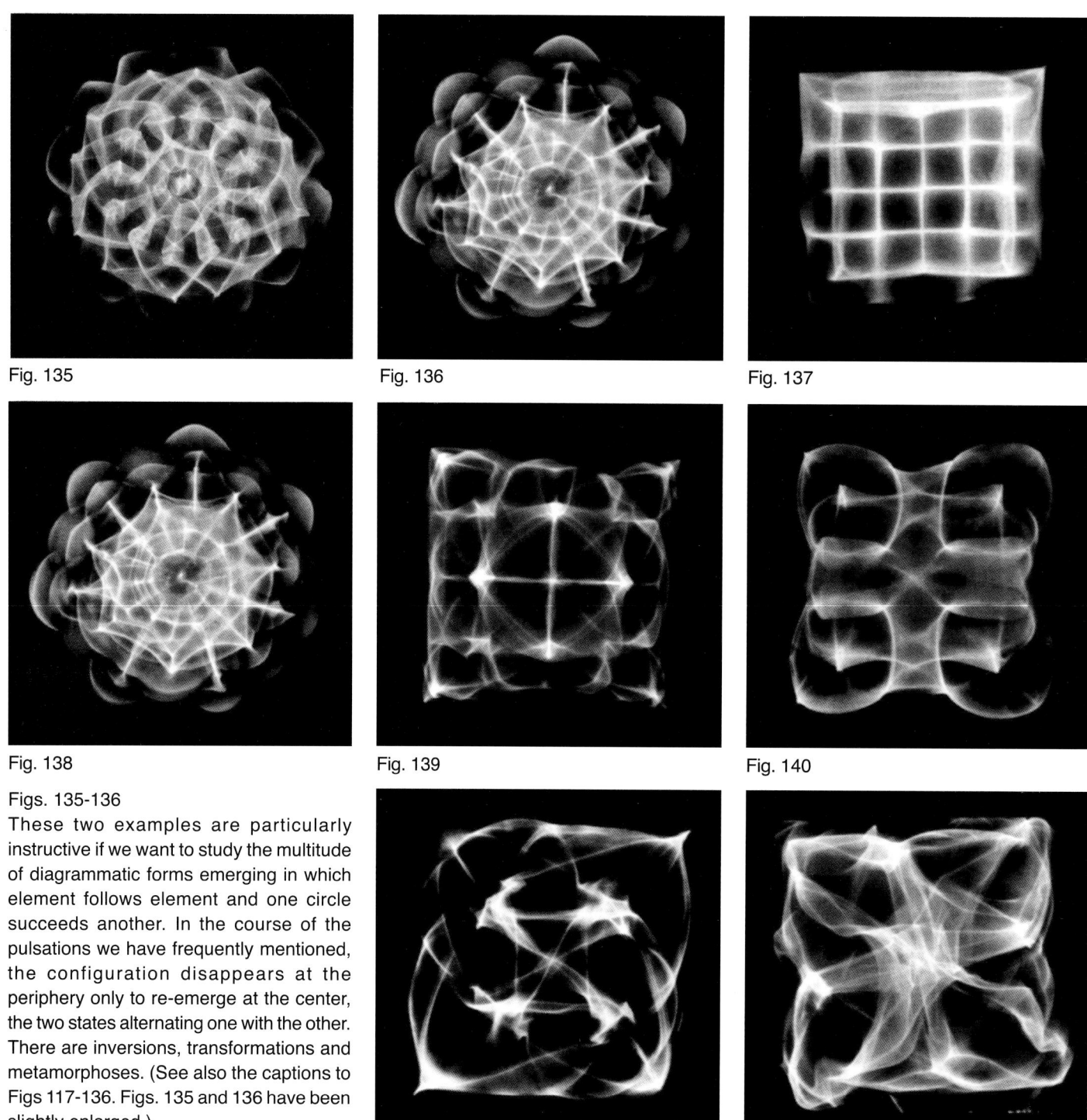

Fig. 135

Fig. 136

Fig. 137

Fig. 138

Fig. 139

Fig. 140

Fig. 141

Fig. 142

Figs. 135-136
These two examples are particularly instructive if we want to study the multitude of diagrammatic forms emerging in which element follows element and one circle succeeds another. In the course of the pulsations we have frequently mentioned, the configuration disappears at the periphery only to re-emerge at the center, the two states alternating one with the other. There are inversions, transformations and metamorphoses. (See also the captions to Figs 117-136. Figs. 135 and 136 have been slightly enlarged.)

Fig. 137-142
First of all we see a basic form, a quadratic lattice. As the amplitude is increased, various figures spring into view, all of which are contained within a square vessel. As we saw in Figs. 129-134, a square container is not an essential condition for the appearance of the tetragonal principle. Here again in these photographs (Figs. 138-142) we have harmonic oscillations. In Figs. 141 and 142 there is a suggestion of rotation. (See also caption to Figs. 117-136. Figs. 137-142 have been slightly enlarged.)

Fig. 143
In this photograph we are looking at the *surface* of the pulsating liquid from the *side*. It must be imagined that, the next instant, the three hummocks will subside and three others will appear between them. The relief changes in time with the rhythm of the exciting frequency. If this process of oscillation were to be viewed from above using the schlieren optical system, a trigonal structure (2x3, as seen in Fig. 128) would be seen. This lively play of plastic forms does not cause turbulence in the harmonic pattern; indeed, it is itself harmonic in character. (Diameter 2 cm.)

for instance, at the periphery, only to re-appear in the center; and in its place new formations of the same order appear at the periphery. These vanish in turn and the original patterns re-appear at the periphery. There is a constant process of shaping and reshaping, an incomparable metamorphic drama. The processes do not impinge on one another: they flow through one another in harmonic "wave curves" and "curve waves", and at the same time concentric to-and-fro motions can be observed.

Anyone viewing the photographs will ask himself: What does this all look like in reality? The interplay of pulsations transpires in three dimensions. If, instead of from above, the vibrating liquid is viewed with the schlieren optical apparatus from the side, an undulating relief is seen. The louder the tone, the more marked this relief becomes. Figs. 143-146 are instantaneous photographs taken from the side. In Fig. 143 three domes can be seen, a moment later (in time with the frequency) they will subside, and in the spaces between them three new ones will appear. Then the process will be repeated, so that the view seen in Fig. 143 will appear again. If, using the schlieren system, this experiment is viewed from above with the camera looking down on the pulsating liquid, a pattern of two times three will be discernible. Similar conditions prevail in the other photographs: Fig. 144 has a quintuple character, Fig. 145 a septuple. Fig.146 shows an intermediate phase but the tetragonal arrangement (2x4) can be clearly recognized.

Let us look at an example showing the process by which vibrations work up into forms. At the same time let the reader remember that the change in the figures is *abrupt*. A series of patterns is shown in Figs. 147-151 appearing in sequence as the vibrations change in quality. First of all there is the basic tone as seen in Fig. 116. The number of concentric rings is determined by the pitch of the tone; the higher the tone the greater the number of rings appearing. The tone is now held at a given frequency, but the amplitude is increased. This, however, must be done with great delicacy. Suddenly, as the crescendo mounts, Fig. 148 appears: a tetragonal figure which, providing conditions are not altered, persists as it pulsates. If we again turn up the volume, Fig. 149, a pentagonal configuration, suddenly flashes into sight. Another change, and Fig. 150, a bilaterally symmetrical pattern springs into view. Fig. 151 is again the result of increasing amplitude; a tetragonal pattern can be discerned. As many as eight of these periodic vibrational patterns have been generated in one and the same experiment. Then, as the amplitude is increased, turbulence begins and the pattern is dispersed. A particularly interesting feature of these experiments is that quantitative intensification and structuration bring forth qualitative phenomena (tetra-, penta-, hexagonal forms, bilateral symmetry, etc.); for the one pattern cannot be drawn over into the other merely by a steady quantitative increase. What we really see in these abruptly appearing patterns

Figs. 144-146
Again a side view. As they pulsate, these reliefs, with their features of mathematical regularity, go through changes which can be followed with the aid of the stroboscope. Fig. 144 shows a pentagonal, Fig. 145 a heptagonal, and Fig. 146 a tetragonal configuration. The lively up-and-down motion of the hummocks at first gives no hint of the strict pattern of pulsation prevailing in the liquid, which can, however, be visualized with the schlieren system. Again and again the observer is made to realize that structure, dynamics and the changes due to pulsation are all subject to the unifying influence of the vibrational effects. (Diameter approx. 2 cm.)

Fig. 144

Fig. 145

Fig. 146

are qualities. And certainly the proposition that Nature makes no leaps *(natura non facit saltus)* does not hold true here; on the contrary here Nature does make [quantum] leaps in an expressive way: quantitative augmentation is accompanied by a corresponding gradation—corresponding discontinuities in which something qualitative appears, *Natura facit saltus!* The process in which quantity and quality are manifested here is also characteristic of another aspect of these harmonic vibrations: it demonstrates the periodic style according to which systems are built up in Nature.

To the processes involving morphology and pulsation must be added flow. All the phenomena are in a process of flow, and these movements proceed in complete harmony with the sequence of events producing the symmetrical figures with their mathematical regularity. Figs. 152 and 153 show the simplest case: a pair of currents is circulating in a bilaterally symmetrical pattern. One current is circulating in a clockwise and the other in an counter-clockwise direction. These are not vortices but *circular* or *ring currents*. (In Fig. 152 a marker dye was used as an indicator. In Fig. 153 lycopodium powder was floated on the current.) The velocity depends on the amplitude. While the excitation proceeds at a constant level, the flow pattern remains unchanged. The question of how the structural patterns behave individually in response to these currents is a particularly interesting one. Fig. 154 provides us with an answer. We see a hexagonal pattern: the brush-like strokes are current pathways made visible with lycopodium powder. The photograph was taken in such a way that the path of these particles recorded itself. Only motion film, of course, reproduces the actual movement, but photographs like Fig. 154 do provide definite evidence of the way in which figure and dynamics harmonize. The flow areas are distributed regularly in a hexagonal pattern. For instance, in the peripheral areas of a hexagonal figure there are six regular pairs of currents, each rotating in a contrary direction. Twelve circular currents can be seen. In the central parts there are additional circular currents, all in a hexagonal pattern. Fig. 155 is taken from a series of experiments with containers of different shapes. A series of different current patterns can be seen. In the central parts it is clear beyond doubt that there are circular currents, or concentric circulation. If the volume of the tone is increased, the indicating particles rush around in a circle. Figs. 116 and 117 can be understood in the light of the structures and flow patterns shown here. The curious lateral branchings are due to the harmonic motions in which the drops of liquid become involved on their introduction. Thus we see how a phenomenon becomes "woven into a whole," only after prolonged observation.

It may be asked whether it is possible to make two tones act upon a liquid "from outside"? Does a regular figure then appear? What is the effect? Here we present three examples of experiments framed to answer this question. These experiments must be carried out with delicate care. Using one tone only, we produce the corresponding structure, and then with the second tone, another corresponding structure. Now the two tones are applied together. It must always be remembered that we really have two tones in action in the vibrating liquid. We must continually think back to the individual tones. And in actual fact it proves possible with two tones, each of which produces a characteristic structure, to bring forth a third figure. In the first example the effects produced by the two tones singly are shown. One tone produces a tetragonal, the other a pentagonal pattern. When the two tones are combined, the bilaterally symmetrical figure seen in Fig. 158 appears. The element on the left-hand side always appears as a mirror image on the right, and vice versa. Again in the second example a pentagonal structure (Fig. 159) and a triangular structure (Fig.160) also produce a bilaterally symmetrical pattern (Fig. 161). And again the elements can be verified on the right and left sides. In the third example Fig. 164 arose from Figs. 162 and 163. It goes without saying, that once more in these experiments the forms produced by the individual notes must first be identified, otherwise there is the risk of being deluded into thinking that a figure is the result of the two notes whereas in actual fact it is due simply to one of the notes predominating over the other.

Now we come to the visualization of vibrations in films of liquid (Figs, 165-170) by utilizing the schlieren system. The structural patterns in the examples selected are of a tetragonal nature. There is a detailed four-sided grid pattern with diagonals and also curved configurations. Once again the whole medium is harmonic in character. These films also pulsate when irradiated by sound; and the corresponding flow patterns also appear there (Figs. 171, 172). The flow paths are indicated, while at the same time the structural pattern shows through. For all the extraordinary complexity of the phenomenon the dominant influence is that of harmonious symmetry and mathematical regularity.

Invariably in these harmonic vibrations the three basic elements of cymatics are in evidence: structure, wave process, dynamics.* They appear as a unit and are wholly inherent in the symmetrical, mathematically ordered harmony. Here, as we have already intimated, the phenomenal categories are sustained in the intrinsic harmonic style. Other categories can be added. The processes pass through regular transformations and at the same time they are pervaded by multiple polarities. Thus the metamorphosis of their morphology and the polarity of their processes and structures can be added to the list of characteristics. It is consistent with this to say that the physical processes intensify themselves. But what does "intensification" mean in this context? It certainly does not mean that we must introduce scales of values and straitjacket our phenomena into them.

*In the first volume, Cymatics, Vol. I, these conditions were described as the basic triadic primal phenomenon.

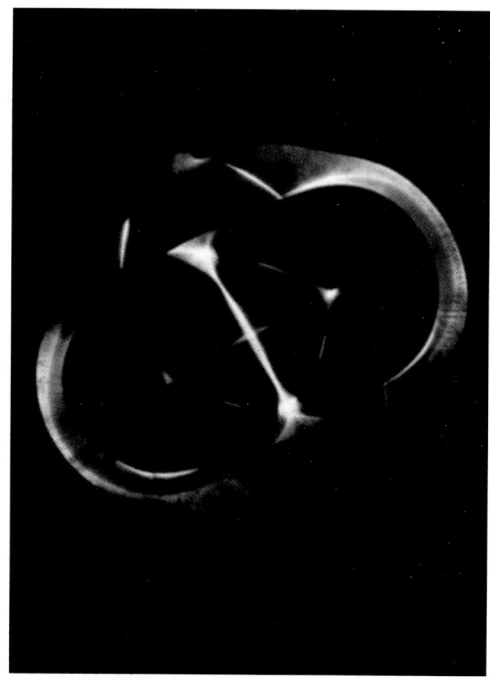

Fig. 147

Figs. 147-151
This series of figures was produced in *one* experiment. While the *same frequency* (the same tone) is maintained, these structures emerge *abruptly* one after the other with increasing amplitude, and in some instances the patterns are quite different (bilaterally symmetrical, tetrogonal, pentagonal). If the amplitude is increased still further, mild turbulences are set up and disrupt the picture. All the harmonic vibration patterns shown (Figs, 117-142) have come about as a result of this abrupt change. Nature really does make leaps.

Fig. 148

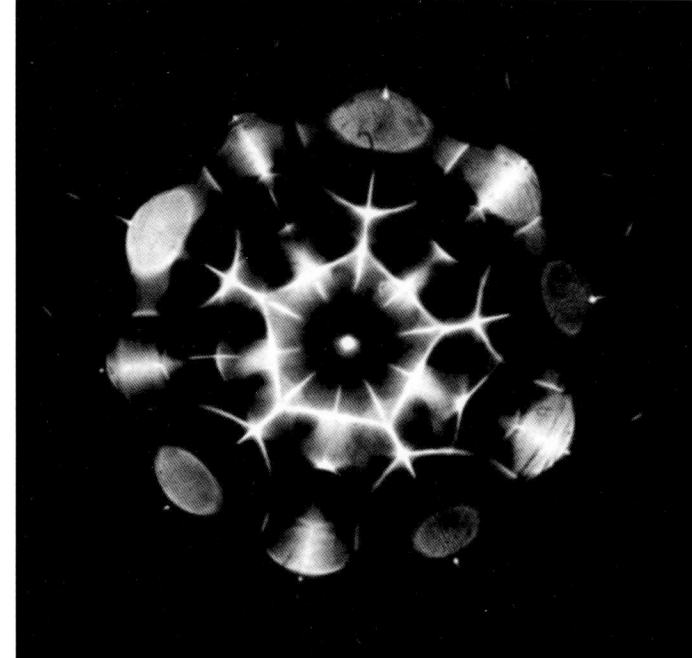

Fig. 149

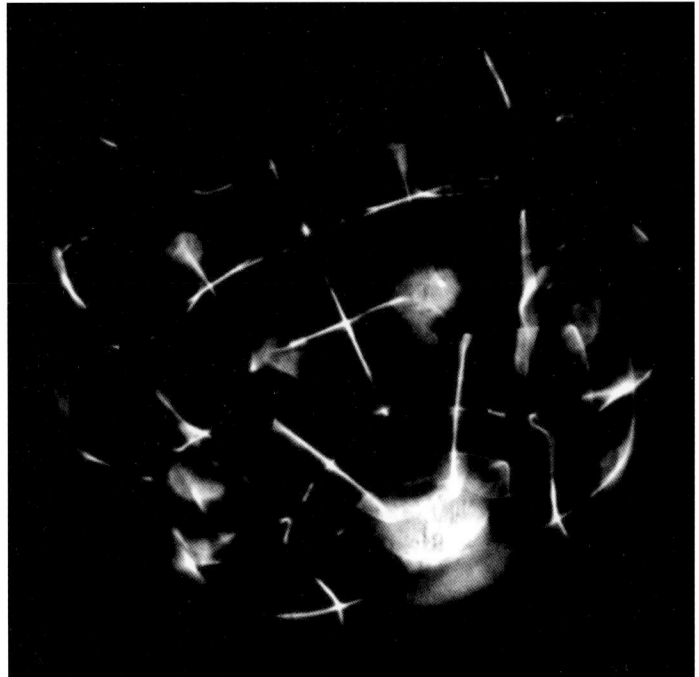

Fig. 150

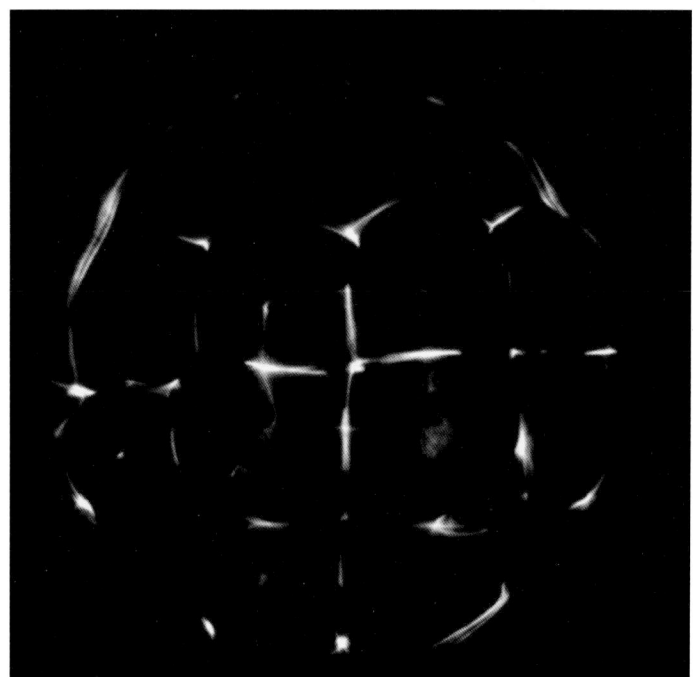

Fig. 151

Figs. 152-154

The currents we have frequently mentioned are illustrated in Figs. 152-155. Figs. 152 and 153 show the simplest case; the liquid is circulating in a flow pattern displaying bilateral symmetry. In Fig. 152 the current paths are visualized with a marker dye, in Fig. 153 by floating particles of powder. These circular currents are rotating in contrary directions. They are a feature of all harmonic figures (Figs. 117-142) but are fitted into the prevailing order, as may be seen in Fig. 154. In this hexagonal figure the current paths are delineated by the floating particles. It will be seen that the currents fit with perfect regularity into the overall pattern. Thus kinetodynamics is added to structure and pulsation. A system is created in which unity is attained through the interplay of the components. However complex things look, they can nevertheless be viewed as a whole in their symmetry and mathematical regularity. A hexagonal figure displays a flow pattern in which, for example, there are 12 circular currents or 6 pairs of currents rotating in contrary directions at the periphery. (Figs. 152 and 153 are slightly enlarged. Fig. 154 is enlarged several times.)

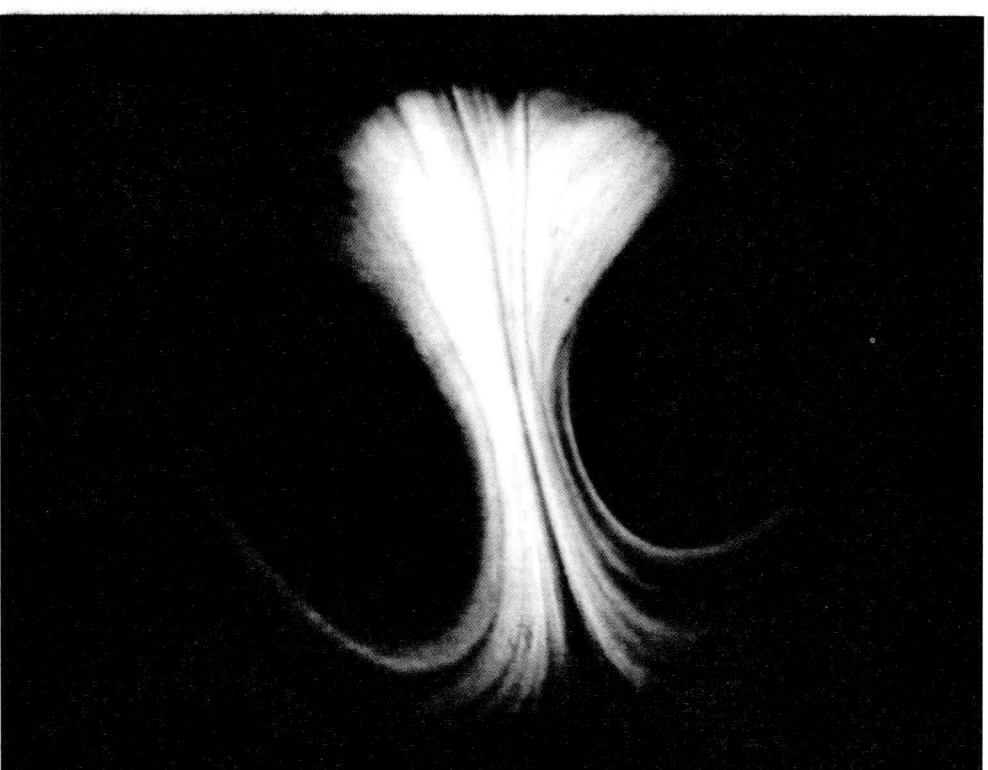

Fig. 152

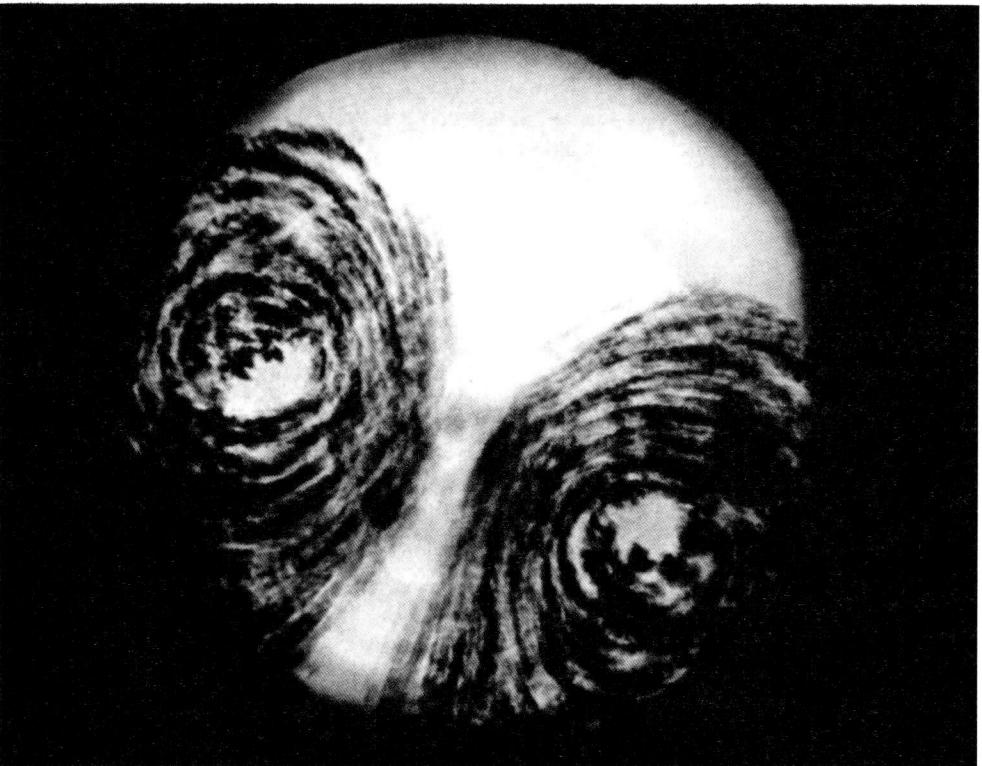

Fig. 153

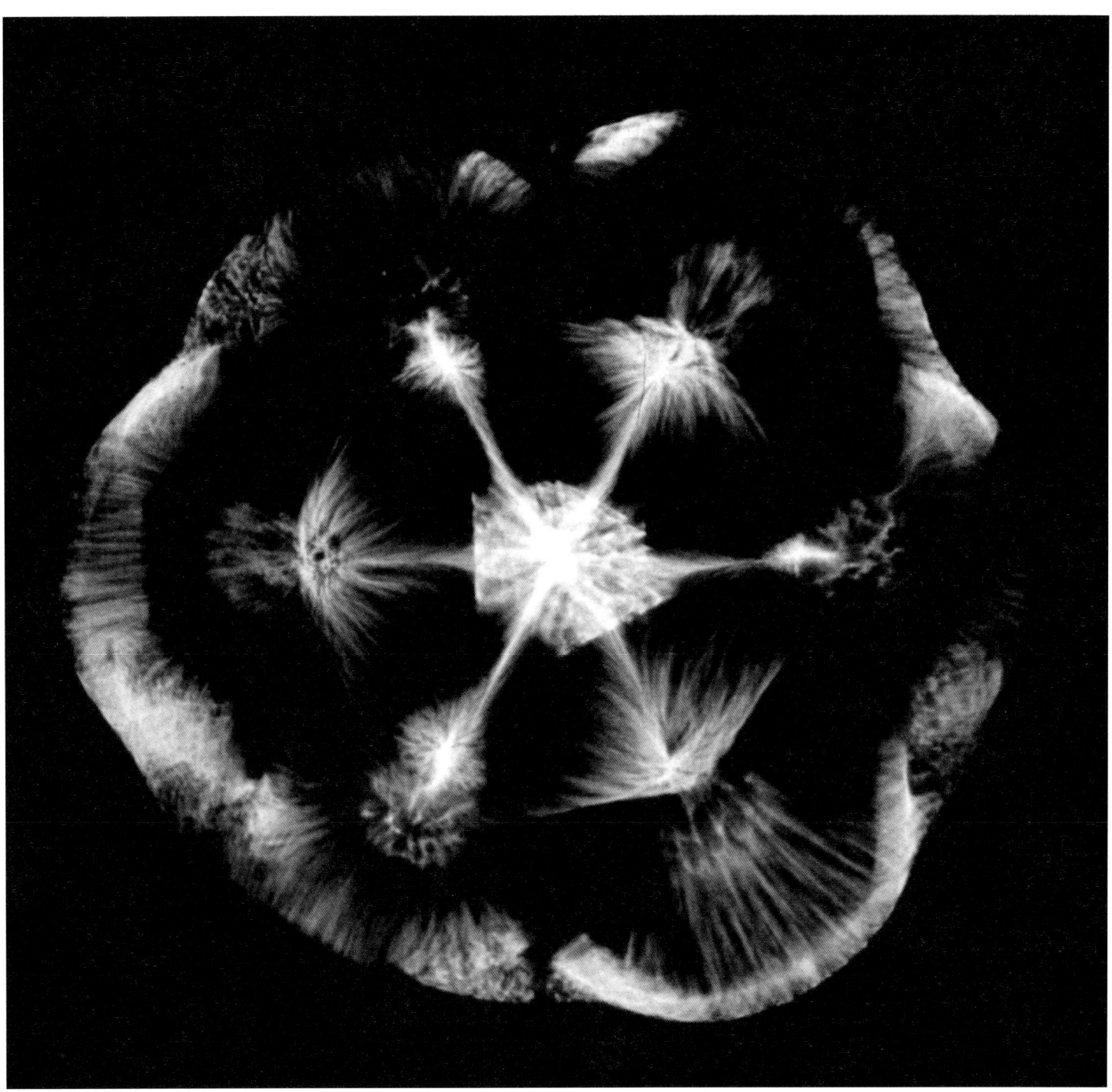
Fig. 154

Fig. 155
Here the liquid excited by sound is contained in a shaped vessel with sloping sides. The flow pattern of the vibrating liquid is shown by the particles strewn on its surface. In the center of the picture there are circular formations which are indicative of the hydrodynamic pattern of circular currents. Their speed depends on the intensity of the tone and is proportional to its amplitude. A "nucleus" is often formed in the center of the flow pattern. Schlieren pictures would also reveal the structure that is present at the same time. (Photo somewhat enlarged.)

Figs. 156-158
If two high-frequency tones, each of which produces a characteristic figure, impinge "from outside" on the liquid, does a third figure appear? This is an interesting question. Figs. 156, 157 and 158 show the result of an experiment designed to answer it. In Fig. 158 a pattern with bilateral symmetry is beginning to take shape as the *resultant* of two tones, which individually can produce a tetragonal (Fig. 156) and a pentagonal (Fig. 157) pattern respectively. Fig. 158 results when the two tones operate *simultaneously*. (Photo somewhat enlarged.)

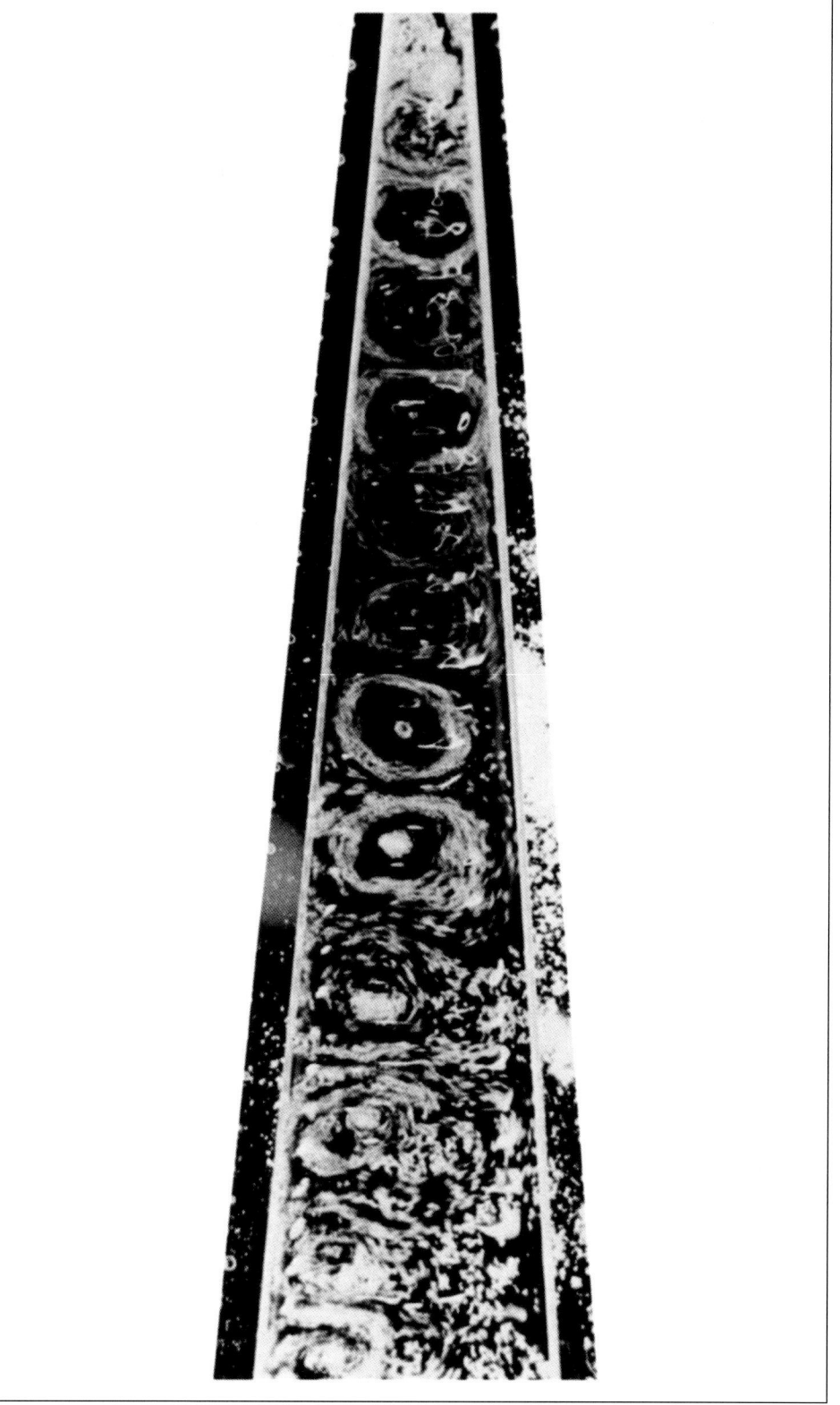

Fig. 155

Fig. 156

Fig. 157

Fig. 158

Fig. 159

237

Fig. 160

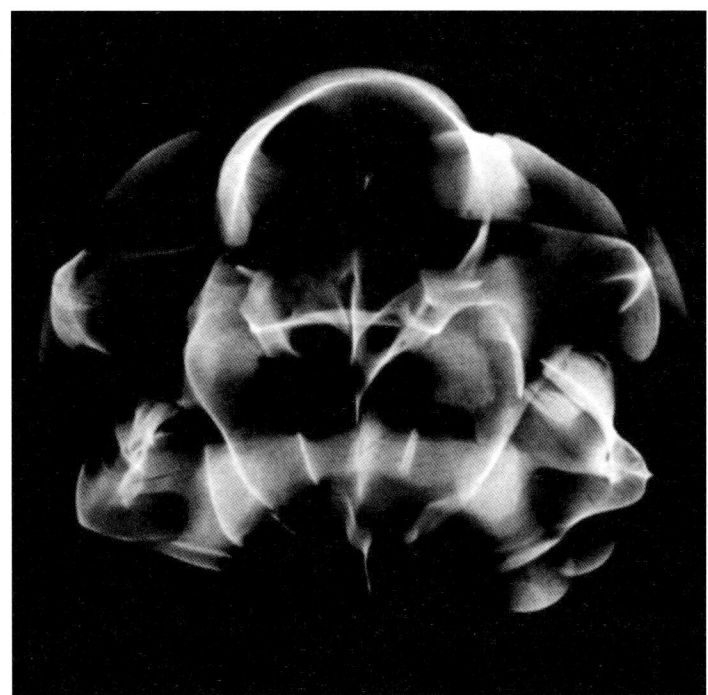

Fig. 161

Fig. 162

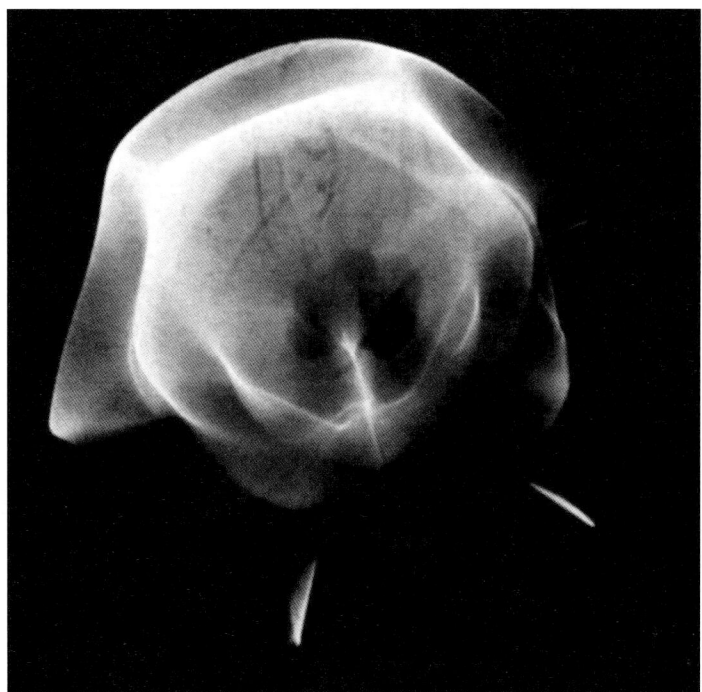

Fig. 163

Fig. 164

Figs. 159-164
Here we see two further examples of the *simultaneous* action of two tones. Fig. 161 is the result of simultaneous irradiation by two tones, which singly produce a pentagonal (Fig. 159) and a trigonal (Fig. 160) pattern respectively. The resultant, seen in Fig. 161, is a configuration which is not present as such in Figs. 159 and 160. It displays bilateral symmetry. Fig. 164 shows the resultant of excitation by two tones which singly produce the patterns of Fig. 162 and Fig. 163. Again it is clear that the resultant displays a third pattern. (Figs. 162 and 163 are slightly enlarged. Fig. 164 has been enlarged several times.)

What really happens is this. The material we obtain through our empiricism (features of time and space, mathematically regular formations, movement, relations, integrations, individuations, transformations, etc.) comes to us in the harmonic oscillations in such a way that, as one outward form yields to another, it also appears to the eye of the understanding mind as an organized whole. The one and the same harmonic quality holds sway in the form, the number, the symmetry, the pulsation, the kinetics, the dynamics, in the time, in the centration, in the transformation, polarity and so forth. This concordance of the entities, this consonance of the various entities (ontic constitutions of existents) relating harmoniously to form a uniform chord in the field of vibrational effects, shall be termed, we suggest, the *primal harmonic phenomenon*. The vibrational effects intensify to produce this primal harmonic phenomenon.

There are a large number of features which we could mention, such as rotary waves which set the whole revolving, and the scintillation of particles under the influence of oscillation. Experiments in which several materials are used at the same time are of particular interest. Fig. 173, for instance, shows turpentine and paraffin subjected to simultaneous excitation. Structural patterns can be seen but at the same time there are signs of bilateral currents. In this way individual substances can be used as indicators. In Figs. 174-176 paraffin and turpentine were again used simultaneously, and hexagonal and tetragonal patterns appeared. Fig. 176 clearly shows the two patterns side by side. The properties of the substances and their viscosity are important factors in achieving these results; but account must also be taken of the topography of the vibrational field generated by the diaphragm. Even though two such formal patterns can be visualized at one and the same time in a wide range of experiments (Figs. 174-176), each retains its own sharply characteristic features. Other conditions again become apparent when foam is vibrated. An extraordinary complexity of features involving structures, currents and pulsations can be seen in the bubble structure. All these different experimental approaches enable us to penetrate ever more deeply into the world of vibrations. At the same time it is obvious that the work is still in a state of flux, or even, one might say, at the very beginning. Here we have begun by presenting, first of all, documentation on sounds and musical tones and their relationships in terms of figures, kinetics and pulsations. What happens in the *succession of tones*, sounds, and musical notes? This is a problem that confronts the research worker, and the sequence of sounds, whether in *melos* or speech, calls for further study. These are fields which still have to be investigated. But first and foremost we are confronted by another set of problems: How do vibrations proceed in natural spheres which are marked by periodic, serial and rhythmic phenomena? Do relationships prevail here which bring these fields of research into the purview of cymatics? The next two chapters are concerned with these questions.

Fig. 165

Fig. 166

Figs. 165-170
If a film of liquid is irradiated by sound, patterns of the kind shown in Figs. 165-170 can be revealed by the schlieren method. A complicated lattice appears comprising squares, diagonals and perpendiculars. Curves, circles and radiating patterns result. But here again, number, proportion and symmetry prevail. In the experiments shown here oil of turpentine was the substance used. Such harmonic patterns can be produced in films of many different liquids by means of vibration. The orderliness prevails down to the smallest corner, e.g. in Figs. 169 and 170, which reproduce details. Cymatic effects of this kind are also produced at the microscopic level and can be clearly recognized even after the original has been magnified several hundred times. We realize that these figures and processes appear in *every* dimension.

Fig. 167

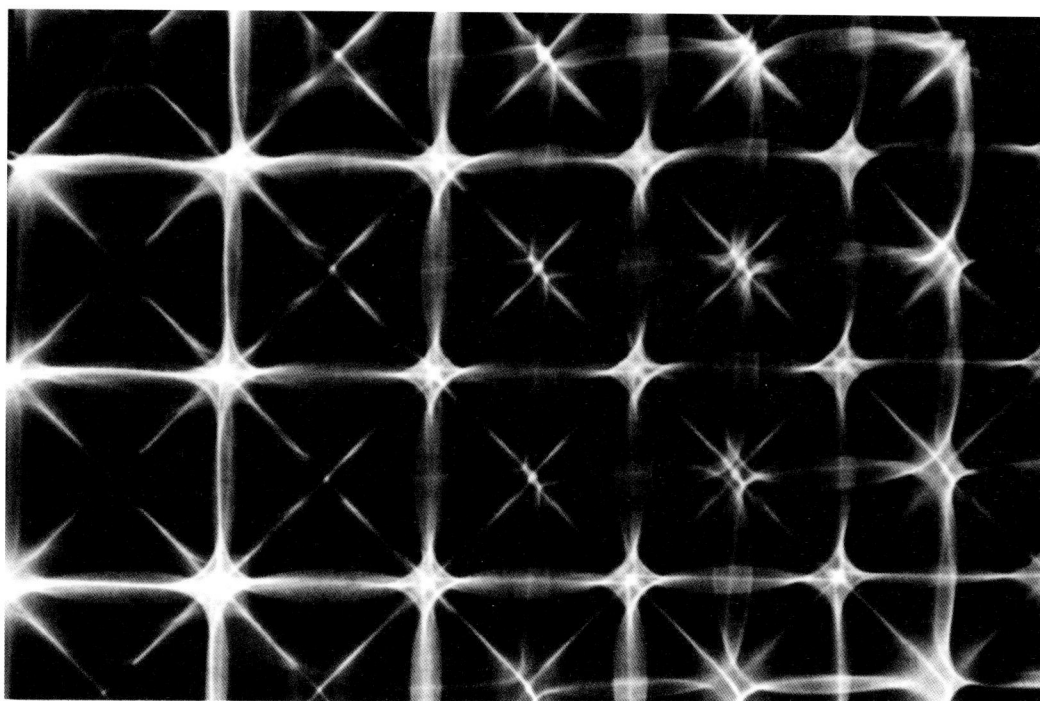
Fig. 168

Fig. 169

Figs. 165-170 (con't.)
All the harmonic vibrations shown here (Figs. 117-170) are macroscopic. If we refer back to Fig. 2, we may now understand what is happening in the curious side branches we observed. The drops of liquid become involved in conditions which we can understand only when they have been revealed to the eye by schlieren methods. Figs. 165-170 show that here, too, currents and pulsations proceed in orderly proportions. (All the figures have been enlarged several times.)

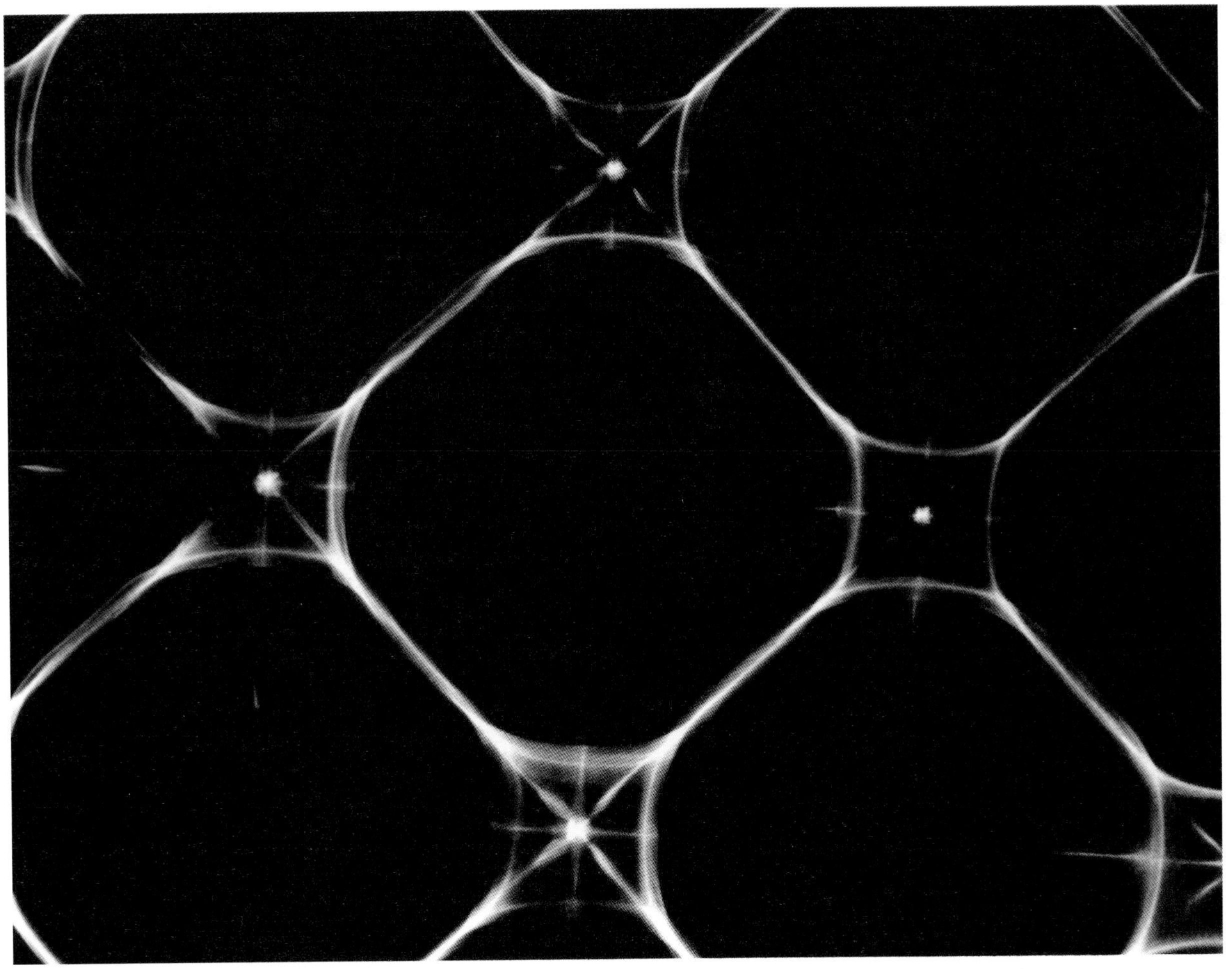

Fig. 170

Fig. 171

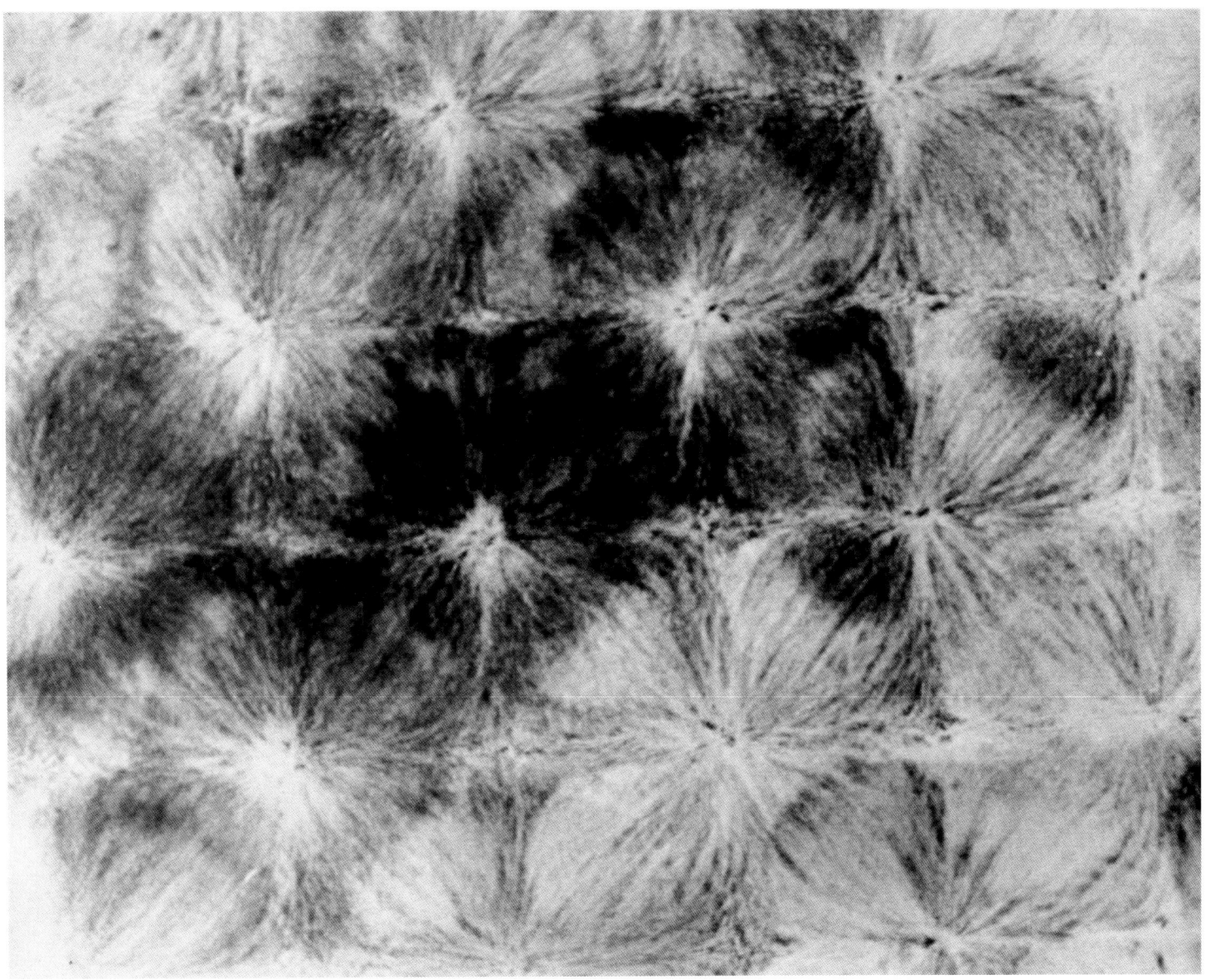

Fig. 172

Figs. 171-172
Lycopodium powder is strewn on vibrating films to visualize the currents. The camera was focused so as to reveal *both* flow paths *and* structures. The tetragonal structure shows through and is interwoven with the pattern of lines representing current pathways. The system of currents is made even more complicated by the fact that the whole structure is pulsating. Once again it should be emphasized that proportion, number and symmetry as expressed (triadically) in figure, wave and dynamics, are the dominant categories of manifestation.

Fig. 173
In some of the experiments several materials were utilized *at the same time*, e.g. paraffin and turpentine in Fig. 173. On the one hand there is a very clearly marked structural formation, and on the other there are characteristic currents, particularly on the periphery, which display a bilateral flow pattern. In this way we can gradually learn through experimentation, to understand and visualize harmonically oscillating systems. Our power of insight thus takes us through the surface appearance to the totalities beyond. As the characteristic behavior of the diverse materials varies according to their different viscosities, vibrational effects can be visualized which would otherwise escape observation. Thus in Fig.173, on the one hand, the structure, and on the other, the flow pattern, can be discerned.

Fig. 174

Figs. 174-176
The topography of the vibrational field, the thickness of the film, and the simultaneous use of different materials all exert an influence on the phenomena, as does the frequency and the amplitude. In Figs. 174-176 we see the curious situation arising when a hexagonal and a tetragonal pattern occur together. Paraffin and turpentine were the materials. The conditions of the experiment can be made more complex by, for example, exciting films of several non-miscible liquids or impressing vibration on various foaming substances. The *multiphase systems* thus created, manifest a great diversity of undulations, currents and pulses, all of which, however, conform to the basic patterns of number, proportion and symmetry. It should be mentioned that besides excitation by plane surfaces (diaphragm, plate) point excitation can also produce harmonic patterns. Hence when multiphase systems are present, specific relationships are set up which display clearly marked morphological, kinetic and periodic features. Conceptual models can thus be obtained which may serve for the investigation of multiphase systems.

Fig. 175

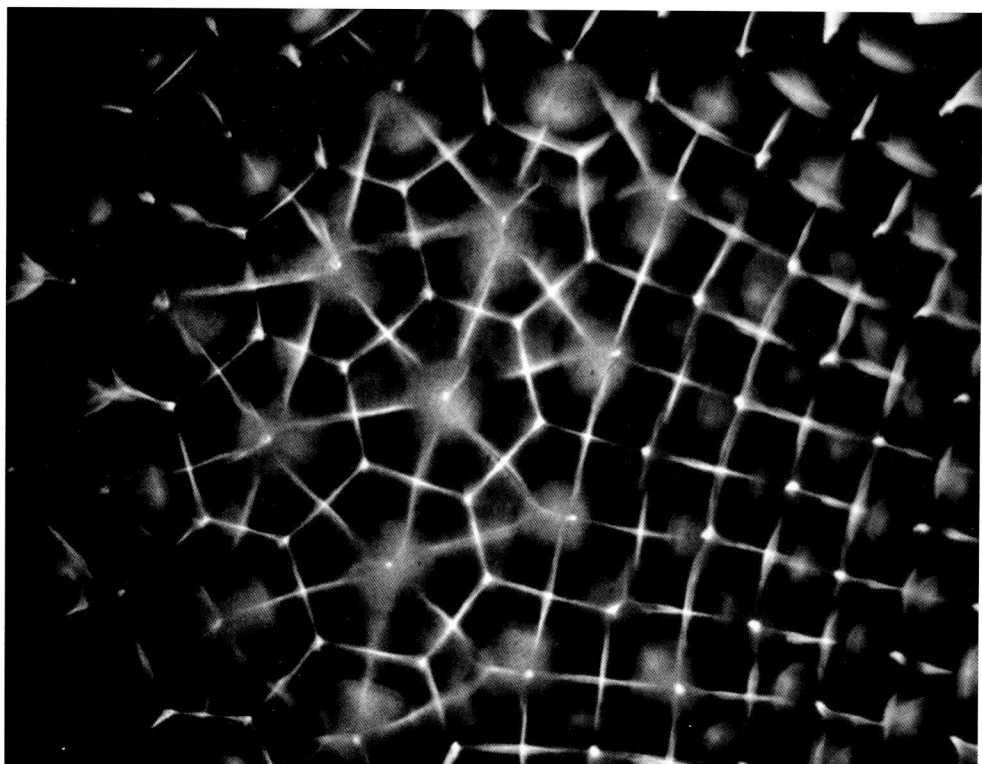

Fig. 176

15

Cymatic Effects in a Wider Context

Wherever vibrations are an essential constituent of a system, cymatic effects must be apparent. We are not in any way concerned with analogies here; we are not saying, for example, that things look similar, or that there is a wider resemblance between this phenomenon and that. If there is in fact to be an actual connection between one set of relationships and another, what we must have is not analogy but homology or identity of nature. We must therefore make observations in one sphere of nature — letting it speak for itself — so that we may ascertain whether vibrational processes, wave phenomena, are an essential part of it. The cymatic effects must, so to speak, put in a fresh appearance. They must declare their presence to us independently in each sphere in question.

But the connections between the different areas of the natural sciences are not only of this kind. The question is whether there are not common features and uniformities of an *essential* nature. In our very first chapter we spoke of the whole and its parts. We came to the conclusion that the part is a modified version of the whole. This is a theme to which we must revert when we discuss the relationship between the various fields. In the conversations Alexander Moszkowski had with Albert Einstein, the problem arose of the extent to which things are investigable. The discussions led to the following question: "Supposing it were possible to find the reason for every property of a grain of sand, would this mean that one had plumbed the *entire* universe? Would there still not be something left unsolved for a complete understanding of the world?" Einstein said that the answer would have to be an unreserved yes. "For a full scientific understanding of what goes on in the grain of sand would be possible only if one knew the exact laws governing events in time and space. These laws—differential equations—would be the most general of all laws from which the essence of all other events must be deducible." The idea was then elaborated by Moszkowski: "This idea can be pursued in a different direction; namely, that any research, however specialized, into the most insignificant of matters, retains a connection with research into the nature of the universe and may be of value to it. If knowledge is held to be perfectible, then any new contribution to it, even the smallest, may be essential and indispensable to the whole." And indeed the world is present in the grain of sand. If we want to acquire a complete knowledge of it, then we must know about gravity, nuclear forces, electricity, magnetism, mechanics, thermodynamics, etc. We have already mentioned the living cell. We must realize that in *each cell* all the hereditary factors are present, that *each cell nucleus* has a complete set of chromosomes (a whole genome). And then, as we have already mentioned, respiration, metabolism, regulations, procreation, plastic and metamorphic potentiality, circulation—in brief, the whole inventory of vital processes—are enacted in the individual cell. The cytologist has to deal with a whole entity. If he keeps his mind open to what the cell has to tell him, he will have evidence enough of its wholeness. Everything that is meant by life can be manifested to him by every single cell. All he has to do is to let the cell speak to him to the fullest extent. Starting from such premises as these, we must find common features, uniformities, and essential affinities in the various regions of nature, but this is tantamount to saying that we must encounter cymatics in diverse areas of observation. It is, so to speak, inherent in the various regions of the universe. But this implies at the same time that it is accessible to research and that our propositions concerning the part and the whole are thus verified by the facts themselves. There can be no question of our imposing our views on anyone. What we have found in the field of acoustic wave phenomena must be encountered in other fields in an equivalent form, and research will, in its own interests, speak the language of cymatics. There must be cymatics in geology (orogenesis), in astrophysics, in meteorology, etc. Now it might be said that we may leave the geologists, astrophysicists, and biologists to uncover these concatenations for themselves. The point is, however, that since vibrational effects are in fact linked together, a knowledge of their phenomenology alerts the eye to their different categories; it provides clues and guiding patterns. Above all it must be stressed once again that a kind of special sense, a cognitive faculty, is acquired with a particular awareness for these connections.

Before taking a look at other areas of research, we will first make a systematic review of what cymatics affords us in the way of phenomenology. It ranges from simple wave formation directly accessible to observation, to complex phenomena in which structures, pulsations, dynamics and kinetics are compounded: it leads to the primal phenomenon (triadics, harmonics).

We have moving and standing waves. Figural elements and currents appear; currents and countercurrents. Steady rotations come into view. Curious diametral influxes occur. Reliefs are formed: attenuations and evacuations on the one hand, up-thrusts and accumulations on the other. Lumps float around when adhesion is reduced; substances liquefy as the result of reduced cohesion. Conglobations and clumpings (Figs. 177, 178) are particularly prominent, and there are also dispersions and expansions (Fig. 179). Turbulences, bilateral vortices and pulsations appear in clouds of powder and spray. There are large-scale circulations. There are transformations in the state of matter, with changes of phase regularly occurring at the same frequency and the same amplitude. There are interferences manifested in pendular movements and in eruptions separated by quiescent phases. There are resonances which may also provoke eruptions. There are to-and-fro effects. Then there is the phenomenon of significant variations in time. The system may be quiescent; and then suddenly and abruptly much happens within a short time. There are phases of the most intense activity, and then phases of stagnation. Widespread tendencies toward repetition are evident. There are correlations. There are symmetrical, harmonic vibrations displaying mathematical regularity (as intervals, musical tones).

If we have before us processes in fields of force which operate side by side, against and with each other, along with currents of energy, streams of matter, upheavals of matter, etc., then there must inevitably be wholesale wave processes due to oscillation. The compendium of characteristics we have just listed will enable us to probe more deeply into such relationships.

The experiments we have conducted with various substances yield some impressive pictures. A wave pattern takes shape and a folded relief comes into being. Transverse formations may result in a network of folds. And then again we see steep wave crests with rearing sides. The formations migrate and their fronts may peter out in circulations. There are also upheavals in the masses themselves. Depending on the vibrating diaphragm, there are accumulations and unfoldings and next to them, attenuations and evacuations. Lumps float around. They dissolve; they separate. Extensions alternate with retractions. Quiescent periods are interrupted by processes of abrupt onset. The edges of the masses reflect the exciting vibration down to the last detail. Bays and promontories, uprisings and subsidences alternate. The whole formation is dominated by patterns of curves and arcs, rotations and vortices. If the substance has solidified, there are characteristic fractures. The individual parts of the masses may overlap each other; they may over-run and under-run each other. Inversions also occur.

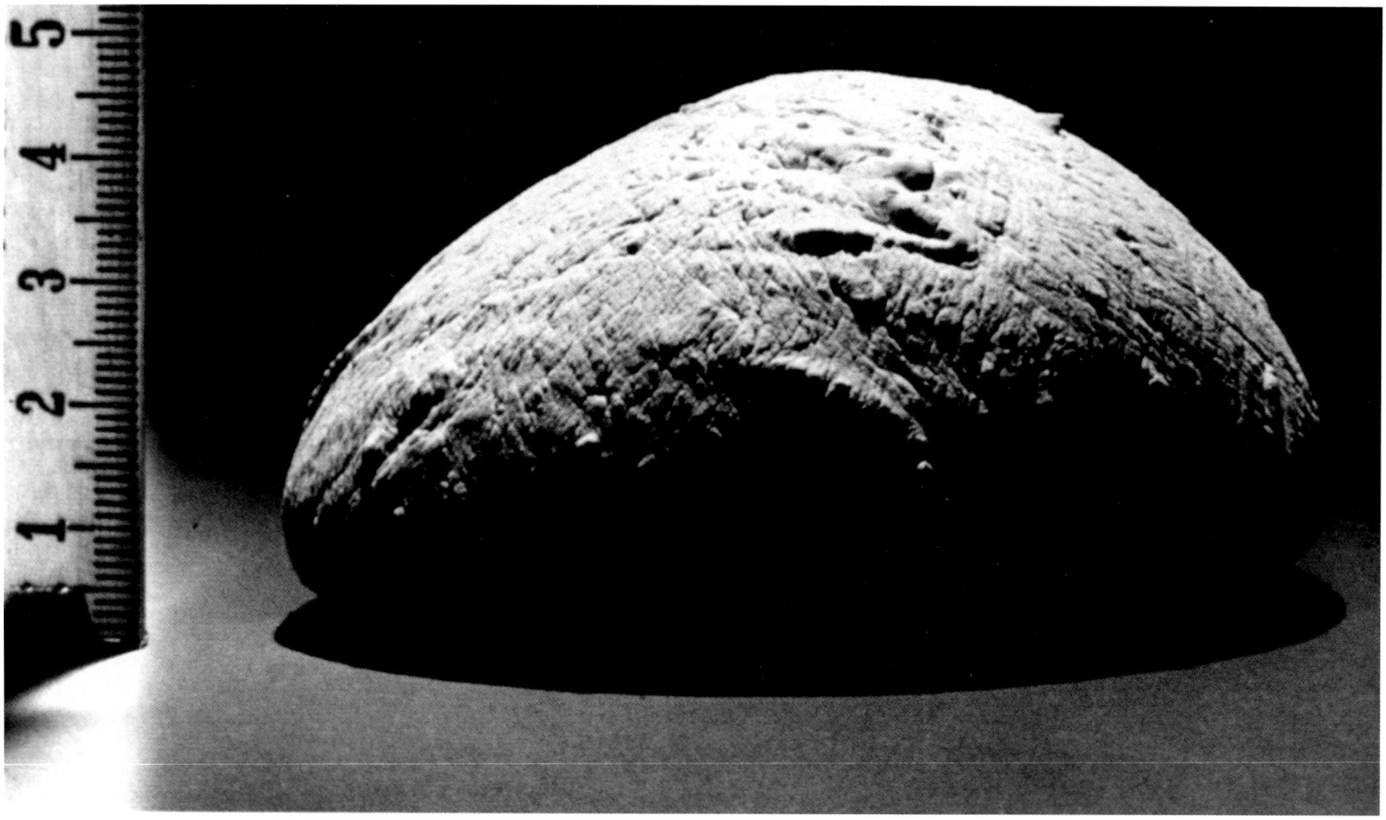

Fig. 177

Figs. 177-178
Conglobation appears again and again as a typical cymatic affect. Masses of powder or viscous pastes are clumped together in circular heaps, spheres, globular clusters, clouds, etc. Fig. 177 shows a ball of plaster which was plasticized by vibration and then solidified. Fig. 178 shows a number of spherules which were formed into a cluster in an oily milieu. The conglobing force of vibration bestows the character of a unified system on these spheroid formations. The results obtained in liquid media by surface tension are here procured by the oscillating effect. A system where unity is conferred and also maintained dynamically by vibration, displays harmonic arrangements (visible, for example, in Figs. 87-90). Wherever fields or currents of energy are excited by vibration, the effects referred to here may be expected. The experiments described may provide conceptual models.

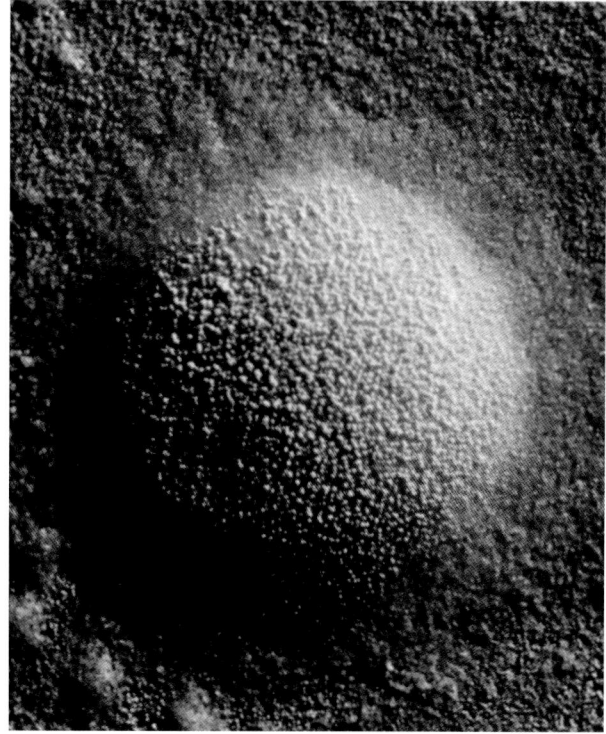

Fig. 178

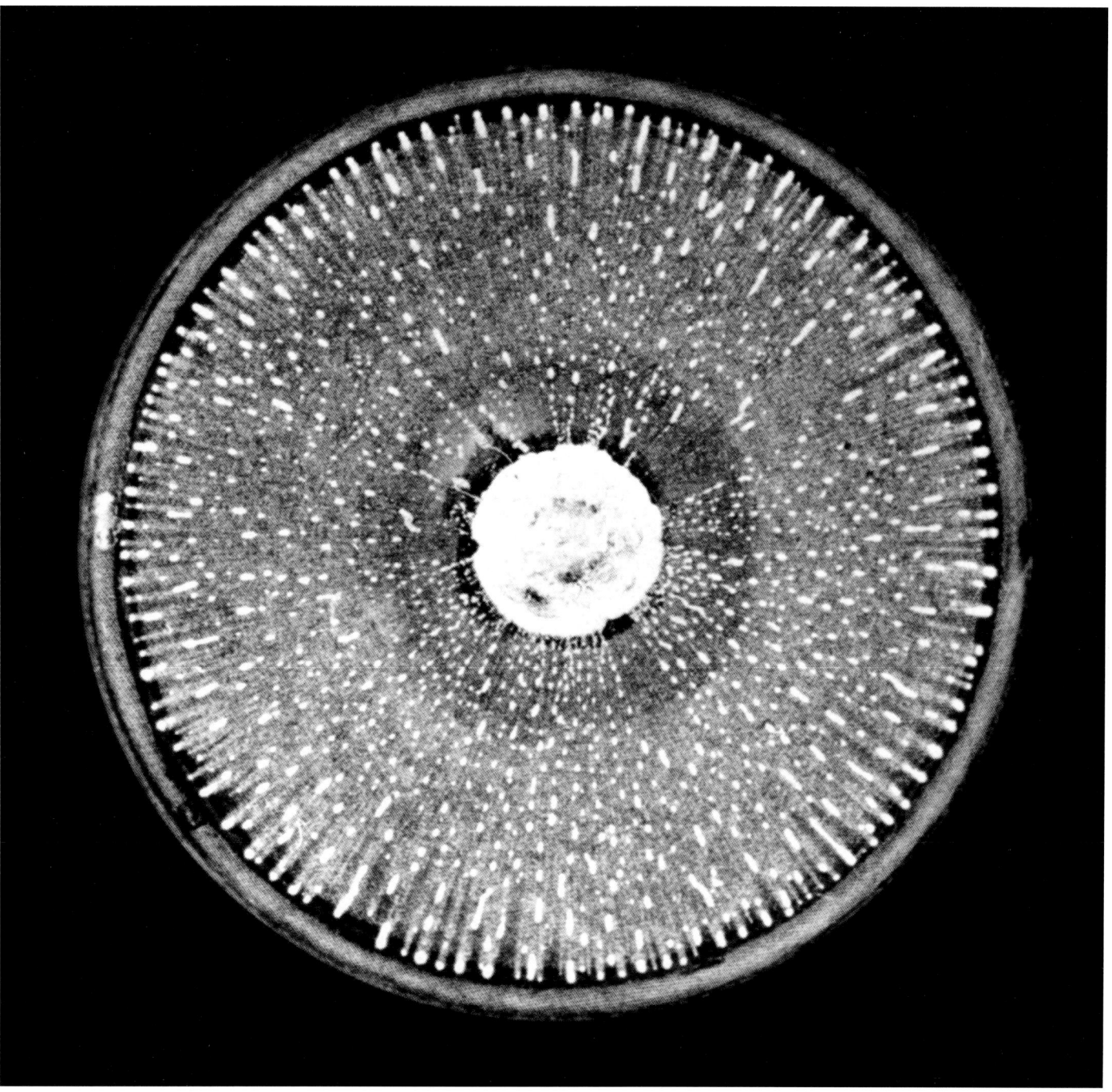

Fig. 179
We are looking down into a funnel, from the bottom of which conglomerations of particles (visible as white streaks) are rushing up to the rim, against the force of gravity. If the amplitude is low, the masses slide down to the bottom because adhesion to the funnel-shaped diaphragm is reduced. Frequency and amplitude can be adjusted so that both processes take place at the same time and the whole vibrational field becomes a single zone of circulation. While one part of the mass expands, another flows to the center. In this way a uniform system of periodic circulation arises. (Photo somewhat reduced.)

The attitude we have taken precludes simply carrying the processes described over into geology; it is not our wish to interpret and explain. Nor do we wish to invoke seismic processes as a *deus ex machina*. Undoubtedly tectonic shocks do produce a change in the behavior of the mass. We need only recall how vibrations can liquefy cement which has not yet set. Eye-witness accounts of earthquakes tell us, for instance, that the masses slipping down a mountainside do not simply pile up on the valley floor but run up the opposite slope as if they were on rollers. Similarly it is certain that some processes can be enacted only under oscillating seismic conditions and, most important of all, that they are briefly enacted, and that active phases are followed by phases of quiescence. To the factors pressure, temperature, incline, sedimentation, etc., must certainly be added the factor "seismic vibration" with a shift in the state of matter.

What we are primarily concerned to do is to look at mountain formation with the eyes of a geologist and to listen to the language he speaks, the concepts he forms, and how he seeks to understand his subject.

To this end let us follow the exposition Hans Georg Wunderlich gives in his book "Wesen und Ursache der Gebirgsbildung", 1966. Wunderlich's descriptions are of particular interest to us because individual processes such as lifting and sinking, extension and compression are viewed together compendiously and conceived as a uniform process. This novel attempt at synthesis, which the author modestly refuses to overvalue, is precisely what we must focus our attention upon. For this exposition emerges step by step out of geological empiricism. To give the reader some idea of these extraordinary ideas we shall quote from the work at some length. It will be seen that the set of quotations we have compiled does not transcend the limits of cymatics; indeed, we find that we are moving in a very specifically cymatic field. Wunderlich conceived mountain formation (orogenesis) as a wave process in the full sense of the word. He draws comparisons between sea waves and mountain waves: they are actually connected together. We quote: "We have at first for the sake of simplicity taken vertical and horizontal movements singly. In actual fact both are concurrent and our next task must be to investigate how the two are combined together."

If we imagine such horizontal and vertical movements as described above as the curve tracings of sinusoidal oscillations, the most diverse combinations are of course conceivable: the two curves may coincide (maximum uprise then roughly corresponds to the maximum movement towards the foreground), move in mirror symmetry (maximum uprise then corresponds to maximum movement towards the background), be shifted one quarter out of phase (maximum vertical movements then occur where the horizontal movement is zero and vice versa) or assume any intermediate position.

If we bear in mind the quantitative sizes of the movements and the directions of the curves (which will not be discussed here in detail for reasons of space), simple reflection will show that the first two of the possibilities mentioned are precluded. They involve an oscillation in each plane of less then 45° to the horizontal, even though the quantities and the signs may vary. Tectonic processes, however, are obviously not restricted to such movements under 45°, and consequently only possibilities 3 and 4 remain for further consideration. The third variant is simply one, if the *simplest*, special case of the fourth possibility: a shift of the curves of the horizontal and vertical movement by a quarter phase yields, given the same numerical values for lifting, sinking, and backward-and-forward movement, circular movement paths, and in the last "general" case, any elliptical movement paths whatever for the mass points. Below we shall take the simpler case of the circular combination of movement sequence, resembling the familiar orbital movement of the water particles of a sea wave. In forward propagation of the wave from left to right (i.e. in the same direction as the orogen front with the vertical and horizontal movement in the preceding sections) the orbital movement is in a clockwise direction.

At the same time, however, it must be remembered that such sea waves are generated by wind movement, starting that is, from the surface of the water. The frontal course of an orogen receives its impulse from the depths; for this reason the direction of the orbital movement is reversed, and is consequently counterclockwise. Whether it is in fact an orbital movement in the strictly circular sense or rather an elliptical pathway is irrelevant for the following discussion, just as the water particles in sea waves do not always describe exactly circular paths. Elliptical pathways arise if the phase shift between vertical and horizontal movements is not exactly a quarter period or if the lifting and sinking, or the forward and backward movement, do not correspond quantitatively; that is to say, that loop-like curves are generated in which the orbital movement is not exactly reproduced in the successive revolutions, but, depending on the amount of the forward and backward movement, is laterally displaced by a variable distance. This process leads to a wandering of the fronts. However, what is important is that we are confronted by a uniform process such as is represented by the wave. "If the forward movement in the direction of the foreground predominates in the horizontal movements during the passage of the front, there follows a zoned displacement of parts of the earth's crust in this direction: i.e. superimposed on the orbital movement there is a clearly defined advance of the crust in the direction of the front movement itself, even though the amount by which it advances is as a rule substantially less. This process might be compared to the formation of water waves on a river flowing in the same direction whereas, of course, in sea waves it is not the water itself but the wavelike impulse which is transmitted further. If, however, forward and backward movements are balanced out on the passage of an orogenic front, each point of the earth's crust will finally return to its point of departure following a true orbital movement when the orogenic cycle is past: i.e. the orogenic contraction is offset by the early orogenic expansion of the crust before and the late to post-orogenic dilatation afterwards. Thus in practice large portions of the earth's crust can be implicated one after the other in a folding process without any actual shrinkage of the earth's crust resulting from the folding contraction; which is an extremely important conclusion for the "space balance" of the earth in view of the ideas prevailing about the contraction of the earth. Whereas it is a well-known fact that the contraction theory is tenable only if all kinds of auxiliary mechanisms be admitted in order to explain the extensive processes of expansion in the earth's crust, the hypothesis of circulation in the above sense provides a simple explanation even for the concurrence of compression and expansion in *one* sequence of movements." In these terms Wunderlich defines the concept of the "orogenic wave", and it does in fact involve an oscillatory process. "In the previous chapter we saw that the periodic functional curves for horizontal and vertical movements, occurring when the front advances, neither coincide exactly nor move in mirror symmetry; there is a phase shift which leads to an orbital movement of the mass points on the passage of the orogenic front. If the phase shift is exactly a quarter period, there is a genuine circular movement, but with larger or smaller phase shifts these mass points may describe any form of elliptical path. In this respect there are parallels to sea waves." From this perspective, mountain-building processes can also be viewed as oscillations. "It might be objected that only sinusoidal oscillations have been employed whereas out in Nature the sequential movements are not likely to show the same regularity; a non-sinusoidal curve would appear to meet the actual case more closely. But this counterargument does not affect the principle. The science of oscillations provides us in Fourier analysis with a tool for reducing any non-sinusoidal to sinusoidal oscillations; the superimposition of two or more sinusoidal curves of differing period, amplitude or wavelength produces any kind of oscillation however complex its character. Our choice of the simplest curve patterns was due to the fact that we wished to keep visualization of the processes uncomplicated. This may be the first step along the path to a complete Fourier analysis of orogenic processes." The orogens migrate in accordance with laws. "There has already been talk for a long time of the migration of folds in the sense of new folds joining the outer edge of an orogen. All these many observations suggest that subsidence, constriction and uprise in the orogen are not fixed to the spot, but are subject to regular shifts of position:

thus vertical and horizontal movements mesh together in the course of the progressive upheaval of the crustal region involved in orogenesis as described above." The same views may be invoked to describe the nappes and the way they slide over the underlying rocks, looking for all the world like surf-riders. "A nappe located in this position may be involved for some considerable time in the displacement of the orogen front without being implicated in the orbital movement of the substrate; just as the surf-rider utilizes a relatively narrow area at the outer edge of the uplift zone where horizontal translation is most favorable and the movement is still directed towards the foreground. And just as the surf-rider has continually new masses of water under his board which carry him forward a distance until, following the orbital movement, they sink into the depths and are replaced by others driven by the same impulse, so the layer underlying the nappe before the orogen front is constantly changing while the impulse always remains the same. The dynamic uplift is all the more effective in that nappes are generally no heavier than the rocks below them, in contrast to the surf-rider on his board. Such a nappe can be regarded neither as a pure thrust mass nor as a simple sliding mass, for upthrust and horizontal impulse are interlinked. Since the nappe is not involved in the orbital movement of its substrate it can escape, at least temporarily, orogenic constriction. It is not the movement of the nappe that stamps its characteristics on its substrate but vice versa, the momentum of the substrate at any given time determining the fate of the nappe it is carrying along." The flow processes in the substrate of the earth's crust are described as turbulences, rotations and convections. There are parallels to the processes in turbulent media. Of course, the time factors involved in this "secular-plastic substrate" are of quite a different order. "But it follows from this view of the problem that the creation of the mountains around the Mediterranean can scarcely be interpreted otherwise than in terms of actively flowing substrates in and under the lithosphere, and that turbulent phenomena in the flowing substrate are of prime importance in the formation of upfolds. Above all, however, it sheds an explanatory light on the way in which mountain chains often peter out abruptly or suddenly change their line of advance, and also on the rapid change in the tectonic planes of alpine orogenesis. Nor are the fronts of the lithospheric circulation in any way fixed in position, although of course, the eddy zones are not displaced at the same speed as the atmospheric cyclonic and anticyclonic eddy systems moving in the general western drift. The secular-plastic circulation of the lithosphere is fixed to a much greater degree since it is gripped in the long-term movement of the earth's crust."

To comprehend geotectonics we make liberal use of a set of concepts comprising flow dynamics, turbulence, rotations, convection cells flowing in different directions, eddies and burbling, etc. Within the limits allowed by the gravimetric and thermal measurements presently available, a comprehensive picture is drawn in which quanta of thermal flux, differences in level in the configuration of the earth, the crustal structure of the earth in terms of its rotary movement, the magnetic field, etc. are all marshalled as elements. We quote Wunderlich again: "While the eddy centers of the thermospheric convection are (apparently) drifting westwards the configuration of these convection cells is also changing at the same time since the flow processes involved *are not* constant and simple (laminar) but, it would seem, secular-turbulent movements with a certain mobility of the stream filaments which are capable of combining in ever new eddy forms. Such shifting patterns in the subcrustal fluctuation bring about permanent changes in the geotectonic configuration of the earth such as appear to have occurred several times in the history of the earth." There is also an affinity between the up-and-down movement of the vortical convection cells and the movement pattern of the atmospheric circulation; in other words we again find a certain parallelism.

It is no part of our plan to initiate a discussion on Wunderlich's theory of orogenesis. This is the province of the geophysicist and geologist. The point we can make as non-geologists is that here we have a field of nature which can be described in terms of the wave processes and phenomena inherent in it. Nor have we any wish to push the parallels to extremes with our comparisons: the cymatics of acoustic waves are no proof of orogenesis, nor orogenesis of cymatics. Still less do we wish to encroach on the geologists' preserves by references to cycles, pulsations, repetitive tendencies and the like. Things may be left to take their course with one set of ideas fructifying the other and our picture of the universe being enriched by the addition of features of oscillation.

Just as we have, in the foregoing, presented geologists and geophysicists with some of the effects of vibration, we should like, below, to bring certain wave processes and their phenomenology to the notice of astrophysicists. Here again it is not our intention to encroach on anybody's preserves and still less to pick holes. There can be no question of formulating hypotheses, of wanting to explain the universe, etc. It is much rather a question of communicating a few facts which might prove suggestive when the phenomenon of oscillation comes under observation. If oscillations and waves appear in cosmic systems, then the corresponding effects may be expected. We shall now describe a series of cymatic phenomena of this kind.

In our experiments with vibrations in various materials and media we constantly find pronounced conglobations when oscillation causes the masses to pile up in circular heaps. They are truly spherical formations. Fig. 177 shows a ball of plaster which has been sculptured into shape by oscillations. In Fig. 178 a pile of tiny plaster spherules has become clustered in an oily substance. Even the finest of particles are formed into a ball. The impression created by the phenomenon is one of uniform formation. If a disturbance is introduced, the whole will be restored to its spherical shape on the removal of the disturbing factor. At the same time the systems may pulsate and, most important of all, convect. If they move in the vibrational field, this change is accomplished in correlation with the other factors involved. The mass flows along as a fluctuating system with spheroid features.

When the amplitudes are great, particles are flung up in clouds. These too, retain their spheroid formation although the most violent turbulence is raging round them. And yet within these storms of particles there are still circulations, eddies and even incipient bilateral symmetrical flow patterns. As the amplitude is changed (modulation) we witness concentrations and expansions of extraordinary violence. A cloud of this kind may suddenly collapse into a dense spherical mass only to burst explosively into a cloud of dust again under the impact of amplitude impulses. Unstable though the turbulences may be, structural tendencies are nevertheless discernible and often take the form of regular patterns.

A remarkable experiment can be performed with a vibrating conical diaphragm. The material at the bottom of the cone rushes up the walls under the influence of vibration. In Fig. 179 it must be imagined that the tiny round masses are rushing upwards and outwards (against the force of gravity). We had already observed an "antigravitational effect" of this kind. In an earlier experiment viscous masses remained in place for the most part when the vibrating substrate was tilted; although they started to slide down when the oscillation was discontinued, they crept back up to their original position on its recommencement (Cymatics, Vol. I).

The behavior of the masses rushing apart (Fig. 179) is remarkable under certain conditions of oscillation. If the tone is loud, they expand; if it is soft, they slip down to the bottom under the effect of gravity and reduced adhesion. The vibration can be adjusted so that these two processes are held exactly in balance. When this happens, the masses stream radially to the periphery while *at the same time* the powder comes flowing back. The whole becomes involved in a remarkable pattern of circulation. The entire process can be recorded on film.

Eruptions in the field of vibration come as a continual surprise. A layer of substance may lie quiescent when suddenly the mass is thrown up high at one point and ejected, whereupon quiet returns. Periodicity can be seen in certain eruptive phenomena. There is also a repetition of the processes on the same spot. This means that there is interference: i.e. reinforcement of the amplitude of traveling waves as they come into phase, and transfer of the excess energy to the indicator particles. A distinction must be made between these processes which occur at characteristic places and times and those which occur suddenly "out of the blue." The latter type is observed when the masses are shifted around. As a result of this, the vibrational characteristics of the system are altered; zones of resonance occur which boost the vibration enormously. The spherical formations illustrated (as uniform globular masses, round heaps of particles, and turbulence clouds) have a pronounced tendency to pulsate. The whole system pulsates regularly and also displays, likewise as part of this tendency, regular structures of the type produced by harmonic vibrations (reminiscent of intervals and musical notes). At the same time the eye should be alert to harmonic frequency ratios in regard both to numerical relationships and also, more particularly, to symmetrical spatial arrangements. This inevitably raises the question: What are the harmonic forms of a pulsating star? An actual harmonically organized polygonal form of pulsation is a possibility to be entertained in this context.

And now we come to rotary effects. There are areas which are in constant rotation. How are they formed? We begin with a mass of particles uniformly distributed upon a steel plate. At certain frequencies they do not form figured patterns (nodal lines) but begin to flow, and the *whole* vibrational field becomes involved in the same process. The traveling waves appear in flow patterns moving more or less in parallel or contrary directions (Fig. 180). These contrary motions nevertheless fall within a larger overall pattern and give rise to rotations in the mass, with the particles flowing into the area of rotation at points diametrically opposite to each other. Fig. 181 illustrates this pattern. A group of areas of rotation can be seen. The sand is flowing into the "arms" of the elliptical or circular heaps of particles, and its constant inflow starts the whole rotating. Fig. 182 is another picture of the same scene. In this photograph a twisting pattern can be seen. The particles in the area of rotation are the earliest arrivals: those flowing in from the arms are the newcomers. After a time all the material from the bridge currents has flowed in and we have the circular and elliptical results seen in Figs. 183, 184, and 185. In the center can be seen a nucleus, which is particularly conspicuous in Figs. 184 and 185. Even though the inflow of particles from diametrically opposed points has ceased, the formation still goes on rotating. If markers are placed in the areas of rotation, the regularity of their revolutions is plain to see. Now, traces of such processes can always be discerned in the vibrational field. Bridges, extensions to vortices, filaments, and more diffuse areas strike the eye. In this context the reports by F. Zwicky* are of particular interest. He describes elements between the galaxies which look like bridges, filaments and extensions, and it is important to note that this author raises the question of oscillation in his discussion.

*Intergalactic Matter, by F. Zwicky, Pasadena, California, "Experientia" Vol. VI, Fsc.12, Pages 441-480, 15. XII (1950); Multiple Galaxies, by F. Zwicky, Erg. D. exakt. Naturwiss. Vol. XXIX, pages, 344-385 (1956).

The morphology of the galactic universe is strongly marked. Its structural pattern is dominated by spiral formations. Bearing in mind the forms produced by oscillation in experiments, it is possible to make out areas in which phenomena exhibit uniform patterns recognizable as halo formations, areas of rotation with diametric inflows, nucleations, disc formations, clusters, intermediate traces as bridges and filaments, and above all, as sequences of events with precise chronological characteristics proceeding from youth to age. Now it would be contrary to logic to argue that, merely because things are so, the universe must necessarily be in oscillation. After describing the phenomena in question, we shall venture a suggestion concerning oscillatory conditions in the universe.

Let us revert to the diametrically arranged area of rotation as seen in Fig. 42 (color plate). Here we have a *constant* "rotary wave" which revolves, one might say, as a diameter. The picture is quite different if there is a spiral motion. This would entail processes in which there are centrifugal effects and masses may be ejected. Hence all rotary phenomena must be examined to decide whether the rotations are constant and due to traveling waves or whether centrifugal forces are involved.

Turbulences are an ever-recurring feature. Although continuous disintegration is typical of turbulences and an impression of complete instability is created, here again there are repetitive tendencies. In fact conditions of turbulence may occur over and over again. At the same time it must be remembered that turbulent media are sensitive to oscillation. ("Sensitive flames" are a familiar example). They may be said quite literally to prepare the way for other vibrational effects.

If we use several frequencies at a time to excite vibration, i.e. we use several crystal oscillators, no sonorous figures of the usual kind appear. Instead only *those* spots are visualized by the marking material where the nodal lines intersect. One often has the impression that areas of concentration are quite irregular in the way they are distributed, whereas in actual fact the events taking place are determined by fields of vibration at certain frequencies. If a number of oscillations are now made to impinge on one another, the picture grows enormously complicated. It looks as if there were no regular pattern at all, as if everything were scattered at random, while in reality the pattern is based on notes, each having a characteristic frequency. This is a point to be borne in mind when observing the extremely complex relations arising in oscillating spaces. The obscure oscillatory pattern in such cases is, as it were, hidden under apparent randomness.

The world of vibrational phenomena presented here is of extraordinary complexity. We are not concerned merely with this or that factor; invariably there are several factors in action at once. Thus we have conglobations, rotations, circulations, currents and structures occurring side by side and also

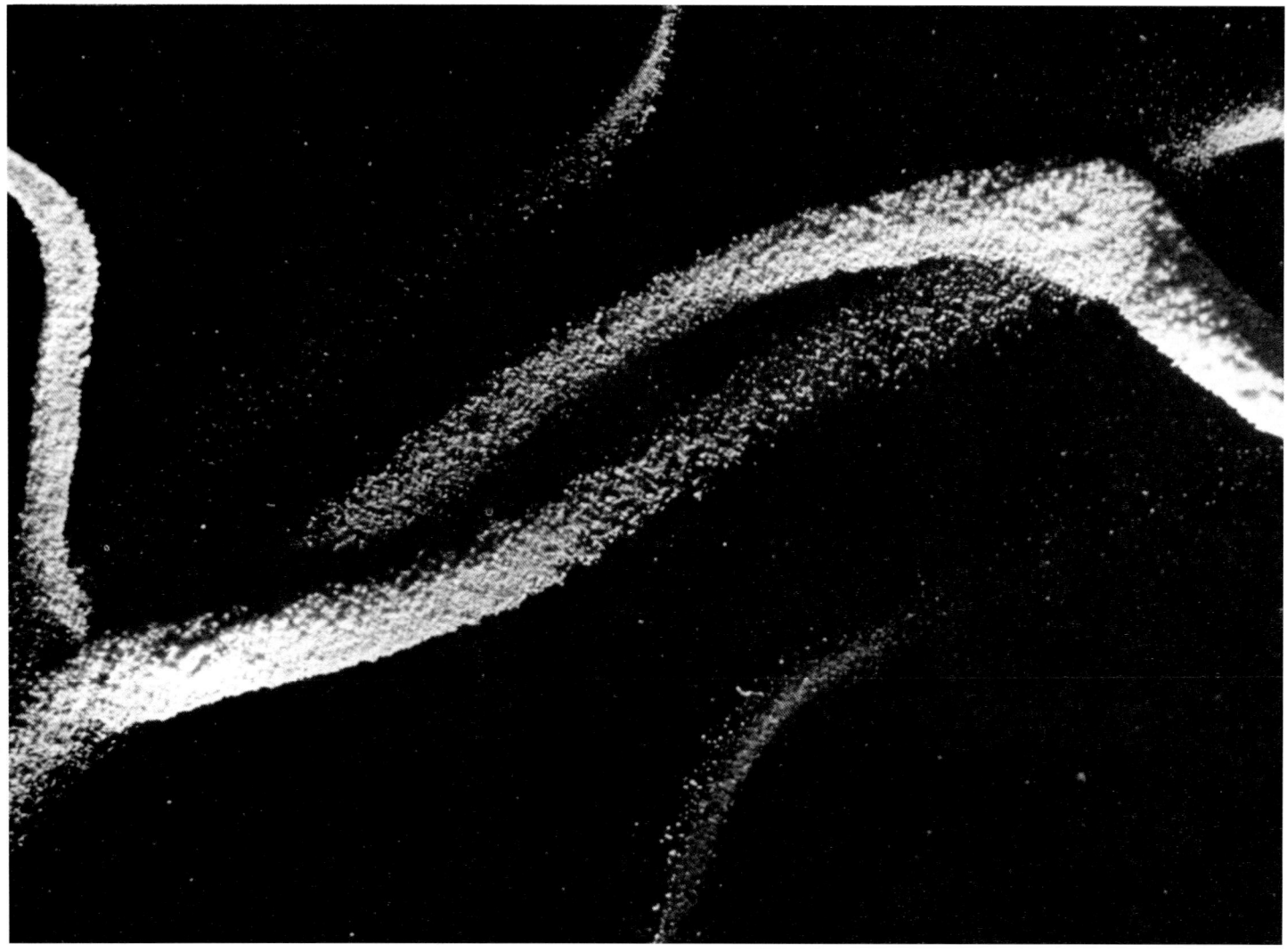

Fig. 180

Figs. 180-182
Since crystal oscillators enable vibrations to be impressed on plates throughout the whole frequency band, continuous waves can be generated and studied. They flow as currents in a characteristic manner, side by side and in contrary directions. These current paths form a system covering the whole vibrational field and they can be visualized by sand strewn upon a vibrating plate. (Fig. 180 reproduces a detail.) Areas of constant rotation arise within these current patterns. They are created out of these currents by the sand flowing into a central area at two diametrically opposed points, as seen in Figs. 181 and 182. Two arms are visible along which the particles flow to join the elliptical or roughly circular area, where they become involved in the rotation process.

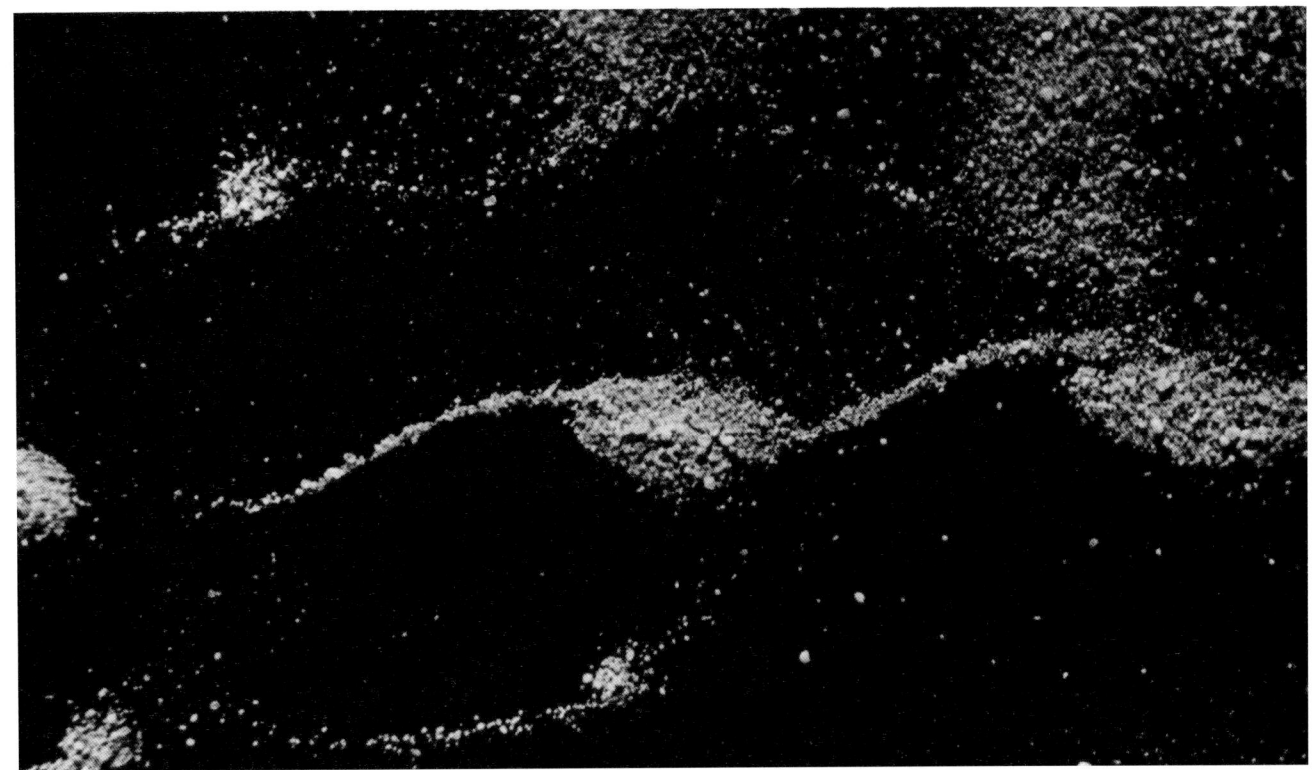

Fig. 181

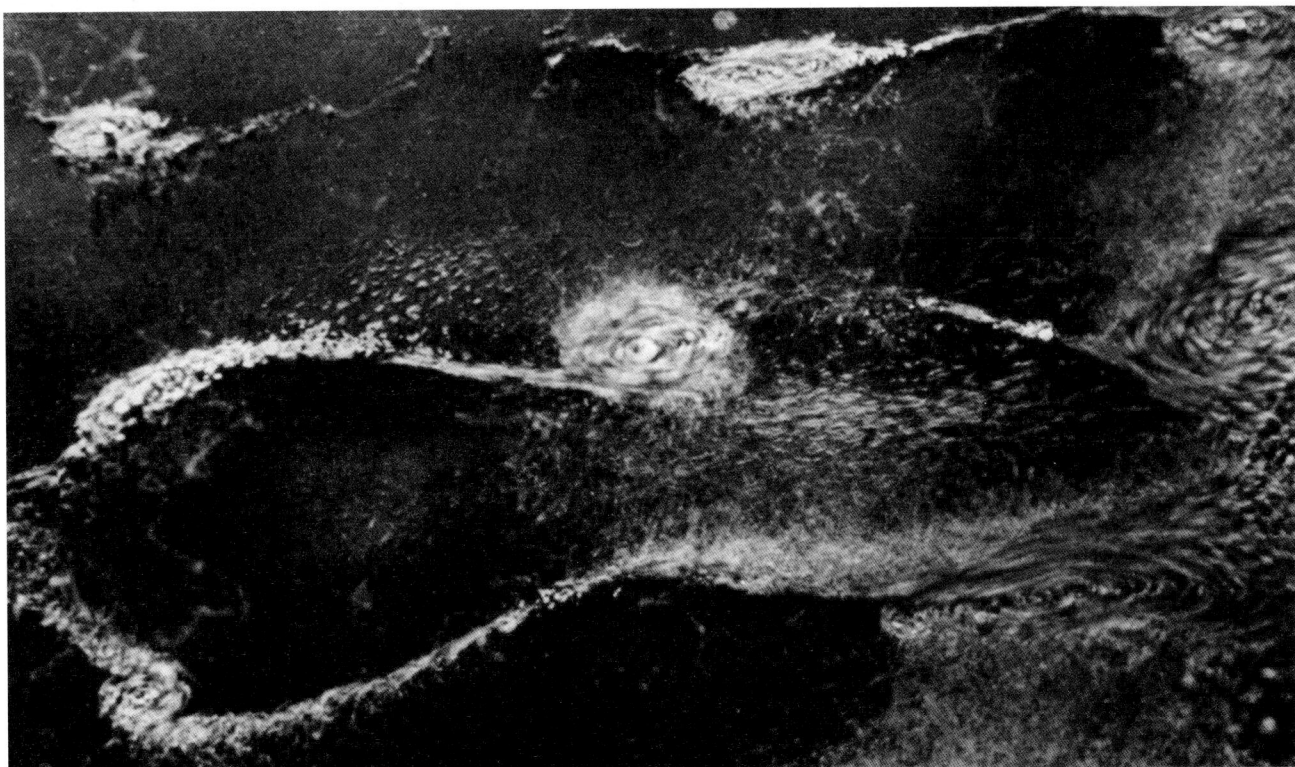

Fig. 182

interpenetrating one another. Their relations are varied and inverted. If the whole now passes into other sets of conditions, new aspects are added to the phenomenology. Magnetic fields are a case in point. If we transfer an experiment involving vibrations in ferromagnetic materials to a magnetic field, the magnetic and the wave phenomena interpenetrate. Currents are diverted or their rate of progress is reduced, new flow paths arise, structural patterns take shape, three-dimensional forms appear, freeze into immobility or go through abrupt changes. Both factors at work here must be borne in mind when observing such experiments. A better idea of what is entailed can be obtained from a closer inspection of Figs. 43-60. The phenomena portrayed here cannot be produced without vibration. Here we have the rudiments of a science of magnetocymatics, just as we have one of magnetohydrodynamics.

Needless to say, we have no apparatus or experimental set-ups out in cosmic space. There can be no question of placing astrophysical constructions on the experiments described here or of foisting them on the universe. Nor does it behoove us as non-astrophysicists to interpret periodic cosmic objects (pulsating stars, pulsars, quasar rhythms, oscillating plasmas, magnetohydrodynamic waves— Alfvén waves— convections, etc.) or even to make an attempt; far from it. All the same, the series of phenomena we have depicted might enable the astrophysicist to see cosmic morphology and dynamics through the eyes of the cymaticist. Currents of energy, expanding substances, and streams of matter may proceed as waves or appear as oscillations *ab initio*, or through contact with other fields of energy or force fields, or as a feature of their characteristic flow pattern. There can be no doubt that such phenomena as these might be considered by the astrophysicist in terms of cymatics. The galactic universe, solar physics, stellar formation, and interstellar media might all be amenable to such an inspection.

Experimental work reveals aspects which prompt us to observe Nature in our environment and stimulate our efforts to understand it. There we find phenomena of which we have already seen equivalents in our experiments with vibrations. All we wish to do is simply to let things speak for themselves in each field of research. Since oscillations are omnipresent in Nature, we must expect to find periodic, rhythmic and cyclic phenomena at virtually every step. We are necessarily surrounded by them every day and hour of our lives. They must be inherent in every sense impression. It is merely a question of adverting to them and the remarks we have made here are intended to bring them to notice. Continuing our description of phenomena along these lines we shall make brief reference to atmospheric phenomena in the troposphere, to clouds, and then to sea waves and their phenomenology. In this way it will become clear that the "realities of our environment" can be experienced in terms of cymatics.

Hardly ever do we look at the sky by day without seeing periodic elements. Even in a cloudless sky the upward and downward movement of spheres of haze can be observed. No sooner are filaments of cloud spun out of the blue sky than an endless pageant of changing forms begins. Shapes, movement, and above all the processes of cloud formation are paraded before us. Let us take a closer look at these processes without concerning ourselves with the physics of droplet formation. Creation and extinction, concentration and attenuation, condensation and dissolution follow each other in an endless sequence. The phases oscillate: there is a rhythmic pulsation which shows up quite clearly as the clouds take shape. One formation is ranged beside another, so that we soon have row upon row of tufts and tuftlets, clumps and clumplets being woven into a pattern. Watching such objects is excellent training for the observer. Let us suppose, for example, we have before us a cloud which we have watched taking shape. After a time we look again. There is the "same" cloud as before, but actually it may, in the meantime, have dissolved, vanished and then taken shape again. It may also happen that we see the "same" cloud continually in the same place. But if we look carefully, we shall see that at one end it is continually dissolving, while at the other end it is continually forming again. Thus it is not a fixed, unchanging object that we are looking at, but rather a process. What persists is the activity, as it were, of the "cloud." The constant cloud, as it appears to be, does not exist. This is why it is useless to take one's bearings on a cloud, to use it, as it were, as a fixed point. We have flown up to clouds to determine their dimensions. It is virtually impossible to discern anything in these formations that might serve as a fixed point of departure. With its convections and vortices the cloud is in a state of constant change and atmospheric flux.

In the cloudscape vaporous masses form up typically in rows. It was Ferdinand Hodler who first thought of introducing such cloud cycles into his paintings. For him they were the great revealers of the all-embracing law he called parallelism. Whatever the type of cloud we look at, the law of repetition is invariably in evidence. There we see cumulus clouds drifting along one behind the other; as they move, they broaden out or they tower up like treetops. Cell after cell wells up on high. Then the clouds arrange themselves in layers again. Vast cloud fields come floating past. Huge waves or delicate ripple patterns take shape. Often various undulations interpenetrate in the most diverse ways. Within the "meshes" of these oscillations, the forms clump together and a dappled pattern of clouds is formed. The zones of concentration form a regular and rhythmic pattern. Whether we traverse the scene horizontally or climb vertically, we invariably find ourselves amidst recurrent formations: one layer after another, one rising mass after another, one wave after another. Wave phenomena are not confined solely to the

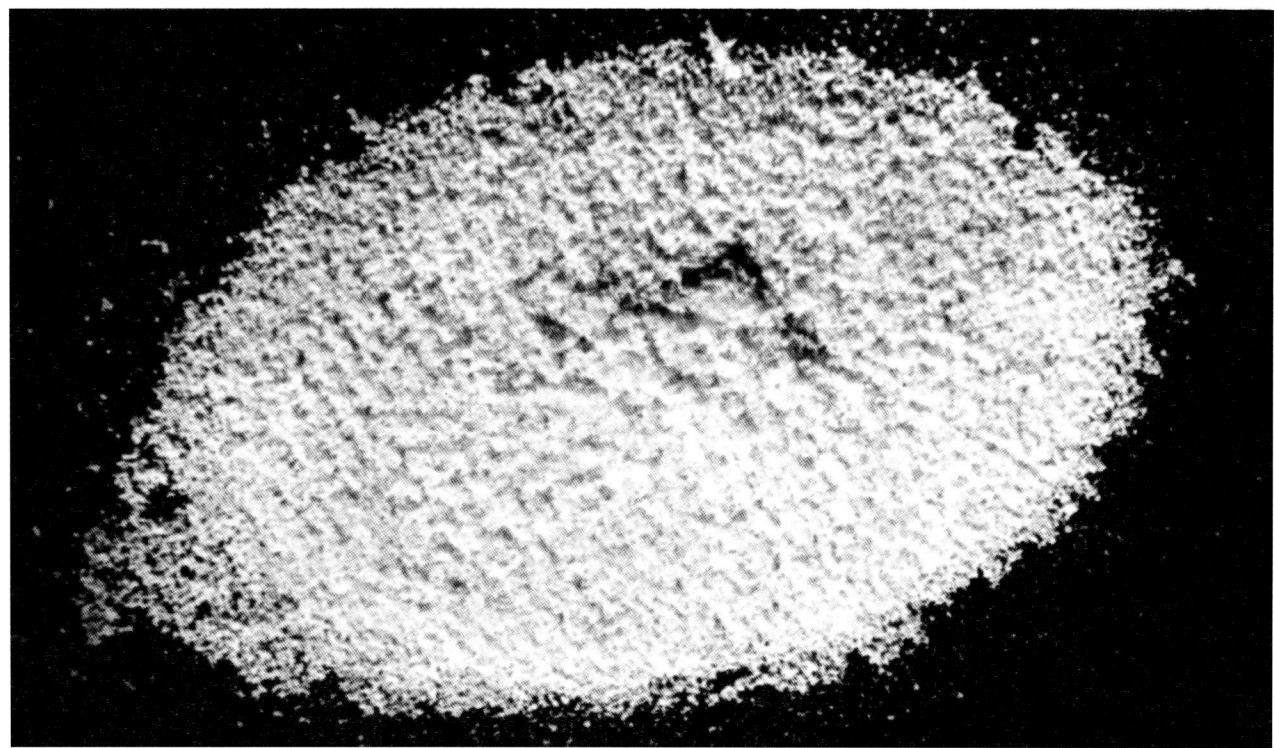

Fig. 183

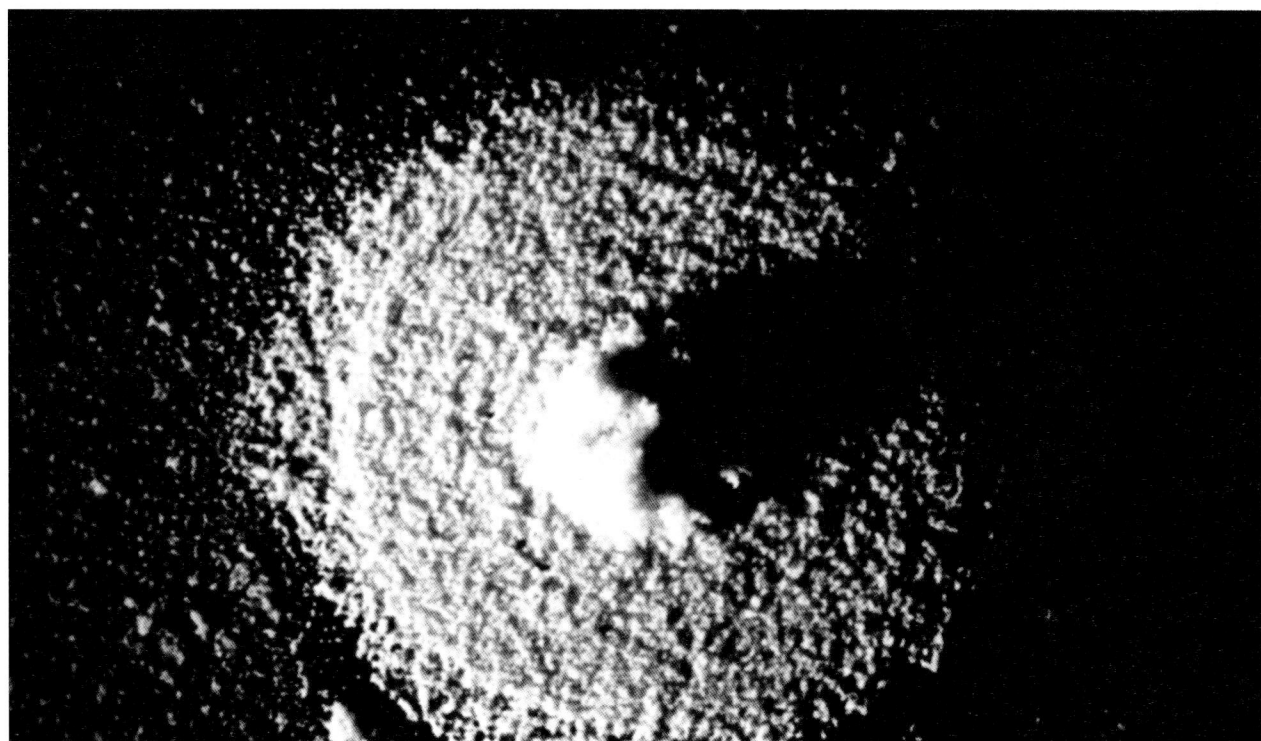

Fig. 184

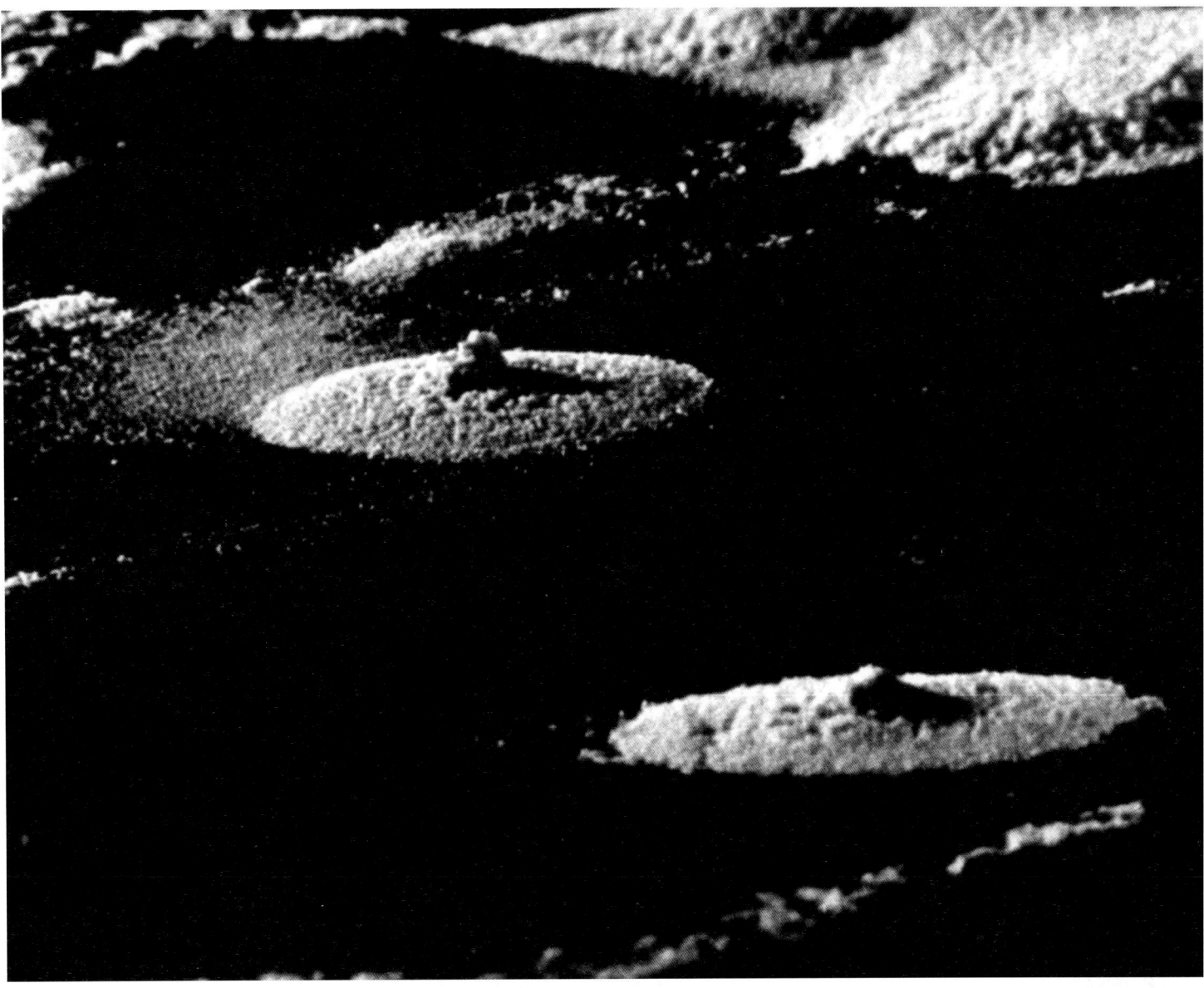

Fig. 185

Figs. 183-185
Here the inflowing process has ceased (Figs. 181, 182). The "arms" have flowed into the discs and the material that formed them continues to rotate there. In the center, a nucleus has taken shape (Figs. 184, 185). The areas of rotation may be elliptical (Fig. 183) or round (Figs. 184, 185). If we look at the oscillatory field upon completing such an experiment, we see, on the one hand, the areas created by rotation, and on the other, in between them, traces of the processes that have transpired: bridges, filaments, etc. The phenomena seen in Figs. 180-185 are a record of a true dynamic morphology manifested as a cymatic effect.

Typus undulatus; indeed, they are almost omnipresent in the atmosphere. We shall deliberately refrain from listing the names of the different types of clouds. Such a nomenclature is, of course, justified and necessary; but what we are concerned with here is the observation of an element which prevails in all cloud formations. Even the "structureless" cloud layer is not lacking in structure. Close examination reveals that here again there is a sequence of concentrations and attenuations. Particularly when seen from an aircraft, this type of cloud also exhibits a differentiated pattern of the repetitive kind. The morphology appearing in all "clouds and waves", in all undulations and interweavings, is also invariably dynamic and kinetic. We are confronted by a whole spectrum ranging from round clumped masses or extended veil-like sheets, through towering mountains of piled, upwelling clouds, to feathery tissues spreading out like wings. But this is not a fixed pattern in which everything is neatly stowed in compartments; everything is in vertical and horizontal motion with currents intermingling. Even in the chaotic-looking pell-mell rush of storm-tattered cloud we can discern eddies and turbulences. And likewise in air masses flowing past each other or clashing frontally, turbulences can be seen in the form of unstable waves. The processes of current and countercurrent are also evident in the same way.

And if we span our vision to take in ever more extensive nexus of atmospheric phenomena, we become aware of vast formations and see islands or "continents" floating along, while in the far distance similar formations can be seen drifting closer. The tendency to repetition is found not only on the regional, but also the supraregional scale. These are everyday phenomena. But if we begin to look at them in terms of the concepts we have been using, we can describe a wavelike field of events permeated by oscillation and striking the eye as configurations or as laminar or turbulent currents in the atmosphere. However small or large our field of vision, we find waves, interpenetrating waves, circulations, pulsations, cyclic layers, serial fronts, periodic creation and extinction, rhythmic morphology, regular laminar flow, and then turbulent, vortical kinetics and dynamics. Whether we are following the mode of origin of a single raindrop or looking down upon the global pattern of vortices through the eyes of the astronauts, it is always this vision of interwoven phenomena that we see, dominated by periodicity and rhythmicity, whatever the changes of the weather. To describe this texture of phenomena in full would call for a whole monograph, the last page of which would never be written, for as we rise from one sphere to the next, we experience thermal, magnetic, chemical and radiant processes, all of which are in rhythm with the alternation of day and night and the cycle of the seasons; and these in turn are subject to the influence of the solar plasma clouds, with the atmosphere and the solar wind interacting. Then there are magnetic storms which cause another whole world of magical phenomenology to be revealed to us in the northern lights. Solar physics and cloud physics are drawn together, and all the spheres (thermosphere, ionosphere, magnetosphere, etc.), however complex their turbulences, bear the imprint of the rhythms of the various thermal, electric, magnetic, chemical, photic, and mechanical forces throbbing, as it were, through space. Starting with the raindrop, our vision broadens to embrace the universe. All these things afford us a graphic idea of what the atmosphere (troposphere) looks like when seen in terms of cymatics.

Our point in giving these descriptions is to sharpen and train the eye toward wave phenomena. For this reason we shall now turn to another aspect of Nature which is particularly enlightening in this respect: sea waves. Here we have cymatics in its native element, for the name of this science comes from the Greek words for wave (*to kyma*, "the wave"; *ta kymatika*, "wave matters"). In turning to a field of natural observation like this, we leave behind the field of pure experimentation and Nature itself becomes the great teacher. The untrained observer will begin by thinking it is a waste of time to take another look at water, since he feels he already knows all there is to know about its motions. After all, the movement of water is an everyday experience and one, it would seem, that calls for no explanation. But once we start to study the phenomenolgy of water, we soon realize that we are incapable of following the currents, waves and undulations involved in its movement. All we see is a bewildering complexity of phenomena. And once again we must admire the accuracy with which Leonardo da Vinci caught the details of water in his drawings. Now, it is reasonable to ask, "What is there that is so important about these wave phenomena?" The answer is that, given the virtual omnipresence of waves and vibrations in the world, it is important that we should be able to take mental possession of them and be aware of them in the specific fields of experience and research, and life in general.

When the observer looks at sea waves and realizes how powerless he is to understand their processes, he turns to the science of oceanography in order to get his bearings. He will find that for centuries leading physicists and mathematicians have been concerned with these problems and have sought to clarify their complexities. An enormous amount of time and effort has been devoted to observation, measurement and calculation. In spite of all that science has achieved, some aspects of wave action still elude comprehension. It seems astonishing to think that the initial generation of waves still raises unsolved problems. Concerning this initial stage we find these observations by Günter Dietrich*: "Calculation and observation both agree that capillary waves adapt themselves in a fraction of a second to the velocity of an incipient wind; the surface of the water responds almost instantly with a delicate shiver while larger gravitational waves only emerge from the rippling of the surface after several seconds. A number of phenomena concerned with the generation of initial waves have been satisfactorily explained by observation of the energetics involved, but the physical processes leading to wave generation are still obscure. This central question remains a subject for hypothesis...." The following experiment may be mentioned in this connection. If a film of water is excited by sound, the surface "crinkles" even if the tone is of small amplitude. This crinkling can be seen if crosswires, or a reticule of parallel wires, can be mirrored on the surface of the water and observed in the reflected light. The images of the wires crinkle as soon as vibration starts (Fig. 186). The crinkling becomes more pronounced as the amplitude increases (Figs. 187, 188). If the amplitude is increased still further, trains of waves are suddenly propagated. The images of the wires vanish and the wave lattice appears (Fig. 189). In Fig. 189 the first signs of waves can be seen. They are not clearly defined because they lie in a different focal plane. The wire image has almost completely vanished. Thus there are two clearly marked phases. First, there is crinkling; and second, a sudden change to wave formation in the proper sense of the word. Being a vibrational pattern, the crinkling is also undular in character and involves up-and-down movement, but wave trains propagated throughout the film do not appear until this abrupt change has taken place. If particles are strewn over the surface of the water, the distinguishing features of the two phases can be clearly seen. In crinkling alone the particles are not involved in a flow pattern whereas such a movement is immediately apparent when wave formation proper, commences. The experiments were performed with vibrations ranging from 45 to 17000 cps. in frequency. Of course, the wind blowing on the surface of the water does not produce a vibrational pattern akin to that generated by sound. But the surface is also crinkled by the wind and when a certain degree of crinkling is exceeded, the heightened state of tension invokes the change to wave formation. Wave formation is preceded by a stage of tension due to crinkling but not involving actual wave generation.

The whole mélange of wave and vibration theory and fluid dynamics must be employed in order to give a quantitative account of the play of the waves. Let us use a phenomenological approach in order to penetrate to the heart

---

* A good introduction to the problems of sea waves is to be found in the monograph "Allgemeine Meereskunde" (General Oceanography) by Günter Dietrich, 1965. We have taken the relevant descriptions, results and information from this work.

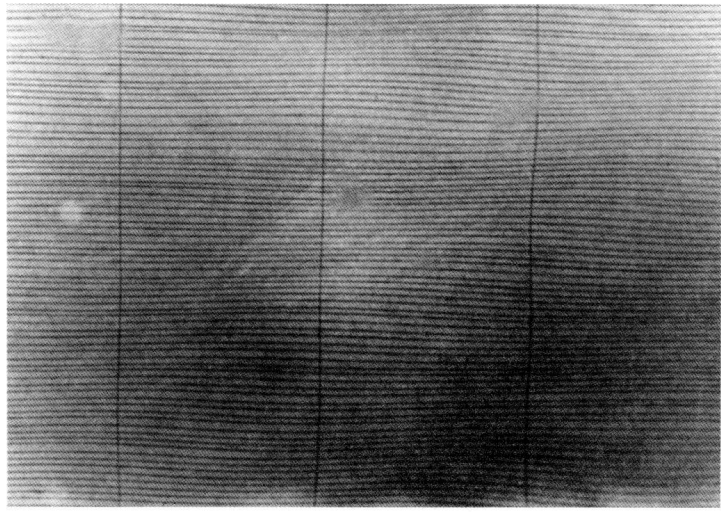
Fig. 186

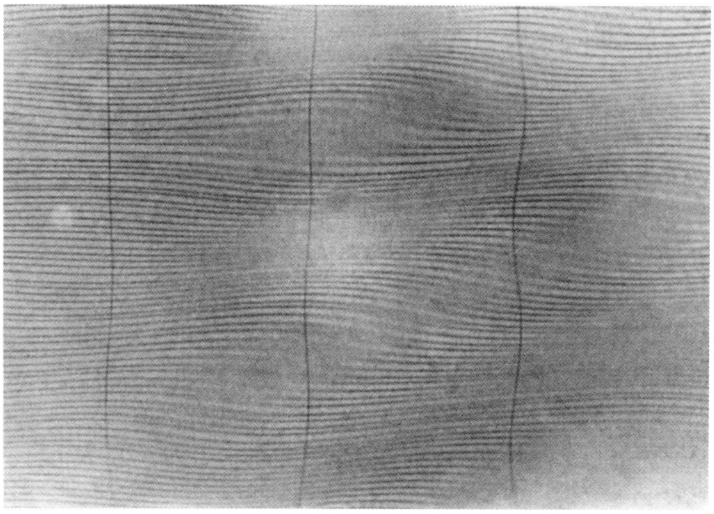

Fig. 187

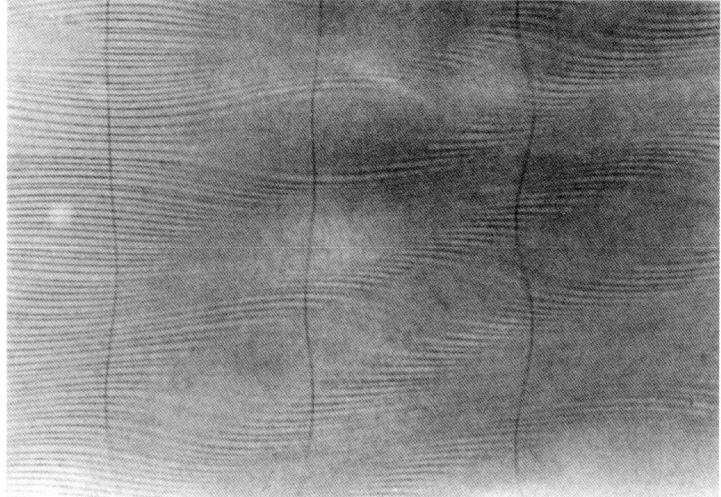
Fig. 188

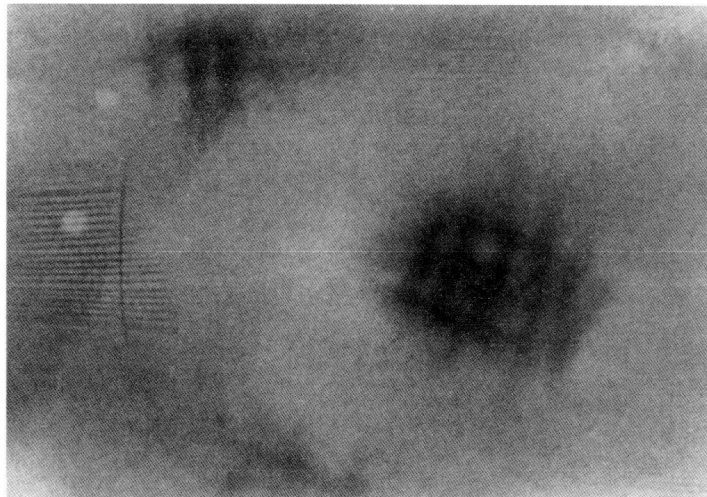
Fig. 189

Fig. 186-189
If a film of water is irradiated by a tone, the surface "crinkles." This crinkling becomes apparent if tensed parallel wires are observed in light reflected off of the surface. In Fig. 186 the vibration has just commenced; it grows more intense with increased amplitude (Figs. 187, 188) and at a certain point changes into true wave formation (Fig. 189). A train of waves on another focal plane can just be made out in Fig. 189; the image of the wires is breaking up. Thus actual wave formation is preceded by a phase of surface crinkling; as the amplitude increases the tension changes abruptly into wave formation and the crinkling vanishes.

of what is happening. What one imagines to be "simply waves" very soon proves to be a rich complexity of formations. The most diverse categories which interpenetrate, overrun and take over from each other, come to light. There are standing and traveling waves, free and constrained vibrations, and surface waves and long waves. Attention focuses mainly on the relations between the atmosphere and the surface of the sea. The wind, being the main generator of waves, must be included. Below we reproduce from the book to which we have referred, a scale of wind and waves which was formerly in practical use and which will give the reader some idea of actual conditions.

Sea and wind		
	No wind	Mirror-like sea.
	Light air	Small scale-like ripple with no crests.
	Slight breeze	Small waves, still short but more marked. The crests look glassy but do not break.
	Gentle breeze	Crests begin to break. Foam mainly transparent, tiny white crests of foam may appear sporadically.
	Moderate breeze	Waves still small but growing in length. White crests appear fairly generally.
	Fresh breeze	Moderate waves having a markedly long form. White crests everywhere. Spray may occur sporadically.
	Strong wind	Large waves begin to form. Crests break and leave behind large patches of white foam. Some spray.
	High wind	Sea rises. White foam from breaking waves begins to stream back in the direction of the wind.
	Fresh gale	Moderately high crests of considerable length. Spindrift begins to blow off the crests. Foam is flung back in well-marked swathes in the direction of the wind.
	Strong gale	High crests, dense swathes of foam in the direction of the wind. The sea begins to "roll." Spray may start to limit visibility.
	Whole gale	Very high waves with long breaking crests. Sea white with foam. Heavy thrust-like "rolling" of the sea. Visibility limited by spray.
	Storm	Exceptionally high waves. The crests of the waves are blown to spray. Visibility greatly reduced by spindrift.
	Hurricane	Air filled with foam and spray. Sea completely white. Visibility severely reduced. Distant view blanked out.

The changes affecting the waves are made plain by this scale. It is not simply a matter of one wave following another. The motion of the waves also varies. "The first thing to be noticed at sea is that not all the waves following each other in succession are of the same height, but that high waves occur in groups. Usually the third or fourth wave is the highest; in between the waves are "blanks" in which the waves are very low. The second striking feature is that although the wave group moves forward, the single wave does not maintain its station in the group but migrates through it. Both phenomena can be explained in terms of the interference of waves of varying lengths" (ibid.). The various types of waves are characterized by shortness or length, and by flatness or steepness in various time sequences. At the same time these wave forms interpenetrate and overrun each other with a highly varied periodicity. There is a complicated periodogram and indeed a fully developed wave spectrum. There are also interference phenomena. "In broad basins standing waves in both a longitudinal and transverse direction are possible. The result of the two interfering oscillations is no longer a to-and-fro vacillation but a turning wave which was termed *amphidromy* by R. A. Harris (1904), who was the first to study such waves." They run as lines of simultaneous high tide like the "spokes" of a seawide wheel which "generally turns left in the northern and right in the southern hemisphere." We shall meet them again in connection with tide waves.

The actual currents of the ocean also enter as a factor in the generation of waves, "Large eddies dominate the pattern of surface currents in the ocean" (ibid.).

The movement of these currents gives rise to turbulence. "In actual fact, movement in the sea is almost without exception disorderly or turbulent. The turbulence of the flow is manifested in the fact that a relatively even movement which might be called the actual ocean drift is subject to short but marked fluctuations in its rate of progress. The direction of flow thus varies" (ibid.). Vast circulations of a horizontal and vertical nature are formed. They are due to the wind, to differences of heat and salinity (thermal and saline circulation). Their course is affected by the rotation of the earth, by friction, and by topographical conditions. To obtain a more definite picture of these complicated conditions we will take a closer look at the Gulf Stream. We see that a natural phenomenon of this kind contains an extraordinary variety of features and that it is really an individualized system with clearly marked characteristics. First of all the velocity of the Gulf Stream is subject to periodic fluctuations connected with the changing speed of the generating wind. Then there are small counterstreams on both sides of the Gulf Stream. And there is also a transverse circulation. As a result of all this, the water mass of the stream is not constant. "Thus for example, near Cape Hatteras, the Gulf Stream no longer bears along tropical water from the region of the trade currents whence it derives its energy, but subtropical water from the Sargasso Sea" (ibid.). This means then, that the Gulf Stream is the same in its appearance, but that the mass of water has changed. The hydrodynamics of the main current also gives rise to "horizontal oscillations which are reminiscent of the meanders of continental rivers. The meanders are propagated with the current and spread until they have completely taken the place of the vortices of the current" (ibid.). As to the stratification of the Gulf Stream, it has been found "that the narrow ribbon of current has an imbricated structure, for it consists of thin layers traveling at high-speeds intercalated with thin layers moving at lower speeds, all the layers being perpendicular to the direction of the stream." The Gulf Stream in fact has a phenomenology of its own with periodic features in its time flow, in the pattern of its currents and in its structure. Characteristics closely resembling those of the Gulf Stream are also to be seen in the Kuroshio current in the North Pacific. The wholeness exhibited by a phenomenon like the Gulf Stream becomes all the more impressive the more one takes its climatic and biological sphere into account. It would be no exaggeration to speak of an "organ of the earth."

The pattern of sea waves and currents we have described above is also influenced by the sun and the moon as tide-generating forces. "Regularly the water rises and falls in a roughly diurnal or semidiurnal rhythm on all the coasts of the world's oceans. And with the same regularity the water washes up to the coast and away again. The rise and fall is called the tide and the to-and-fro motion of the water is known as the tidal current. Tides and tidal currents are two different phenomena of one and the same process" (ibid.). In these phenomena, which are significantly periodic, astronomy and hydrodynamics act in association; this joint action is reflected in oscillations. At the same time the amphidromies, to which reference has been made, develop as rotating waves. They are formed by tidal forces under the influence of friction and the rotation of the earth. They are true oscillating systems. It is interesting to study the way in which these oscillatory processes in the ocean act together with those of bays and bordering seas. There is interplay between the oscillations of the oceans and those of bays and adjacent seas. Free oscillations, co-oscillational tides and independent tides manifesting themselves in equivalent formations of nodes and antinodes show up as resonance zones. "The results reveal that the period of the free oscillations, which is determined by the depth of water and the effect of the earth's rotation, has a decisive influence on the elevation of the tidal range. At certain critical depths the free oscillations are in resonance with the forced oscillations deriving from the tide-generating forces" (ibid.). The oscillatory pattern thus produced, embraces the oceans of the world and the waters communicating with them.

The phenomenon known as *seiches*, which again involves oscillations in bodies of water, is also deserving of mention. "Standing waves as free oscillations occur in all natural bodies of water. F. A. Forel, a Swiss physician, was the first to recognize them as such in the Lake of Geneva. The term 'seiches', which had been used locally since time immemorial for variations in the level of the lake, has been generally adopted to describe free oscillations in more or less enclosed bodies of water. Since F. A. Forel wrote his classic monograph on the seiches of Lake Geneva in 1895, a number of lakes and bays have been studied and oscillations involving one or more nodes have been identified. The seiche theory has enabled oscillations of short periods responsible for indentations in the water-level recordings of tidal movements to be identified as the natural oscillations of small basins, and periods of many hours as the characteristic oscillations of whole areas of sea. The causes of seiches have often been studied. Just like long progressive waves, they must also be due to extraneous forces. They occur particularly, for example, when the wind has forced water up to one end of a bay and, on dropping, causes the mass of water to vacillate to-and-fro. Seiches may also be due to variations in atmospheric pressure, particularly when the period of pressure fluctuation coincides with the natural period of the oscillating system, under which conditions the seiches are intensified by resonance" (ibid.).

The oscillatory pattern becomes particularly marked when "in the tides of shallow seas there are, besides the astronomical tides, what are known as shallow-water tides appearing rather like overtones in acoustics. Their angular velocity is either an integral multiple of the arguments of the astronomical tides, in which case they are known as overtides, or there are combinations of various astronomical tides, known as combination tides" (ibid.).

The internal waves which form on the boundary between two layers of water are also of interest. Here again there are short traveling internal waves and long traveling internal waves of a tidal character, and also natural oscillations which appear as "cellular waves or stability oscillations." Once more resonance may also figure as a factor in wave generation.

These brief references and observations, for which the monograph we mentioned was a handy source of information, will, it is hoped, enable the reader to form a graphic picture of sea waves in all their complexity. It is a grandiose picture that we see: it ranges from the orbital paths of water particles on the smallest scale through wave formations of every kind, through interferences, turbulences, circulations, resonances, to the vast influences of sun and moon, and thus to oscillatory processes which embrace the oceans of the world. Nature creates this pattern of oscillations and waves on a global scale of cymatics.

In the preceding chapter we saw pictures of figures which, generated by sound and strictly ordered according to proportions, number and symmetry, spring into existence as the amplitude is modulated. Are these conditions reflected in their essential features in other dimensions of nature? Looking at the "pictures" wave mechanics gives us of the electron of the Hydrogen atom in its various states of excitement (with the surface probability of the electron) we might answer the question in the affirmative. These "clouds of electrons" such as the atomic physicist describes, are oscillatory systems ordered according to proportions, number and symmetry. Thus these phenomena, which have long been familiar to the atomic physicist, also fall within the province of harmonics and may be observed as such*. If we consider other atomic levels, there must, if the levels involved are of a simple numerical nature, again be *electron-space-time configurations*, although they will invariably be highly complex.

---

*The objection that the atom involves something "totally different" from acoustic vibrations does not hold if we remember that essentially the *same features* (proportion, symmetry, oscillations) are apparent in a great diversity of systems, and in all of them we are dealing with three-dimensional wave systems. The oscillations (periodic system, harmonics or overtones) occur at intervals expressible as quantities. (The fundamental frequency multiplied twice, three times, four times, etc. gives the "overtones." These are natural and originally quite independent of "our" music.) The different levels to which the oscillations rise by leaps now acquire a qualitative character: the appropriate element in the case of the atom [quanta]; the appropriate interval in music. Needless to say, a vibrating piano string is not an electron shell. Yet the processes *qua* waves in terms of proportion, number and symmetry have close affinities categorically and phenomenologically, even though in atoms, by reason of their four different variables of state (which, however, also have different levels of a numerical quantitative nature) the complexity is enormous. We have no ulterior motive in taking a look at all three phenomena together. We do not plan, say, to introduce the Pythagorean harmony of the spheres as a *deus ex machina*. What we are trying to do is to grasp the actual features wave harmonics have in common wherever they occur. Moreover, the criticism that wave-mechanical "pictures" are, of course, not pictures but only loci of probabilities, is not valid because the orbitals of the electrons nevertheless obey laws which are reflected in such spatial representations of position. There is not just any number of probabilities: there are the probabilities which exist on the basis of arrangements involving number, proportion and symmetry. We might perhaps say that certain patterns of regularity in these various regions such as harmonic overtones and the elements of the periodic system, highlight features these vibrational systems have in common. If the tones are built up into periodic series or periodic revolutions (as say, in a spiral) we find the "same" tones occurring again and again in the octaves. Whenever the frequency has been doubled and we arrive an octave higher, we perceive the "same" tone only an octave higher. By the way, exactly the same kind of pattern appears when we look at the vertical columns of the periodic table. The essential thing is that these oscillatory systems are manifested in terms of number, proportion and symmetry. It is therefore logical to assume that the harmonic principle will also be encountered by research in nuclear physics. Such a statement is not made without due and careful thought. Indeed one may be sure that it requires an enormous feat of imagination to conceive a complicated hierarchical oscillatory model reflecting the harmonic figuration and dynamics of the elementary particles. Here again we will respect the boundaries we have set to our work but nevertheless enter a claim that the criteria of proportion, number and symmetry in oscillatory systems also holds true in nuclear physics.

The periodic system of the elements shows these ascending relationships in an orderly numerical sequence. If these systems of number, proportion and symmetry (the atoms of the periodic system) are studied in terms of their ascending and descending genetic levels, we find a pattern similar to that presented by intervals, musical sounds, chords, and harmonic frequency spectra considered as a stepped series of vibrations. If the periodic system is involved in genetic processes, as it is in fact in star formation and the biography of stars, we have in genesis on the atomic scale a "score" which is played in cosmic systems. These processes would in fact, as "cosmic music", be the equivalent of music and its domains. It is not speculative philosophy that we have inferred here, but the real signature of these wave systems, however different each may appear to our perception. We have no desire to encroach on the preserves of atomic physicists. But, without overstepping the bounds we have prescribed, once more we would point to the criteria of vibrational systems involving number, proportion and symmetry.

The scope of our vision is extended when we turn to molecular dimensions. Here again number, proportion, symmetry and oscillation predominate, as described by chemistry. In crystals, in the solid phase, we recognize systems exhibiting these harmonic features. Although the imagination is called upon to perform prodigious feats, and boggles at first, research in physics shows beyond doubt that the configurations of minerals are in fact ordered according to number, proportion and symmetry *through and through*. Here again, the harmonic categories can be verified. Even in the province of chemistry, proportions, symmetries and numbers (particularly in mass ratios) appear in such a way that we can find equivalents in intervals, tones, chords and harmonic frequency spectra. (In music we take one note and add others to it, thus obtaining intervals, notes, chords, and polyphony. At the same time we would emphasize that the harmonic levels of oscillation under consideration here are *natural* systems or harmonic overtones).

It is not the purpose of this book to pursue this line of thought any farther. It is mentioned because cymatics brings the imagination to the inside of things and sets the powers of concrete thinking tasks of supreme difficulty. Paradoxical though it may seem, it is the purpose of these tasks, and part of their solution, to "hear" the systems of Nature, whether in the cosmic or the mineral-chemical sphere; i.e. to perceive them in a way equivalent to the perception of sound in acoustic space.

We cannot go further in our observation of harmonics without turning our attention to organic nature. The next section will be devoted to biological systems.

16

The Biological Aspect

Throughout the animal and vegetable kingdom Nature creates in rhythms, periods, cycles, frequencies, reduplications, serial phenomena, sequences, etc. This is the style in which natural structures are built and it is ubiquitous. If we take a few examples, we shall see that this is the all-dominant mode of appearance. Let us look at histology, the science which deals with the structure of tissue. The very origin of the word tissue (Latin *texere* = to weave) is a significant comment on the prevailing conditions: cells are arrayed in rows, one pattern following another, wherever we look. The intercellular structures take the form of frameworks, networks, grids, families of elements continually repeated and following each other in regular sequence, forming a woof and weft whether looked at with the naked eye, through the light microscope, or through the electron microscope.

If we pass the individual organs under review, we shall find this rhythmicity of formation wherever we look. To start with the animals, let us look first at the integument, the skin. Apart from the cell tissue as such with its vast array of repeating elements, we find all the skin structures, scales, feathers, and hairs arranged in serial patterns, in regular formations, in rows and tracts, etc. Each structure is in turn a serial product. A bird's feather in which the pterylae are arranged serially, one after another, is a good example. Turning to the muscle tissues, we do not find compact homogeneous masses but organized fasciculi of fibers with the elements arranged in rows, one bundle next to another. Series of fibers of this kind continue in sinews which irradiate into the ligaments and bones. Everywhere there are fibrillae, lamellae, and folia which develop into the spatial frameworks of the sinew, ligament and bone organization. In the fields of the sensory cells, in the layers of the ganglion cells, and in the immensely complex communications between these systems, we still find that this principle of periodic seriality prevails. The same holds true in the digestive organs; for instance in the tracts of intestinal villae. And then there are organs such as the liver, the pancreas, kidneys, etc. which represent "tissue" in the true sense of the word, comprising as they do *one* basic element (the liver cell, the pancreas cell, the nephron or renal unit) which are regularly repeated. In a word, wherever we turn, we invariably find the style of structure we have described. It might be argued that these things are as they are and there is no reason for their being otherwise. But, whether we take them for granted or not, whether they are commonplace or not, we *must*, if we want to characterize Nature's mode of construction, state a case for this mode of periodicity which is ubiquitous in Nature's creations. In cell division in particular, in mitosis, the process of the regular repetition of polar phases occurs as a function of space and time. Here we have a strict cycle of well-characterized phases, which is strictly repeated in terms of space and time and also of number and proportion. We may speak of biological periods and biological oscillations in the proper sense of these words. Needless to say, there are other substrates and also other media and complicated chemisms. The basic features, however, remain the same:

Repetitions
regular repetitions
regular repetitions of polar phases
regular repetitions of polar phases as a function of time (cycle)
regular repetitions of polar phases as a cyclic function in time
and as a spatial function.

We can, as it were, build up such events in steps: in this sense and only in this sense can we speak of biological oscillations in the proper sense of the word.

However, the regular and consistent repetition of basic elements is not restricted to the major organ systems (integument and nervous, supporting, digestive, and procreative systems, etc.): we also find segments occurring serially, as it were, as elements of "style" in the general structural principles of organisms. The metamerism or fundamental division into segments of vertebrates, arthropods (insects, crabs, etc.) and a number of worms appears as a regular repetition articulating all the systems such as vessels, skeleton, muscles, nerves, etc. and incorporating them in the segmentation. Look at an earthworm, or a snake whose vertebral column may comprise hundreds of vertebrae!

But it is not only the structural elements which show a repetitive periodic character; functions also proceed rhythmically, in regular cycles and serial processes. This is exemplified by the pulsations of the heart, the respirations, the oscillations in contracted skeletal muscle, the autonomic rhythmicity of the intestinal musculature, and the serial action currents of the nervous pathways, etc. We see that these periodic processes occur generally in vast numbers and what is more, that this style of construction is dominant and ubiquitous. The remarks we have made about the animal kingdom also apply to the vegetable kingdom. Here we have a virtually endless array of repetitive elements: node follows node, leaf comes after leaf, shoot after shoot, and bud after bud, to name but a few examples amongst the higher plants. Here again the questing eye can range far and wide in morphology and physiology: wherever we look we can describe what we see in terms of periodicities and rhythmicities. And here again the prevalent style of Nature's structures is such that we may speak of periodic morphology and periodic physiology in the broadest possible sense. Now since these two fields form, of course, a living unit, we may simply speak of the essentially periodic character of the animal and vegetable kingdoms together. Let us place on record what we mean when we say this: when Nature creates anything, it creates it in this periodic style. This characterization does not, however, say anything about the essential nature of the various steps and stages. The phases of vegetation, of blossoming, of fruiting, for instance, or the systems of digestion, respiration or nervous regulation, etc. must be seen in the context of metamorphoses, developmental stages and functional cycles. But whatever appears, whether it be in this phase of development or on that level of organization, whatever it is that comes to light, it has rhythmicity and periodicity writ large all over it.

But this is by no means the end of the regular formations in organic nature. Quite the contrary, throughout the two kingdoms of organisms we have formations and organizations where the elements are repeated regularly in number, form, and proportions and in symmetrical and centered patterns. Here we have the bilaterality of the arthropods and vertebrates, the molluscs and worms, the radial symmetry of the Coelenterata (corals, medusae), and also the echinoderms (starfish, sea-lilies, sea-urchins). Here symmetry, number, dimensions and proportions are the ubiquitous keynote of the organization. How do such things come into being? The embryogeny of these organisms enacts these processes for us. If we take a close look at these embryologies, we shall see the first intimations of multiplication in aggregations of cells on the germ; they appear in groups and number 2, 3 or 4. They are arranged in a plane or directed towards one point. And thus, little by little, a miraculously organized animal takes shape. From all the vast number of embryogenies, let us take that of the anthozoa, the corals. On the gastrula-like basic form there appear two ingrowths, followed by two more, making four, and then another four, making eight in all, which develop into the septae or internal walls. In this way octocorallia or alcyonaria develop. In other groups two more pairs are formed, giving rise to hexacorallia or zoantheria. What we have now are regular architectural designs. And these arrangements are reflected in the internal structure, in the layout of the muscles, the digestive organs, and the feelers (tentacles). At the same time bilateral symmetry prevails throughout the basic design. The sixfold character is doubled up; 12, 24, etc.

Let us look now at the embryogenesis of the starfish. In the larva of the starfish there are two primordia which are pentagonal in shape and located on either side of the larva's stomach. These two pentacled organs unite, "lie up" one against the other and produce the young starfish.* Thus an organism, as it were, arises in the organism. The starfish arises from one "part" of the larva. A pentagonal creature emerges from a non-pentagonal basic form. One embryogenesis after another could be described in this way. Thus not only serial regularities arise but also organisms displaying symmetry of number, shape and proportion. And here again we find this type of formation occurring consistently throughout the organism. Even unicellular creatures show regularity of form and configuration. Let us look, for example, at the radiolaria with their skeletons of silica, of which Ernst Haeckel discovered and described hundreds of species. He illustrated a selection of them in his "Kunstformen der Natur." The mind boggles when we look at these space lattices, star patterns, bell-shaped networks, etc. Here the wildest improbabilities become visible fact. For millions of years, billions upon billions of these and similar creatures have been floating in the sea in thousands of varieties. Each of these microscopic creatures is a miraculously wrought texture in which harmonic characteristics may be plainly seen.

Phenomena of the same kind arise in the vegetable kingdom. Regular configurations meet our eyes as we pass from the algae (diatoms) to the highest form of flowering plants, and all display symmetry of form, number and proportion. If we look at roses, lilies, primroses, cruciferae, ranunculaceae, papilionaceae, labiatae, orchids, etc., we see patterns displaying perfect harmony. How can we comprehend these symmetrical relationships? To give ourselves a lead, let us compile a list of the regularities we have observed in the physical field under the influence of vibration:

---

*Here we follow the descriptions of Korschelt and Heider, *Vergleichende Entwicklungsgeschichte der Tiere.*

1. Polygonally pulsating drops (Figs. 73-79).

2. Three-dimensional formations regularly pulsating (soap bubbles, Figs. 80-86).

3. Heaps of particles (spores of the club moss) which form symmetrical patterns, uniformity being conferred by the conglobing effect of the vibration (Figs. 87-89).

4. A paste forms regular segments under the action of sound. Vibration can also cause it to clump into a homogeneous round body in which the material circulates (Fig. 90).

5. With the electron beam we can exhibit regular pathways with two frequencies. If the frequencies stand in certain interval-like ratios to each other, the electron beam describes paths displaying symmetry of number, form and proportion (Figs. 91-104).

6. Similarly the mechanical pendulum produces patterns of orderliness with its oscillations (Figs. 110-112).

7. Rendered visible by the schlieren method, liquid systems such as drops, liquids in containers, and films, produce perfectly harmonic figures when subjected to vibration, the patterns changing abruptly one after the other as the amplitude is modulated (Figs. 118-142).

8. If two frequencies are made to impinge on one and the same liquid system, a figure appears which is the resultant of these actions. Each frequency produces a definite figure of its own. The resultant is a "third" figure (Figs. 156-164).*

Now it is beyond doubt that, *where organization is concerned,* the harmonic figures of physics are in fact essentially similar to the harmonic patterns of organic nature. Here we have, say, a pentagonal vibration pattern (Fig. 122) exhibiting a quintuple configuration in a number of circles, each with its own distinctive form. Here we see alternation and equidistance such as we find, for example, in the higher flowers. When we look at *these* regular patterns,

---

*It might be mentioned in this connection that we must not lose sight of the rotary waves that appear. If such a vibrating system performs a translatory movement, we find that the wave movement follows a screw-like path (e.g. a spiral).

what we actually see is identity of configuration. However, we must not lose sight of the fact that in the one case these have to do with physical oscillations and in the other case with biological formations whose peculiar features are determined by both the biological substrate and the biological process.

Have these two categories of phenomena anything to do with each other? Can we obtain a firmer mental grasp of them in concrete terms?

First of all, we can establish the following points in our minds.

In the first place we have the certain experience that harmonic systems such as we have visualized in our experiments arise from oscillations in the form of intervals and harmonic frequencies. That is indisputable.

Secondly, we are familiar with the style of nature which is characterized throughout by rhythmicities and periodicities, so that we can speak of biological periodicity, and even of biological oscillation in the strict sense; i.e. regular repetition of polar phases as a function of time and space.

Thirdly, we have knowledge of the interplay of factors in the organic world.

We have:
Antagonisms and synergisms.
Inhibition and excitation.
Damping and stimulation.
Suppression and liberation, etc.

We know, however, that the interaction of these factors goes much farther. We know that the processes are adjusted "to one another", that there is a delicate interplay of regulatory factors governing the way they are followed out and that they are often correlated with and proportional to each other.

Fourthly, we can say: if biological processes on and in the biological substrate proceed in an interval-like manner, there must be a corresponding pattern in this field of operation. If biological rhythms operate as generative factors at the interval-like frequencies appropriate to them, then harmonic patterns must be necessarily forthcoming.

Fifthly, it follows that if harmonic configurations appear in organic nature (morphological and physiological), then what we see before us is the result of the rhythms, intervals and frequencies of the generative factors. In other words: harmonic phenomena appear where the generative factors operate within a harmonic order.

What is more, it is impossible to imagine that formations which display perfect symmetry of number and proportion should not arise from origins in which a similar pattern is inherent. This style as a process of becoming must be an essential feature of the genetic field. *That* this is in fact the case is certain. *How* Nature proceeds in these matters, that is the question. It is not enough for us to feel sure that harmonics goes with harmonics. We want to have an insight into the way Nature works.

In this connection we have the repeated impression that Nature wishes to show us its method of working, that it wishes to remove the blinders from our eyes. Take animal pigmentation as an example. How improbable a phenomenon is the zebra's striped pattern! And yet zoologists are already speaking of wavelengths and interferences in the zebra coat. Patterns of stripes, transverse or longitudinal, or a combination of the two, are common enough in the animal kingdom. Such patterns abound among the birds, insects and fishes. It is particularly interesting to compare the markings of the tiger with the pigmentation of the leopard and jaguar. In the tiger we have a more or less serial pattern. In the leopard there are characteristic groups known as "rosettes." Brehm has the following to say about the markings of the leopard: "Towards the sides the ground color lightens down to the belly and the inner side of the limbs, which may be pure white. On this basic color there are a number of black spots which may be solid or in the form of rosettes. There are solid spots on the head, the throat and chest, and the extremities. They may also appear on the shoulders and hams, particularly in the types where the rosettes are small; in the types where they are large, they are usually scattered profusely over these parts, too. The back, the side of the body, the belly and the inner side of the limbs are covered with rosettes. These vary greatly in size, and an attempt has been made to separate animals with small spots (leopards) from those with large ones (panthers), but the fact that the spots may vary in size from a walnut to patches of up to 3 inches in diameter has made this impossible.

"The annular spots are made up of a number of small black circles, imperfect in shape, moon-like spots and dots united in a broken ring, resulting in real rosettes with as many as eight small spots, rarely only two, and usually five to seven. The area enclosed by these rosettes is usually more vividly colored than the ground color of the coat. It is usually free from black, but there are exceptions among both African and Asiatic panthers in which black dots also appear inside the largest rosettes near the back. But this is never so regular or pronounced as in the jaguar. The rosettes continue past the base of the tail but become solid dots nearer the pure white tip. In some forms the tail is only spotted, in others the spots form rings before the tip but are not closed on the underside of the tail.

"Frequently the spots are arranged in rows, mainly on the back where, particularly in the rear half, there are often two longitudinal rows comprising elongated rectangular spots accompanied by two parallel rows of rosettes, whereas the other rosettes run from the top front down to the bottom rear in diagonal lines which are not particularly conspicuous. The spots are also patterned, mainly in longitudinal rows, on the head and at the sides of the throat, and there are one or two transverse rows on the breast." There are also similar patterns of pigmentation in the jaguar and the snow leopard. We mention these animal markings in particular because they are virtually an ocular demonstration of the problem confronting us. Here is one line of approach we can follow in our study of the origin of regular design in organic nature. We can plunge ever deeper in our observation of Nature and seek to read the riddle of its forms and processes. We can explore this or that periodic process and discover the rhythmicity of this or that structural motif. What we want to do is, as it were, to learn to "hear" the *process* that blossoms in flowers, to "hear" embryology in its manifestations and to apprehend the inwardness of the process.

A second line of approach likewise leading into the harmonic operations of Nature opens up when we plan an actual research project. Molecular biology provides us with some vital clues. In this field of research we come to know whole worlds. In the nucleus of the cells are the chromosomes. These contain in rows the genes which play a crucial role in the processes of development: they are thought to be the carriers of hereditary features. Many of the activities of these genes have been elucidated. There are regulator and structural genes, etc. The point of interest to us is that these genes interact in a variety of ways. They excite and provoke, they impart "information"; but on the other hand they have an inhibitory and blocking action. In molecular biology there are pointers to the necessary existence of a delicate and finely adjusted concert of actions. This is a sector of observation on which our attention is sharply focused. It is obvious that the impulses of these regulating, controlling, directing genes proceed in rhythms: i.e. that they are of a periodic nature. But what are the rhythms? What are the intervals? What are the chords and frequency spectra? How is the score of this gene orchestra played? It can be taken as absolutely certain that the admirable methods of microbiology will shed more and more light on the activities of the genes. Our contribution to the discussion concerns the criterion of periodicity and the ratio (interval) between the periods involved. The object will be interrogated in terms of biochemistry, bioelectrics, biodynamics, biostructure, etc. and its rhythms spelled out. This is one line of advance. The next concerns the materials and processes which are originated by the activities of the genes and their messengers. Their periods must likewise be recorded. But this brings us ever closer to the primordia themselves which, apart from having the physical capacity to respond to outside influences, create the impression of being *unities* or at least *homogeneous systems.* Hence they can, as systems, be generated by rhythms.

Setting up the program for a biological research project is by nature only a "first fitting." But at least an indication can be given where the important seams have to be. If research, working along this line of advance, is successful in discovering a pattern of frequency in biology, then it has reached the preliminaries to configuration, the processes which precede the appearance of structure. These are divested of mystery if they prove in fact to be the harmoniously ordered operations of frequency. A path can then be imagined leading via "oscillating spheres" to the harmonically formed configuration. (By oscillating spheres we understand the sequence of observational fronts.) An objector might question the necessity for us to perform such complicated experiments involving a complex methodology in order to explain the harmony and symmetry of natural events. Organic entities displaying features of regularity, number and proportion must, he may say, be originated in one way or another by generative factors which are themselves harmoniously organized. How could anything perfectly harmonious arise out of the unharmonious? Hence there must be harmony of some kind in the genetic process. Again—our objector might continue—you feel quite assured that the whole is also dominant in the individual parts. Then, on this view, harmony must also prevail in the generative sphere of organisms. To this it may be replied: if one holds that harmony is at work even in embryology in the generative phase, the actual processes must nevertheless be *seen* in their frequencies and intervals.

So many questions concerning the effects produced by these rhythmicities and periodicities remain unanswered. There is really no alternative but to set about a research project which might be defined in these terms: How do the generative factors act to determine the embryogenesis of organisms which present harmonic forms throughout? Since the formative style of Nature in all its aspects is rhythmic, periodic and cyclic, how do living intervals, sounds and frequency spectra act in embryology? Time and effort should be devoted to experiments to enable these generative realities to be seen and recognized.

17

Historical Review / Methodological Preview

As was pointed out time and again by philosophers and historians in discussions with the author following his inquiries about harmonic vibrations, harmonics were the main basis of early philosophies. Indeed these ancient doctrines consisted essentially of harmonics. The basic tenet of the Pythagorean philosophy was that arithmetic, geometry, astronomy, music and ethics were all to be explained in terms of numbers and their ratios ("everything is number"). Erich Frank has now convincingly demonstrated in his book "Plato und die sogenannten Pythagoreer" (Plato and the So-Called Pythagoreans) that Plato was in touch with the Pythagoreans of Southern Italy, particularly with Archytas, who was active there in the first half of the 4th century BC as a physicist and mathematician (statesman and military commander). Plato visited him and learned about his discoveries and investigations. Few though the surviving fragments by Archytas may be, they suffice to show that he discovered the numerical ratios of the intervals of the vibrating portions of strings. This knowledge was of supreme importance to Plato and under its influence he created systems of order in which the numbers 1, 2, 3, and 4 (as "ideal numbers") became nothing short of world-creating principles.

A more familiar knowledge of Plato's views reveals that everything in his world view is arranged according to measure and number, proportion and mathematical form. The Platonic dialogue of most importance for us in this respect is, of course, the "Timaeus", for it is here that Timaeus the Pythagorean describes the creation of the world and tells us in detail how Nature and man came into existence. He explains how the world (earth, water, air and fire) is built up from triangles and how such triangles are formed into regular solids which are assigned to the different states in which matter exists. To give some idea of what this Platonic world is like, we will quote a few sentences from the "Timaeus": "In the first place, then, as is evident to all, fire and earth and water and air are bodies. And every sort of body possesses solidity, and every solid must necessarily be contained in planes; and every plane rectilinear figure is composed of triangles; and all triangles are originally of two kinds, both of which are made up of one right and two acute angles; one of them has at either end of the base the half of a divided right angle, having equal sides, while in the other the right angle is divided into unequal parts, having unequal sides." (What we have, then, is a triangle with two equal sides subtending a right angle, and a right-angled triangle with unequal sides). "Now of the two triangles, the isosceles has one form only; the scalene or unequal-sided has an infinite number." Equilateral triangles are then considered alone. "...and four equilateral triangles, if put together, make out of every three plane angles one solid angle, and out of the combination of these four angles arises the first solid form."

This is the tetrahedron or the three-sided pyramid. In a similar way octahedrons, hexahedrons (cubes), icosahedrons and dodecahedrons are evolved. These five regular solids are assigned to the elements for appropriate reasons: the cube to earth, the tetrahedron to fire, the octahedron to air, the icosahedron to water, and the dodecahedron to the universe. The book then describes how these elements act together and are transformed and the many varieties of fire, earth, air and water are mentioned.

"We must imagine all these (particles) to be so small that no single particle of any of the four kinds is seen by us on account of their smallness: but when many of them are collected together, their aggregates are seen."

Now it does not matter at this juncture whether we find these ideas right or wrong, childish or inspired. What is important is the impression which such facts create — an impression of the tremendous orderliness of things. With this in mind, let us turn now to look at the scale which Plato constructed. It was, Plato tells us, in accordance with this scale that the creator fashioned the soul of the world. Let us say at the outset that the reader will at first understand little or nothing of what he is told. For centuries some of the best minds have labored on commentaries regarding this scale, but space will not allow us to go into them here. First of all let us quote the words in which Timaeus describes the creation of the soul of the world, bearing in mind, however, that these are things which are impossible to grasp on first acquaintance, let alone experience as realities. We quote: "He made the soul (sc. of the world) in origin and excellence prior to and older than the body, to be the ruler and mistress, of whom the body was to be the subject. And he made her out of the following elements and on this wise: Out of the indivisible and unchangeable, and also out of that which is divisible and has to do with material bodies, he compounded a third and intermediate kind or essence, partaking of the nature of the *same* and the *other*, and this compound he placed accordingly in a mean between the indivisible, and the divisible and material. He took the three elements of the *same*, the *other*, and the *essence*, and mingled them into one form, compressing by force the reluctant and unsociable nature of the *other* into the *same*. When he had mingled them with the essence and out of three made one, he again divided this whole into as many portions as was fitting, each portion being a compound of the *same*, the *other*, and the *essence*. And he proceeded to divide after this manner: First of all, he took away one part of the whole (1), and then he separated a second part which was double the first (2), and then he took away a third part which was half as much again as the second and three times as much as the first (3), and then he took a fourth part which was twice as much as the second (4), and a fifth part which was three times the third (9), and a sixth part which was eight times the first (8), and a seventh part which was twenty-seven times the first (27). After this he filled up the double intervals (i.e. between 1, 2, 4, 8) and triple (i.e. between 1, 3, 9, 27), cutting off yet other portions from the mixture and placing them in the intervals, so that in each interval there were two kinds of means, the one exceeding and exceeded by equal parts of its extremes, the other being that kind of mean which exceeds and is exceeded by an equal number. Where there were intervals of 3/2, and 4/3 and of 9/8 made by the connecting terms in the former intervals, he filled up all the intervals of 4/3 with the interval of 9/8, leaving a fraction over; and the interval which this fraction expressed was in the ratio of 256 to 243. And thus the whole mixture out of which he cut these portions was all exhausted by him. This entire compound he divided lengthwise into two parts, which he joined to one another at the center like the letter X, and bent them into a circular form, connecting them with themselves and each other at the point opposite to their original meeting-point; and, comprehending them in a uniform revolution upon the same axis, he made the one the outer and the other the inner circle. Now the motion of the outer circle he called the motion of the *same*, and the motion of the inner circle the motion of the *other* or *diverse*. The motion of the *same* he carried around by the side to the right, and the motion of the *diverse* diagonally to the left. And he gave dominion to the motion of the *same* and the *like*, for that he left single and undivided; but the inner motion he divided in six places and made seven unequal circles having their ratios of two and three, three of each, and bade the orbits proceed in a direction opposite to one another; and three he made to move with equal swiftness, and the remaining four to move with unequal swiftness to the three and to one another, but in due proportion."

These then are the words Plato puts into the mouth of the Pythagorean Timaeus to describe the creation of the world soul in accordance with the musical intervals, at the same time relating the numbers of the intervals to the distances of the planets. The numbers of the intervals are also the measures of the planetary spheres and of the cosmos. This musical scale represents at one and the same time the mathematical structure of the soul of the world and also the essence of the universe modeled on it. Now, these relationships are not meant as allegories, but as realities. Man, too, is fitted into these orderly systems. "Of all movements, that is best which a body initiates itself, for it is most closely related to the movement of thought and the universe.... First, then, the gods, imitating the spherical shape of the universe, enclosed the two divine courses in a spherical body that, namely, which we now term the head, being the most divine part of us and the lord of all that is in us."

"And the motions which are naturally akin to the divine principle within us are the thoughts and revolutions of the universe. These each man should follow, and correct the courses of the head which were corrupted at our birth, and by learning the harmonies and revolutions of the universe, should assimilate the thinking being to the thought, renewing his original nature, and having assimilated them should attain to that perfect life which the gods have set before mankind, both for the present and the future."

According to this description, the way the universe is ordered is reflected in our thought, in our minds. Birth causes these orbits to be shaken and perturbed. By investigating these harmonies we can restore in our heads the ordered systems of the universe. Plato's picture of the world can inspire and animate our minds and fertilize our thinking. The impression we receive has undoubted grandeur. Yet, if we ponder over these things and try to project ourselves into Plato's mind, it must be admitted that we cannot make them part of our living experience and that our thinking on the nature of the world cannot be creatively advanced in this way. It is quite possible to fashion an ethical platonism and live according to its categories; but of course, such a paste-and-scissors Plato is no longer Plato. In him these tremendous visions and vast organized systems sprang from the plenitude of his mental life. But if we try to think in terms of the harmony of the spheres, correct though everything may be in its proportions, these things become insubstantial; indeed, so far as the Plato that lived and taught at the Athens Academy is concerned, ghostly. One may know his Plato inside out, yet he will not think creatively in these proportions and numbers, in these cosmic revolutions which are, at one and the same time, the good, the true and the beautiful.

Granted the importance of our relationship to the great thinkers of earlier ages, let us take another example: Heraclitus. On a careful reading of the "Fragments of Heraclitus," we again have the feeling that we are on familiar ground. His soaring visions of the opposites are vividly present in our minds, for in our experiments with vibrations we have witnessed this constant interchange between polarities. The crest of the wave one moment has become the trough the next. What is now the periphery in a pulsating body becomes the center. A hexagonal shape loses its hexagonality and then has it re-imprinted by pulsation. And yet what is involved in all these is a unified process, a unit: the wave phenomenon. This idea is inherent in the world picture of Heraclitus. Here are some quotations from the Fragments: "Unite whole and part, agreement and disagreement, accordant and disaccordant; from all comes one and from one all."

"Into the same river you could not step twice, for other and still other waters are flowing."

"Into the same river we both step and do not step. We both are and are not."

"One and the same thing exists in us, living and dead, awake and asleep, waking and sleeping, young and old. For these several states are transmutations of each other."

"All events happen in the form of opposition and all things are in a constant process of transformation... and the world arises from fire and dissolves again into fire, in certain periods, in constant change, to all eternity. But this happens according to Fate."

The events of the world take place through the interaction of these opposites but in the interplay between them measure and harmony arise.

Heraclitus:
"This world, the same for all, neither any of the gods nor any man has made, but it always was, and is, and shall be, an ever-living fire, kindled in due measure, and in due measure extinguished."

"...that the world now dissolves in fire and is now formed again from fire in certain measures."

"Of the opposed forces those that lead to the creation of things are called war and strife, and those that lead to the world fire, harmony and peace, and the change between them 'the way up and down', and the world is formed accordingly. For, by condensing, fire becomes moist, and condensing further, becomes water. But water, when it is solid, changes to earth. And that is 'the way down'. And again the earth dissolves, and out of it comes water, and out of that the rest. That is to say, he attributes almost everything to the exhalation of the sea. But that is the 'way upward'."

"The way upward and downward are one and the same."

"For there could be no harmony if there were no high notes and low notes..."

"The opposites are united, and out of the opposed (tones) comes forth the fairest harmony, and by strife all things arise."

"Nature also strives for the opposites and, out of *them* and not the identical, produces harmony..."

"They do not understand how that which separates unites with itself. It is a harmony of oppositions, as in the case of the bow and the lyre."

Whereas we have sensed that everything so far has hung together, we are confronted with a difficult problem when we come to a word central to Heraclitus's conception of world creation: Logos. In his Fragments the word first appears in this description.

"To this universal Logos which I unfold, although it always exists, men make themselves insensible both before they have heard it and when they have heard it for the first time...."

"... Although the Logos is common, the majority of people live as if they had an understanding of their own."

"The Logos which shapes things according to the up and down of opposites is Fate."

Heraclitus explains the Logos, which permeates the whole universe, as the essence of Fate. "This is the ethereal material, the primal seed of the creation of the universe and the circulation of things, to which a certain measure is set."

Here we are like Faust sitting in his study after the Easter walk: Who is the Logos? A number of translations come to mind; world intelligence, world law, truth, meaning, power, etc. Even if we can bring ourselves to accept the view that in truth "the Word" may be meant here: Where does this get us? That is precisely our difficulty today: we cannot understand by "the Word" anything which is capable of creating, guiding, or directing the world; we do not see how sounds can be forces which can shape the world. Whereas with Heraclitus there were points where we felt we were on common ground, with the Logos we find ourselves confronting the incomprehensible.

It is not our wish to discourage anybody from reading the Greek thinkers by what we have said above. To the contrary, what we are trying to do is to see them as they were. And when we do so, of course, we realize that we no longer share their capacity for vital comprehension nor their creative conception of ideas. We are remote from their sources. We see them like fair landscapes bathed in the golden light of evening, but their sun is no longer alive in our contemporary hearts. Although almost all our instruments of thought are derived from these epochs, although we are still fed and nourished by them, we are no longer part of them. We experience them only as shadowy images and we must take care not to become ghosts.

To give some further idea of this outlook, of this world feeling in ancient times, let us turn to Aristoxenos. He was an Aristotelean (active in Athens in 350 BC) and the little of his extensive work still surviving deals with harmonics, rhythmics, melodics. Let us take as an example what Aristoxenos says in connection with rhythm. "We must imagine two different natures, that of rhythm and that of the material to which rhythm is applied (rhythmizomenon), having the same relations to one another as a plan has to the object that is planned. For as the body takes shape and form in different ways if its parts are placed in various positions and postures, either all or some of them, so each of the individual rhythmizomena takes a number of forms depending, not on its own formative power, but on that of the rhythm. For if one and the same spoken text is separated into divisions of time differing from one another, it undergoes changes which correspond to changes in the nature of the rhythm itself. This is the case with melody and whatever else can be rhythmically shaped by such a rhythm consisting of divisions of time.

"But here we must revert to the explanatory example we have cited and try to have a better understanding of each of the two parts we have named, rhythm and the rhythmizomenon. For, on the one hand, in none of the rhythmizomena is any of the forms the same, form being a certain arrangement of the parts of the body which comes about by each of them taking a certain position and posture, after which the form is also named. And, on the other hand, the rhythm is not one and the same with any of the rhythmizomena but something which orders the rhythmizomenon and accordingly shapes the divisions of time in this way or that...."

Thus rhythm is one thing and the element to be "rhythmized" is another. Examples of materials to be rhythmized are: melody, speech, the human form. For Aristoxenos, rhythm is what gives shape to speech, melody and the human form — in poetry, in music, and in the dance. These appear individually or combined in a work of art — for example when a poetic text is spoken, sung and represented in dance. It is clear that Aristoxenos holds rhythm to be a creative force. It is truly and essentially existent and is embodied formatively in the material (in speech, melody, and the dance).

The purpose of these aphoristic references is to allow us to see these ancient world pictures and systems through modern eyes. Of course, to do justice to the matter, we should have to expand our remarks into a monograph. Our brief sketches of the thought of Archytas, the Pythagoreans, Plato, Heraclitus, and Aristoxenos shed nothing more than side-lights on currents which have joined to form the great tide of human thought. On the one hand Pythagoras and Heraclitus, and indeed Plato too, point back to ancient Egypt, and even to Chaldaea, while on the other, these trends of thought continued to run through many minds down to Giordano Bruno and to Johannes Kepler. All Kepler's investigations were inspired by the vision he had on July 9, 1595, when he saw the five regular bodies inscribed in the planetary spheres. He bent his whole mind to verifying this vision by the application of exact numbers and it was in this way that he formulated his three laws of planetary motion.

But one should read Kepler for oneself. There we find an attitude of mind which we can admire and respect, but it can no longer inspire creative power in us.

Whether we regard these splendid visions of universal harmony and a cosmic order of periods as fanciful speculations, as abstruse number mysticism, or as childish errors, is something each may be left to decide for himself. But in looking back upon the great world vision of Plato or Heraclitus, we realize that these systems of number, proportion and symmetry can no longer be living realities for us. They are out of tune with our mental constitution. If we look, for example, at the figures of Giordano Bruno in which he represented the contents of the universe, we simply lack the capacity to enter imaginatively into such structures. And even if we can comprehend this view and apply and use it, the old traditions no longer afford us a full and satisfactory sustenance for the mind. We are at liberty, of course, to construct a world picture out of the relationship of number and proportion as the ancients did, and we may even find a certain satisfaction in so doing. But these systems of harmonies will not bring us into a creative and evolutive relationship with the world. How can we conceive such a relationship?

The visual arts, painting first and foremost, are the most significant expression of the currents and trends of an epoch. In our own day we are witnessing a tremendous struggle between form and dynamic. We find constructivism of the utmost sophistication which soars into a mathematical empyrean where it freezes into immobility. Then there is also kinetic art where everything is in motion, rushes around, is impelled electromagnetically into space, and resolves itself into an agitated mass, into a storm of molecules. And there is every shade in between. The banks of this turbulent stream of schools and experiments are piled with contorted scrap and macabre rubbish. Yet, transcending all aesthetic evaluation or conventional condemnation, processes can be described which intimate a feverish effort somehow and somewhere to establish bearings in the world. Why these remarks in a book on observations in physics? The point is this. What finds typical expression in the visual arts is also to be found in the natural sciences. Around us, before us and within us there are mysteries discernible as configurations, processes, figures and movements, but all ill-defined, seen as in a glass darkly. It is useless here to plead disinterest; these mysteries accompany us through life, and indeed we ourselves *are* these mysteries. Nor can we claim to have kept our balance in this baffling maelstrom. For it is precisely the elements essential to such an equilibrium that are missing. We snatch after traditions, seek salvation in the greatness bequeathed by the past, and all the while are swept along by the cataract of our times.

Where does cymatics come in here? If we can comprehend the wholeness of vibration or oscillation, and grasp the totalities in which it is manifested, then we have caught hold of reality. We have an instrument for bringing clarity to our view of the physical nature of the world. In cymatic phenomenology it is the *whole* phenomenon upon which we concentrate as we follow Nature unswervingly with eyes, ears, and brain. In this enterprise we bring our mental powers into play; we train and tutor our cognitive faculties. Again and again we have stressed the importance of developing faculties of observation, perception and insight. It is not a question, then, of ever more sophisticated apparatus and experimental designs but rather of a cognitive capacity which is continually unfolding. How do we know that this cognitive capacity can develop into perceptive faculties? Through our processes of conceptual and imaginative thinking we perform on phenomena what Goethe called "the highest operations of the spirit." We are not debating the limits of knowledge as a parade-ground exercise. He who sets such limits describes himself*; they are valid for him.

What will happen if we pursue these mysteries, pressing ever further with our investigations, life itself will show. What should our procedure be? By letting sounds, noises, musical tones elicit their effects we have discovered perfect systems of order comprising numbers, proportions and symmetries. These systems are not rigid figures; they pulsate, flow and undergo transformation. They weave textures out of their polarities and metamorphoses. They grow in intensity to become phenomena manifesting "everything" in patterns where orderliness prevails in spite of all the kinetics. Should this now be taken as a recipe for a "symmetrical" view of life and the world? Certainly not. More and more we realize that we are only at the start. Our ridge walk is only just beginning. Now we are confronted by *sequences* of tones, *sequences* of sounds, melody, language, speech. In what succession do their configurations occur? How do periods and rhythms follow each other in series which, again, represent wholes?

As scientists we cannot taboo the mysteries confronting us, and forbid ourselves to investigate them. And so we ask ourselves: What is the nature of speech? What strange rhythms are there in the spoken word? What is it that pulses in the measures of verse? Yes, we ask ourselves— without for one moment reneging as scientists— where does a Beethoven *Appassionata* or a Mozart *Jupiter Symphony* come from, what is the source of Goethe's language...etc.?

---

*In this connection it is interesting to compare what Kant said in "The Critique of Judgment" in the second part (Critique of Teleological Judgment, 1790) on man's cognitive capacity vis-a-vis organized existence and what Goethe said in his "Attempt to explain the Metamorphosis of Plants," also published in 1790. Space forbids us to quote the passages in question here. Nevertheless the reader's attention should be drawn to this dramatic situation in the history of thought. The confrontation helps bring out more clearly the truth of the dictum: "He who sets limits to knowledge describes himself and his own cognitive position."

It is not aesthetic descriptions, explanations and cultural commentaries that are required here. Indeed, in our view, when an *Appassionata* is written, the systems of order, the proportions of the melodies, the symmetrical variations, dynamics and configurations all belong to one and the same cosmos. We do not wish to analyze and pick to pieces such a precious work. To the contrary, the work must take us back to its *Urgrund*, to its primal cause; we enrich our experience, we train and exercise our powers, we perform our "cognitive gymnastics" and see where we get to, what the mysteries hold in store for us. So it is not to the dissecting table that we go but to the generative, creative, ever-active primal cause. But what do we mean by primal cause? Have we not simply succeeded in re-obscuring the issue? Far from it.

In German the prefix *"ur"* is used when we wish to designate something from which much is derived, which comprises and holds much in itself (*Urtier*, *Urpflanze*, *Urphanomen* = protozoa, protoplant, proto-phenomenon, etc.). How do we reach such a primal cause of vibration? This is where the human larynx, the human organ of speech invites attention. It is omnipotent in the sense that in its frequency band it can represent and generate everything, including — within its range — the whole of cymatics, with its figures, circular currents, turbulences, harmonics, etc. It is not simply a matter of studying its anatomy, its curious physiology, and its origin in and metamorphosis from the gill basket of the fish, although this takes us straight into the mystery of this organ; it is more a question of making ourselves thoroughly familiar with its activity. Will anything come of this? Is there a point to it? That, of course, is what we do not know in advance. This is an open question at first.

We proceed according to the stated method. *Methodos* means actually "to go after," and hence "look into." We will look into the larynx, which potentially contains the whole range of cymatics within its capacity. It is therefore also a proto-organ, or rather — to use the category appropriate here — it is truly and really the primordial Word. What is this primordial Word? That is what we bring our minds to bear on as if upon a mystery, what we seek to approach *methodically* in cymatics, what we will dedicate ourselves to, using all our powers of sight and hearing and with modern science as our basis. One thing is clear: we are not in pursuit of a phantom, rather we are directing our perceptive powers to the organ of speech and also to the organ of hearing which is closely bound up with it, both of which are invested with an almost all-pervasive enigmatic quality.

In our research we move towards a creative world, towards a world-creating power. That in itself—in virtue of its creativity—provides for the investigator, for the artist, and for every truly living man, an element in which he can breathe, live, fashion and work. We have no hesitations or doubts. What will come of it? Will the mysteries be solved? Shall we come through alive? How will this adventure end?

We *are*, in reality, this mystery; in it we *become*; it is not that man simply *is*, he is *becoming* all the time with an ever fuller and clearer consciousness.* If he looks back at the great minds of the past, he does so in independence and self-assurance. But in admiring and respecting those visions of a world of harmonies he feels that a responsive chord has been struck, for he carries in his heart the new cosmos as the mystery of the primordial Word seeking revelation.

That, then, is a preview of the methodology to be applied in further cymatic research.

---

*Those who have a vital understanding of the idea of development, cannot agree that there are already human beings; instead they will say that men will for the first time *"become"* men.

Closing Comments from the Publisher
by Jeff Volk

If you are one of the rare readers who has made it this far, I mean actually having read this book from the beginning up to this point— then I offer you my respectful congratulations... and a suggestion. Go back and review the final chapters of each volume: *The Basic Triadic Phenomena*, at the end of Vol. I, and the final chapter of Vol. II, *Historical Review / Methodological Preview*, (and the two chapters before that if you're really into wave phenomena!)

In other words, this is not a book to be read once, nor even several times. It is something to be assimilated, gradually, over time. Jenny's studies of animating "the dust of the earth" into lifelike, flowing forms, with an invisible force that is largely imperceptible to our gross senses— feels like an alchemical blend of magic and mysticism. But upon close observation, one discovers that the causal principles are as precise as Pythagoras' theorems; they are consistent, they are coherent, and they are replicable.

Jenny's *Cymatics* is a very deliberate study, a visual representation of the Creative Principle— and as such, it is always changing, expanding, evolving. Unlike the inert substances that we've observed as the subjects of these experiments, we too are also comprised, even *animated*, by a force that is not restricted to these principles, yet which still functions impeccably within their domain.

As I mentioned in my opening comments at the very beginning of this edition, I view cymatics as a living metaphor, a representation in physical phenomena of certain principles that have universal application throughout the entirety of manifest creation. A metaphor is a way of telling a story, an aphorism that translates an abstract understanding (for example, $E=MC^2$), into concrete terms— something tangible and familiar, to which we can more easily relate. As Ted Gioia mentioned in his Foreword, Pythagoras is reputed to have instructed his students, "This stone is frozen music."

This was undoubtedly more accessible to his students than Einstein's abstract assertion would have ever been. Likewise, but even more so, do these cymatic images convey a profound Truth, in a direct and immediate way that resonates deeply within our consciousness. It does so, because it illustrates, in splendid detail and in real time, and right before our very eyes— the invisible principles that animate our universal playground. Yes, matter is energy, but it remains inert until acted upon by some "other" form of energy, something qualitatively distinct, yet intimately familiar to each of us. It's surely far more expansive than we can comprehend, but we can sense it directly…through resonance.

In Jenny's experiments, that universal energy is stepped down, so it is perceivable to us as sound. In the Vedas— the oldest extant texts in our four-millennia blip of written history— this dynamic interaction between energy and form is referred to by the Sanskrit terms, *Purusha* (the animating force) and *Prakriti* (structure or form).

I first saw these concepts articulated in Ayurvedic teachings, though I now believe they arose from a timeless tradition known as the Light and Sound teachings (*Surat Shabda Yoga*), which are continually unfolding to this day.

While cymatics is a physical study of wave phenomena, (energy perceived as sound interacting with matter, with its composite of structure and form), it may also be recognized as a representation of the "Formless Emanation," (*Purusha*), made manifest through the medium of *Prakriti*, just as we are. Might that be why it appeals to us so intensely; why we resonate with it so effortlessly?

So, when the time is right, I invite you to continue your journey through the myriad expressions of cymatics, by perusing the Commentaries written by noted cymatics enthusiasts, that concludes this edition.

These personal reflections provide a synopsis ranging from rigorous analysis to playful curiosity, creating a wealth of detailed research that has tantalized our awe-struck eyes over the past couple of decades since we first published the composite two-volume edition of Jenny's *Cymatics,* in 2001. But for now, please, take a well-earned respite to enjoy a bit of a palate cleanser.

AND SOMETIMES, HAPPINESS

*The syncopated surging of the surf
as it beats relentlessly against the shore
pounding out the rhythms of the day—
a metronome just slightly out of sync,
attack a little faster than decay.*

*The sure, incoming tide brings its array
building up its kingdom grain by grain,
storing up vast treasures
from untold distant lands
that promise lasting pleasure
cast in swiftly shifting sands.*

*Hoarding countless measures
as its rhythms shape the day—
crafting sculpted castles
and then washing them away.*

*© 2024 Jeff Volk
CymaticSource.com*

~~~~~~~ ❋ ~~~~~~~

I'd like to close this section with a taste of something truly profound— (a sound insight, perhaps?)…
from One who listens devotedly, perpetually, to the *Anhat Shabda*, the unstruck sound.

*The world in which you live has not been projected **onto** you but **by** you.*
~ Sri Gary Olsen

Index

A

Abraham, Ralph 2, 312, 322-324
acoustics 6, 25, 32, 144, 275, 314-315, 323
acoustic vibration 29, 143, 276
adhesion 102, 103, 142, 156, 163, 178-180, 184, 187, 251, 253, 259
Adlington, Jacob Lee 331-332
alcohol 134, 219, 220
Allgemeine Meereskunde 270
alternation 30, 224, 268, 283
Alfvén Waves 264
amoeboid behavior 170
AMI 750, AMI 850 318-319, 321
amphidromy 273
amplitude modulation 50, 135
animal markings 286
animal pigmentation 285
annular formation 117
annular wave 112
Anthroposophy 12-13, 322, 335
antigravitational effect 39, 156, 259
antinodes 32, 39-40, 82-83, 274
Appassionata 297, 298
architecture, architectural forms 207
Archytas 289, 295
Aristotelian metaphysics 221
Aristoxenos 294-295
astronomical tides 275
astrophysics 250
atomic theory 134
axial symmetry 38

B

Bach 23, 75, 77, 313
Beaulieu, N.D., Ph.D., John 2, 6, 11, 15-20, 342-344, 351
Beethoven 297
bell 39, 173, 282, 285
benzine 219, 220
bilateral circulation 147, 148
bilateral currents 240
bilateral symmetry 54, 71, 149, 159, 170, 216, 218, 229, 231-232, 234, 236, 239, 282
biological rhythms 284
biological systems 192, 277
blood cells 312-313, 318
bone lamellae 96
brass plate 38
Brehm 285
Brownian molecular movements 192
Bruno, Giordano 295-296
butterfly wings 96

C

capillary interspaces 142
capillary waves 270
cardiology 135, 309, 314
catalysis 97, 135
cellular waves 275
changes of phase 140, 151, 251

chemical periodicity 97
chemical reactions 31, 41, 95, 97, 100
chemistry 29, 40, 192, 277, 286
Chladni, Ernst F.P. 308, 323, 339, 351
Chladni Figures/Plates 22, 35, 37, 308, 323, 331, 339
chromosomes 250, 286
circular currents 172, 231, 234, 236, 298
circulation 28-29, 35, 72, 80-81, 84, 106, 111, 113, 115-117, 123, 129, 132, 135, 140, 147-149, 157, 159, 163, 170, 175-176, 179-180, 231, 250-251, 253, 256-259, 261, 268, 273, 275, 294,
cloud(s) 29, 82-83, 251-252, 259-260, 264-265, 268-269, 275
cloud formation 265, 268
club moss 40, 43, 80, 175, 283
cochlea 53
cohesion 99, 142, 156, 169, 178-181, 184, 187, 251
colloid chemistry 40
combination tides 275
concentric 47, 55, 95-96, 105, 108, 113, 122, 175, 216-217, 229, 231
concentric circles 108, 216
concentric circulation 231
concentric/radial patterns 47
configuring field 145
conglobation 179, 180, 201, 251-252, 258, 261
consciousness 17, 23, 32, 299-300, 324-326, 333, 335, 341-342
continuous waves 47, 262
convection 257-258, 264-265
copper 46
correlation 84, 140, 142, 170, 172, 251, 259
cosmic music 277
crispation 308, 324
Critique Of Judgment 297, 306
Cromwell, Mandara 3, 315, 316-321, 336, 346
Crowe, MMT, MT-BC, Barbara J. 2, 11, 351
crystal(s) 29, 32-33, 35, 38, 41-42, 46, 52-53, 96-97, 112, 119, 132-134, 191-192, 195, 261, 262, 277
crystal oscillator(s) 33, 35, 38, 41-42, 46, 52-53, 261-262
curvilinear patterns 96
CymaGlyphs 310, 315, 321
CymaScope 308-315
cymatherapy 315, 317-319

D

da Vinci, Leonardo 12, 269, 323, 336
deaf 74, 75
de Broglie, L 29
Debye 132
dehiscence 98-101, 169, 189
dendritic (formations) 98-101, 113, 119, 169, 189
diatoms 282
die Akustik 32
Dietrich, Günter 270
diffraction patterns 29
discharge paths 189, 192
discontinuity 192, 210
dolphin(s) 310, 311, 338
dynamokinetic morphology 184

E

ear 32, 46, 53, 77, 81, 117, 123 219
echolocation 310-311
eddies 48, 53-54, 82, 106-107, 124, 257-259, 268, 273
eddy formation 54
egg white 220
Eichelbeck, Reinhard 347
Einführung in die Atomphysik 132-133
Einführung in die Physik 134
Einstein, Albert 19, 249, 300, 342-343
electric leakage discharges 189-190
electron 28, 143, 207-208, 210, 212, 275-276, 279, 283
electronic shell 276
electron beam 143, 207-208, 210, 212, 283
embryogenesis 282, 287
embryogeny 281
energy medicine 19-20
epithelium 169, 170
equidistance 224
eruption 84, 176, 259
ether 219-220

F

Faraday, Michael 308-309, 311-312, 323-324, 336
ferromagnetic substances 142
films of liquid 217, 232
Finkelnburg, Wolfgang 132
flow(s) 37, 54, 99, 106, 114, 117-118, 123, 155, 161, 163, 185, 199, 226, 253
flow dynamics 257
flow patterns 72, 75, 88, 199, 231-232, 259, 260
flower 108, 226, 282-283, 286
fluid dynamics 270
fluorescent liquids 51
forced oscillations 274
Forel, F. A. 274
Fourier analysis 256
Fragments of Heraclitus 292
Frank, Erich 289
free oscillations 274
frequency modulation 83, 136
funnel 253

G

galaxies 142, 260
genes, genetic 135, 141, 277, 285-287
genome 250
geology 132, 250, 254
geotectonics 257
Gilbert Ph.D., Robert J. 3
Gioia, Ted 10, 21-24, 300, 324
glass 46, 95, 99, 169, 170, 172, 296
glycerin 48, 58, 60, 62, 220
Goethe 12, 26, 79, 108, 115, 140-141, 207, 297, 340, 344
Goetheanum 12
Gothic tracery 207
gravitational waves 270
guitar 38
Gulf Stream 273-274

H

Haeckel, Ernst 282
harmonic figures, forms 234, 260, 283, 287
harmonic frequency ratios 260
harmonic oscillations 1, 138, 219, 228, 240
harmonic overtones, phenomena 133, 143, 191, 207, 212-213, 217, 219, 221-222, 224, 228-229, 231-232, 234, 240-241, 243, 247, 251-252, 260, 276-277, 282-284, 286-287, 289
harmonic phenomena in fluids 143
harmonic vibrations 145-146, 199, 212, 217, 224, 231, 232, 243, 251, 260, 289
harmony of the spheres 276, 292
Harris, R.A. 273
hearing 32, 54, 74-75, 298
heat 29, 41, 133, 273
Hendus 133
heptagonal figure 105, 207-208, 213, 221, 230
Heraclitus 292-296
Heureka Exhibition 347
hexagonal (regularity) 34, 48, 82, 105, 197, 200, 221-222, 225, 229, 231, 234, 240, 247, 293
histology 29, 132, 279
Hodler, Ferdinand 265
hollow organs 170
honeycomb(s) 48, 59
Hooke, Robert 22, 336
Hughes, Margaret Watts 22, 336
hydrodynamic phenomena 51, 53-54, 67, 72, 103
hydrophysics 132

I

imbricate 48
influent streams 37
interference(s) 35, 37, 48, 61, 80, 83-84, 136, 141, 163, 251, 259, 273, 275, 285
interfering waves 108, 111
Intergalactic Matter 260, 306
internal waves 275
interpenetrating waves 268
intervals 143, 189, 207, 210, 220, 251, 260, 276-277, 284, 286-287, 289, 291-292
iron filings 102-103, 178-180, 182, 192-193

J

Jupiter Symphony 23, 75, 78, 297

K

Kanazawa, Kenichi 21-22
Kant 297
kaolin paste 111
Kelemen, Gabriel 336, 346
Kepler, Johannes 295
kinetic 31, 47-48, 83-84, 116, 124, 131, 133, 135, 195, 210, 247, 268, 296
kinetic-dynamic processes 83, 131
kinetodynamics 234
Korschelt and Heider 282
Kunstformen der Natur 282
kymatika 31, 269

L

labile waves 62, 64
lamellae 96, 199, 279
laminar flow 54, 84, 268
larynx 143, 298
Lauterwasser, Alexander 331, 336, 339-341, 351
lattice(s) 28-29, 33, 40, 48, 60, 63, 97, 107, 114, 145-146, 160, 228, 241, 270, 282
Lawlor, Robert 2
Leeds, Joshua 301, 336-338
lemniscate 54, 72
Lichtenberg figure(s) 32, 189-192
Liesegang, Raphael Eduard 95
Liesegang ring(s) 95-97, 99-100, 189, 357
Linton, Rachael 333-335
liquids 38, 48, 50-51, 59, 67, 104, 132-134, 145, 191-192, 217, 219, 241, 247, 283
Lissajous figure(s) 47-48, 60, 207, 210, 212
lithosphere 257
Logos 294
lycopodium powder 38-39, 40, 79-80, 83-85, 93, 175-176, 199, 201, 205, 231, 245

M

magnetic field 102-103, 142, 179-182, 184-185, 210, 212, 258, 264
magnetocymatics 264
magnetohydrodynamics 264
Manners, Peter Guy 315-321, 336
marginal zones 47, 51, 54
mass ratios 277
mathematical arrangement 143, 201
McKusick, Eileen Day 2
mechanical pendulum 143, 207, 210, 213, 283
Meehan, Jodina 327-329
membrane 19, 58, 97, 112, 220, 317, 337-338, 309-311, 348
mercury 54, 104-106, 108, 133, 199
metal foil 169-170, 172, 197, 199
metamorphosis 47, 51, 132, 232, 297-298
metamorphosis of plants 297
meteorology 250
Metzner, Jim 3, 330
mineral, mineralogy 29, 40, 132, 277
minor seventh 210, 212
mitosis 142, 280
molecular biology 142, 286
morphogenesis 224
morphology 29, 100, 108, 132, 159, 172, 180, 184, 189, 190-191, 193, 224, 231-232, 261, 264, 267-268, 281, 314, 341
Moszkowski, Alexander 249
mountain waves 254
Mozart 12, 23, 75, 78, 297, 313
multiphase systems 247
Multiple Galaxies 260
music 10-11, 15, 19-21, 23-24, 75, 77, 136-137, 143, 207, 276-277, 289, 295, 300, 312-313, 323, 327, 336, 341, 343, 349
musical intervals 292
musical notes, tones 240, 251, 297
musical scale 292

N

nappe 257
networks 40, 48, 98-99, 279, 282
nodal lines 32, 38-40, 46, 50, 55, 82-83, 260-261
northern lights 269
nuclear physics 29, 276
nucleus 29, 108, 236, 250, 260, 267, 286

O

oceanography 269-270
octave 210, 276
Olsen, Sri Gary 5, 301
orbital movement 255-257
orogenesis 250, 254, 257-258
orogenic front, cycle, wave 256
oscillation 46, 60, 98, 100, 107, 121, 165, 167, 197, 212, 220, 246, 252, 254, 261, 264-265, 274-275, 287
oscillatory pattern 220, 261, 274-275
oscillatory ratios 210
oscillogram(s) 207, 210, 220
oscillograph 207-208, 210, 212, 220
overtides 275
overtones 275-277

P

paraffin 220, 240, 246-247
parallel/diagonal patterns 47
parallel waves 159
parallelism 258, 265
pattern 37-39, 41, 47, 107, 216, 219, 251, 258-259, 260-261, 264-265, 268, 270, 273-277, 279, 283-285, 287
pentagonal 105, 197, 199-201, 213, 221-222, 229-230, 232, 236, 239, 282-283
Perez-Martinez, M.D., David 2, 325-326, 342-344
periodic circulation 253
periodic figures 207
periodic precipitation 95-97, 99, 357
periodic system of the elements 277
periodic table 276
periodicity 28-31, 96-97, 100-101, 108, 113, 121, 131-132, 135-136, 165, 191, 259, 268, 273, 280-281, 284, 286
phase 33, 48, 80, 121, 140, 147, 151, 170, 177, 197, 199, 200, 207, 210, 212-213, 220-221, 229, 251, 254-256, 259, 271, 277, 281, 287
phenomenology 15, 18, 20, 22, 29, 31, 46, 54, 83, 97-98, 335, 342-344, 107, 121, 132, 139-140, 142-143, 157, 163, 221, 250, 258, 264, 269-270, 274, 276, 297, 325, 343
photoelasticity 29, 39, 174
physics 6, 29-30, 132, 195, 221, 224, 250, 264-265, 269, 276-277, 283, 296
physiology 29, 97, 132, 135, 281, 298
piezoelectric effect 32, 33
planar symmetry 38
plant 169-170
plasma 29-30, 108, 184, 269
plaster 160, 252, 258
Plato 12, 25, 289-291, 292, 295-296

Plato und die sogenannten Pythagoreer 289
Pohl 134
polar phases 28, 192, 195, 224, 280, 284
polygonal pattern 197, 224
Popp, Fritz A. 340
precipitations 41, 95-96, 99, 357
primal harmonic phenomenon 240
psychoacoustics 338
pulsation 28, 50, 84, 129-130, 140, 143, 148, 180, 197, 199, 201, 216, 224, 230-231, 234, 240, 260, 265, 293
pure metals 133
Pythagoras 23, 295, 300, 336
Pythagorean 276, 289, 291, 295, 306

Q

quanta 258, 276
quantum 3, 19, 25, 231, 309, 341-342, 344
quartz sand 35-36, 37, 53, 57, 82-83
quasi-crystalline "structure" of liquids 191

R

radial 47, 55, 80, 86, 111, 113, 116-118, 129, 179, 221, 224, 281
radial circulation 80, 111, 113, 117
radiating forms 224
radii 221
radiolaria 282
rectangular 47, 48, 286
Reichel 31
Reid, John Stuart 25, 308-315, 317-318, 336-337, 320
resonance 136, 141, 163, 251, 260, 274-275
respiration 28, 135-136, 142, 250, 281
rhythm 31, 75, 97, 101, 199, 229, 268, 274, 295
rhythmizomena 295
rock formations 96
Ross, Ivy 349
rotary currents 54
rotary waves 177, 226, 240, 283
rotating waves 274
rotation 2-3, 5, 19, 23, 29, 35-37, 44, 47-48, 53-54, 70-71, 83- 84, 103, 106, 113, 132, 134, 175-176, 228, 251, 257, 260-262, 267, 273-274, 300, 310, 316, 318, 320-322, 332-333, 349-340, 344
rotational currents 54
rotational effect 35-36, 44, 47, 83

S

saline circulation 273
salt 95-97, 99, 151, 153, 156-158, 195
Sauter 133
schlieren method 146, 173, 217, 219, 241, 283,
scintillation 240
sea wave 254-256, 264, 269-270, 274-275
seiches 274-275
sensitive flames 106, 261
serial patterns 279
Shalin, Dr. Leonard 343
sine wave 220
sinusoidal oscillations 254, 256

sinusoidal tone 220
Smoluchowski, M. v. 192
soap bubbles 197, 199, 201, 283
solar physics 30, 264, 269
sonorous figure(s) 32-35, 38-42, 46-47, 49-53, 55, 57, 83, 88, 105-106, 110, 192, 261
soul 290-292
sound 7, 29-30, 38, 54, 72, 74-76, 82, 99, 103-107, 112, 116, 135, 141, 143, 145, 165, 180, 184, 193, 200, 216-217, 219-221, 224, 232, 236, 241, 270, 275, 277, 283
sound healing 19-20, 326, 342, 344
Soundflower Experience, The 3, 320-321
speech 74-75, 135-136, 143, 240, 295, 297-298
spherules 80, 82, 84, 252, 258
spiral current 165
spiral formation(s) 81, 261
standing waves 48, 251, 273-274
Stanford, Nigel 349
steel 33-34, 35-38, 41-42, 45-46, 49-50, 52-53, 55, 82-83, 260
Steiner, Rudolf 7, 12-13, 26, 322, 336, 346
stroboscope 39, 80, 82, 103, 105, 111, 118, 147, 155, 165, 167, 173, 220, 224, 230
stroboscopic effect 220
stroboscopic light 39
Stuten, Christiaan 6, 12-14, 22, 348
surface tension 104, 108, 165, 167, 169-170, 197, 199, 201, 205, 252
symmetry 38, 54, 69-71, 143, 159, 170, 201, 205, 207-208, 210, 212-213, 216, 218-219, 221-222, 224, 226, 229, 232, 234, 236, 239, 240-241, 245, 247, 254, 256, 275-277, 281-283, 285, 287, 296
synesthesia 349
systematic circulation 175
systems theory 15-16, 20

T

ta kymatika, to kyma 31, 269
tectonic planes 257
Technorama Science Museum, Winterthur, CH. 347
tetragonal (forms) 105, 199, 221, 224, 226, 228-230, 232, 236, 240, 245, 247
The Critique of Judgment 297
thermal 31, 135, 257-258, 268-269, 273
thermal circulation 273
thermal flux 258
Tierlandschaften 13
Timaeus 25, 289-291
Times Square CymaScope display 314
Toccata & Fugue in D minor 23, 77
Tomatis, Dr. Alfred 337
tone 34, 38-39, 41, 50, 53, 58, 71, 75, 80-83, 87, 91, 93, 98, 105, 113-114, 143, 145, 148, 151, 155-156, 175, 184, 199, 201, 217, 220, 224, 229, 231-232, 236, 259, 270-271, 276
tonoscope 74-77, 135, 144
transverse circulation 273
traveling waves 136, 260-261, 272
triadic (primal) phenomenon 124, 131-132, 134, 136, 232

triangular 105, 224, 232
trigonal 197, 199, 221-222, 224, 226, 229, 239
troposphere 264, 269
turbulence 48, 51, 82, 106-107, 134, 147, 165, 167, 192, 229, 257, 259-261, 273
Turczan, Lachlan 349
turpentine 220, 240-241, 246-247

U

ultra-sonic 48
ultraviolet light 51
undulation 28, 50, 117, 147-149, 155, 162, 247, 265, 268-269
Urgrund 298

V

van der Waals forces 134
Vergleichende Entwicklungsgeschichte der Tiere 282
vertebrates 28, 280-281
vibrational medicine 3, 19, 342
violin 32, 39
viscous paste 117, 147, 149, 157-158, 179-180, 201, 205
vocal figures 75
voice 74, 75-77, 143
Volk, Jeff 2, 5, 10-11, 25-26, 137, 300-301, 317, 333, 336, 346
von Békésy, G. 53
vortices 47, 51, 53-54, 67-72, 82, 98-99, 104-108, 145, 231, 251, 260, 265, 268, 273
vowel 74-75
vowel figures 74

W

Waller, Dr. Mary D. 308
water 48, 82-83, 133-134, 151, 153, 156, 165, 167, 170, 172, 197, 199, 217, 219-220, 255-257, 269-271, 273-275, 289-290, 293
water jet 165
wave crests 117, 121, 251
wave curls 51
wave field 39, 48, 51, 82, 116
wave harmonics 276
wave lattices 40, 145-146, 270
wave mechanics 275
wave train 147
Wesen und Ursache der Gebirgsbildung 254
wood 38, 46
Wunderlich, Hans Georg 254

X

X-ray 133-134
X-ray diffraction 133

Z

Zenneck 35
Zickendraht 107
zones of resonance 260
Zwicky, F. 260

Index Notes

CIRCULATION
bilateral circulation 147, 148
bilateral currents 240
changes of phase 140, 151, 251
circular currents 172, 231, 234, 236, 298
circulation 28-29, 35, 72, 80-81, 84, 106, 111, 113, 115-117, 123, 129, 132, 135, 140, 147-149, 157, 159, 163, 170, 175-176, 179-180, 231, 250-251, 253, 256-259, 261, 268, 273, 275, 294
concentric circulation 231
convection 257-258, 264-265
copper 46
eddies 48, 53-54, 82, 106-107, 124, 257-259, 268, 273
Gulf Stream 273-274
kinetic-dynamic processes 83, 131
laminar flow 54, 84, 268
orbital movement 255-257
periodic circulation 253
radial circulation 80, 111, 113, 117
rotary currents 54
rotation 29, 35-37, 44, 47-48, 53-54, 70-71, 83-84, 103, 106, 113, 132, 134, 175-176, 228, 251, 257, 260-262, 267, 273-274
rotational currents 54
rotational effect 35-36, 44, 47, 83
saline circulation 273
seiches 274-275
spiral current 165
systematic circulation 175
thermal circulation 273
transverse circulation 273
traveling waves 136, 260-261, 272
turbulence 48, 51, 82, 106-107, 134, 147, 165, 167, 192, 229, 257, 259-261, 273
undulation 28, 50, 117, 147-149, 155, 162, 247, 265, 268-269
vortices 47, 51, 53-54, 67-72, 82, 98-99, 104-108, 145, 231, 251, 260, 265, 268, 273

FORMS/FORMATIONS
annular formation 117
cloud formation 265, 268
conglobation 179, 180, 201, 251-252, 258, 261
dendritic (formations) 98-101, 113, 119, 169, 189
eddy formation 54
harmonic figures, forms 234, 260, 283, 287
periodic figures 207
radiating forms 224
rhythmizomena 295
rock formations 96
sonorous figure(s) 32-35, 38-42, 46-47, 49-53, 55, 57, 83, 88, 105-106, 110, 192, 261
spherules 80, 82, 84, 252, 258
spiral formation(s) 81, 261
vocal figures 75
vowel figures 74

HARMONICS
harmonic frequency ratios 260
harmonic oscillations 1, 138, 219, 228, 240
harmonic overtones, phenomena 133, 143, 191, 207, 212-213, 217, 219, 221-222, 224, 228-229, 231-232, 234, 240-241, 243, 247, 251-252, 260, 276-277, 282-284, 286-287, 289
harmonic phenomena in fluids 143
harmonic vibrations 145-146, 199, 212, 217, 224, 231, 232, 243, 251, 260, 289
harmony of the spheres 276, 292
primal harmonic phenomenon 240
wave harmonics 276

MUSIC
<u>Appassionata</u> 297, 298
Bach 75, 77
Beethoven 297
cosmic music 277
Jupiter Symphony 23, 75, 78, 297
minor seventh 210, 212
Mozart 12, 23, 75, 78, 297, 313
music 75, 77, 136, 143, 207, 276-277, 289, 295
musical intervals 292
musical notes, tones 240, 251, 297
musical scale 292
octave 210, 276
overtones 275-277
psychoacoustics 338
<u>Toccata & Fugue in D minor</u> 23, 77
violin 32, 39

PATTERNS
alternation 30, 224, 268, 283
bilateral symmetry 54, 71, 149, 159, 170, 216, 218, 229, 231-232, 234, 236, 239, 282
concentric 47, 55, 95-96, 105, 108, 113, 122, 175, 216-217, 229, 231
concentric circles 108, 216
concentric/radial patterns 47
flow patterns 72, 75, 88, 199, 231-232, 259, 260
heptagonal figure 105, 207-208, 213, 221, 230
hexagonal (regularity) 34, 48, 82, 105, 197, 200, 221-222, 225, 229, 231, 234, 240, 247, 293
honeycomb(s) 48, 59
imbricate 48
interference(s) 35, 37, 48, 61, 80, 83-84, 136, 141, 163, 251, 259, 273, 275, 285
lattice(s) 28-29, 33, 40, 48, 60, 63. 97, 107, 114, 145-146, 160, 228, 241, 270, 282
Lichtenberg figure(s) 32, 189-192
Liesegang ring(s) 95-97, 99-100, 189
Lissajous figure(s) 47-48, 60, 207, 210, 212
mathematical arrangement 143, 201
networks 40, 48, 98-99, 279, 282
nodal lines 32, 38-40, 46, 50, 55, 82-83, 260-261
oscillatory pattern 220, 261, 274-275
parallel/diagonal patterns 47
pentagonal 105, 197, 199-201, 213, 221-222, 229-230, 232, 236, 239, 282-283
radial 47, 55, 80, 86, 111, 113, 116-118, 129, 179, 221, 224, 281
rectangular 47, 48, 286
serial patterns 279
tetragonal (forms) 105, 199, 221, 224, 226, 228-230, 232, 236, 240, 245, 247
triangular 105, 224, 232
trigonal 197, 199, 221-222, 224, 226, 229, 239

SYMMETRY
axial 38
bilateral 54, 71, 149, 159, 170, 216, 218, 229, 231-232, 234, 236, 239, 282
planar 38
radial 281
symmetry 38, 54, 69-71, 143, 159, 170, 201, 205, 207-208, 210, 212-213, 216, 218-219, 221-222, 224, 226, 229, 232, 234, 236, 239, 240-241, 245, 247, 254, 256, 275-277, 281-283, 285, 287, 296

WAVES
Alfvén Waves 264
amphidromy 273
annular wave 112
capillary waves 270
cellular waves 275
continuous waves 47, 262
gravitational waves 270
interfering waves 108, 111
internal waves 275
labile waves 62, 64
mountain waves 254
orogenic wave 256
parallel waves 159
rotary waves 177, 226, 240, 283
rotating waves 274
sea wave 254-256, 264, 269-270, 274-275
sine wave 220
standing waves 48, 251, 273-274
traveling waves 136, 260-261, 272
wave crests 117, 121, 251
wave curls 51
wave field 39, 48, 51, 82, 116
wave harmonics 276
wave lattices 40, 145-146, 270
wave mechanics 275
wave train 147

References

This section lists the various references cited in both volumes of Dr. Jenny's original *Cymatics* editions for which some bibliographic information was provided. They are listed here preceded by the page number of the current edition. Other authors referenced in the work may be found in the Index.

The first entry listed below cites two quotations selected by the publisher to book-end this edition. Further explanation appears below, at the end of this list.

P. 5 and P. 301 Sri Gary Olsen, MasterPath

P. 31 Reichel, (1960)

P. 32 Chladni, E. F. P., die Akustik, (1802)

P. 107 Zickendraht, (1932)

P. 132-133 Finkelnburg, Wolfgang, Einführung in die Atomphysik, (1954)

P. 134 Pohl, Einführung in die Physik, (1959)

P. 192 Smoluchowski, M. v., (1916)

P. 254 Wunderlich, Hans Georg, Wesen und Ursache der Gebirgsbildung, (1966)

P. 260 Zwicky, F. Intergalactic Matter, "Experientia" Vol. VI, Fsc. 12, Pp. 441-480, 15. XII, Pasadena, CA, (1950) Multiple Galaxies, Naturwiss. Vol. XXIX, Pp. 344-385, (1956)

P. 270 Dietrich, Günter, Allgemeine Meereskunde, (General Oceanography), (1965)

P. 273 Harris, R. A., (1904)

P. 274 Forel, F. A., (1895)

P. 282 Haeckel, Ernst, Kunstformen der Natur

P. 282 Korschelt and Heider, Vergleichende Entwicklungsgeschichte der Tiere

P. 289 Frank, Erich, Plato und die sogenannten Pythagoreer, (Plato and the So-Called Pythagoreans)

P. 289-291 Plato, Timaeus, (Platonic Dialogues)

P. 292 Heraclitus, Fragments of Heraclitus

P. 297 Kant, "Critique Of Teleological Judgment", Critique Of Judgment, (1790)

P. 297 Goethe, "Attempts to Explain the Metamorphosis of Plants", (1790)

Sri Gary Olsen is the spiritual guide and director of the MasterPath, a contemporary articulation of the ancient Teachings of Light and Sound, adapted for the Western culture and founded by Sri Gary in 1987. These spiritual dictates uphold the sanctity of the Original Teachings while moderating traditional Eastern vows and elaborate rituals to lead the soul back to its own innate Mastership. The publisher finds many of these same universal principles reflected in the science of cymatics and has selected a couple of quotations from Sri Gary's extensive writings for further contemplation. More detailed information may be found at http://www.MasterPath.org.

Commentaries
on Hans Jenny's *Cymatics*

Prominent contributors describe how
he catalyzed the science of cymatics

Jenny's Legacy in the Annals of Science
by John Stuart Reid

It is not an overstatement for me to say that Hans Jenny single-handedly kick-started cymatic science. I know this to be true because it was Jenny's work that inspired me in 1997 to conduct a cymatics experiment in the Great Pyramid, which led me to create the CymaScope, a cymatics instrument now in the hands of scientists in many countries. The story of my Egyptian research[1] is now well known to many people and is not the focus of this review. Instead, I wish to explore Jenny's vision for cymatics in terms of its manifestation in the world today. His legacy lives on, not only in the homes of thousands of enthusiastic amateur experimenters who are fascinated by cymatic phenomena, but also in many professional laboratories that apply cymatic principles, pioneered by Jenny, to shed light on scientific mysteries; mysteries that may have forever remained unsolved without cymatic science.

Hans Jenny (b 1904 d 1972) was well-read in science, and he found inspiration in the work of Ernst Chladni (b 1756 d 1827) and Michael Faraday (b 1791 d 1867), both of whom made important contributions to vibrational phenomena, Chladni with his sound figures, known today as Chladni Figures[2] and Faraday with what he termed, "crispation" experiments.[3] Jenny's own experimental research began with vibration of particulate matter, such as lycopodium powder on metal plates, and his work then extended into ferrofluids (cymatic-magnetic phenomena), mercury, oils, and water.

Two other notable researchers in this field of study are worthy of mention, one prior to Jenny and one contemporary with him.

Margaret Watts Hughes

Dr. Mary D. Waller

Margaret Watts Hughes (b 1848 d 1907) was a Welsh singer who developed an acoustic sound visualization device that she named the Eidophone, a direct predecessor of Jenny's Tonoscope, and with which she documented a series of extraordinary natural forms created by singing into her device, all beautifully illustrated in her book, *The Eidophone Voice Figures*.[4] She was the first person to show cymatic forms above a musical staff, illustrating how they related to the traditional musical notes on the staff. Dr. Mary D. Waller was an English scientist and professor of Physics at the Royal Free Hospital Medical School in London, and a contemporary of Jenny. Waller worked extensively with Chladni plates, making them "sing" by applying "dry ice", something she discovered from an ice cream vendor when the frozen carbon dioxide accidentally touched his bicycle bell! Waller was directly influenced by Chladni's work and by Faraday's experiments, and her book, *Chladni Figures, A Study in Symmetry*,[5] provides many wonderful insights

---

[1] https://cymascope.com/egyptology/
[2] Original work: *Entdeckungen über die Theorie des Klanges (Discoveries in the Theory of Sound)*.
[3] Faraday, *The Life*, James Hamiltion, Harper Collins, page 235. ISBN: 0-00-716376-2
[4] https://soundmadevisible.com/product/the-eidophone-voice-figures/
[5] *Chladni Figures, A Study in Symmetry*, Mary D. Waller, G. Bell & Sons Ltd.

into cymatic phenomena. An insightful quote on its title page from Michael Faraday, reads: "It is quite comfortable to me to find that experiment need not quail before mathematics, but is quite competent to rival it in discovery." Both Faraday and Jenny were experienced experimentalists and fascinated by vibrational phenomena, but not comfortable with mathematical modeling of their own discoveries.

We can imagine that Jenny would have been justly proud, perhaps even surprised, at the sheer number of sciences to which his cymatic principles have been applied in the half-century since his passing:

Asteroseismology (the study of sounds emitted by stars, via modulations of starlight), Biology, Botany, Cardiology, Electromyography, Electroencephalography, Geometry, Hematology, Materials technology (film deposition), Musicology, Neurophysiology, Oceanography, Oncology, Ornithology, Physics, Phonology, Quantum Physics, Sound Therapy, Water Science, and Zoology. The following are some examples of the many ways that Jenny's imagination, dedicated research, and extraordinary powers of observation have led directly to new discoveries in science.

We begin with a brief overview of the CymaScope instrument, which in turn was based on Jenny's Tonoscope invention.

Jenny was the first to suggest that such a device may one day assist deaf people in their acquisition of speech, and although his Tonoscope device was never widely employed for this purpose, the CymaScope instrument has found application in phonology, the study and organisation of speech sounds in languages, as I will soon describe. Jenny's electroacoustic

An early CymaScope prototype, particulate matter instrument

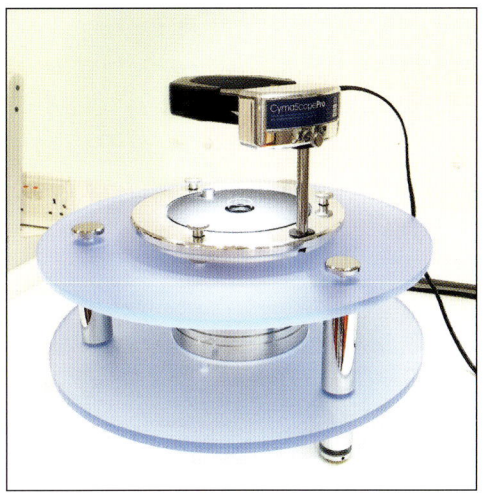
The CymaScope Pro, water-based instrument

Tonoscope consisted of a transducer-excited sand-strewn membrane, stretched across a resonating chamber. Early iterations of the CymaScope instrument employed a similar principle, except the membrane tension could be carefully controlled by a series of spring gauges, having recognized that the geometry of the cymatic pattern that formed on the membrane was partially influenced by the membrane tension.

However, it soon became evident that particulate matter could not reveal sufficient detail due to its two-dimensional nature, and experiments proceeded with electro-mechanical vibration of fluids, which carried the potential to reveal not only a far greater degree of detail but also quasi-3D imagery, since the injected sound organizes all the molecules of the liquid, both on the surface and in the subsurface. Pure water proved to exhibit the optimal response to imposed vibrations, offering an extremely fast transient response to sounds as short-lived as a few milliseconds.

The other main difference between Jenny's Tonoscope and the CymaScope concerns the neutralization of resonant properties inherent in the electromechanical drive system. In Jenny's Tonoscope such resonances naturally 'colored' the resulting imagery by giving preference to frequencies with which the drive system and resonating chamber were structurally resonant. To create a flat response, resulting in CymaGlyphs (sound images) that represent an accurate model of the injected sound, the CymaScope employs electronic filtering, to compensate for the resonant characteristics of the drive system. The result is a new type of analog scientific instrument that makes sound visible. The diameter of the water area is important because the wavelengths of the injected sound are automatically compressed in a ratio determined by the density of air versus the density of water. Only a certain bandwidth of frequencies will optimally generate wave patterns in a given diameter of water. It is for that reason that I designed three models of the CymaScope instrument to cover different portions of the sonic spectrum.

Dolphin language

One of the first applications of the CymaScope concerned making visible dolphin echolocation sounds, a study that was published in the Journal of Marine Science.[6] Jack Kassewitz, of SpeakDolphin.com, a dolphin research organization, reached out to me in 2009 and asked if it would be possible to image dolphin echolocation sounds. We worked for two years on this project, and in 2011 we succeeded in making visible a range of submerged plastic objects upon which a research dolphin had been tasked to echolocate. As Jack held each object under the water the reflected echolocation sounds were picked up by a hydrophone. The signals were then processed by high-specification audio equipment and sent to me as a series of sound files. I began by injecting each audio file into the CymaScope instrument and was overjoyed to see each object recreated in the instrument's water-filled cuvette. A plastic flowerpot, a cube, a cross, and even a child's rubber duck became members of this historic research initiative. When imaging the plastic duck sound file, I was concerned that the "duck" had no head; I saw only its oval-shaped bottom. But Jack subsequently explained that he had held the head while submerging the plastic duck underwater and that the dolphin "saw" only its bottom with its echolocation sense! Our press release was published worldwide, including in our local newspaper which titled the story, "Dr Doolittle, the scientist who speaks with Dolphins." The street A-board, sported the headline: "Scientist could help us talk to dolphins."

These awe-inspiring results confirmed that dolphins can see with sound, since their echolocation sounds, reflected from the submerged objects, contain holographic data that is used to recreate the object on a membrane. In the case of the CymaScope instrument, it is the sensitive surface of medical-grade

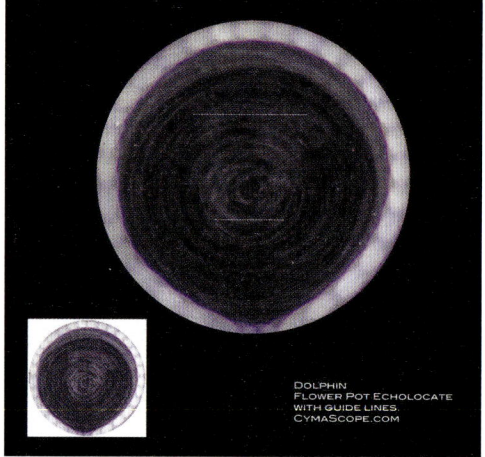

The historic image of a flowerpot, seen by the dolphin and imaged by the CymaScope instrument. The main image includes dotted guidelines to show the extent of the faint imprint of the flowerpot and the image even shows part of the hand holding the flowerpot.

John Stuart Reid beside the local newspaper A-board

---

[6] *A Phenomenon Discovered While Imaging Dolphin Echolocations Sounds.* Jack Kassewitz et al. Journal of Marine Science. DOI: 10.4172/2155-9910.1000202

water that functions as the membrane, whereas in the dolphin, it is the Tectorial membrane in its left and right cochlea that functions as the vibrating membrane. The imprinted image is then "read" by cilia in the Organ of Corti, similar to assembling a picture in pixels on a computer screen. This led us to hypothesize that dolphins have evolved a sono-pictorial form of sight and communication, helping them to navigate in murky waters and to send and receive pictorial information as a form of language. For example, a scout dolphin, circling the pod of dolphins, sees a shark coming toward their pod, takes a "photo" of it, and then beams that image to its cohorts, alerting them to potential danger.

A few years later, in 2015, we captured the image of a submerged man, Jim McDonough, a colleague of Jack Kassewitz. The research dolphin had actually been tasked to echolocate on Jim's face, but instead it imaged the whole of his body, and when we made the sound file visible cymascopically, we saw two frames in which Jim's arm was seen to move a little, suggesting that dolphins see in moving sound pictures, with each returning click pulse from the dolphin's echolocation beam representing an individual "frame" of the movie. The press release, titled "What the Dolphin Saw" went viral. Of course, we can never truly see what a dolphin actually sees with its echolocation sense, but we can now approximate its sono-pictorial sense with cymatic technology.

Cymatic physics

In 2017, I assisted Rupert Sheldrake and his son, Merlin, with aspects of their seminal study, concerning sonic frequencies in the range from 50Hz to 200Hz, made visible by the CymaScope instrument. Published by the Water Journal [7] and titled: "Determinants of Faraday Wave-Patterns in Water Samples Oscillated Vertically at a Range of Frequencies from 50-200 Hz", it is considered by many scientists to be a landmark paper. It extends the study of Faraday Waves (now more commonly known as cymatic patterns) far beyond other papers on this subject, using novel techniques that permitted precise control of the acoustic energy entering the CymaScope's visualising cuvette and providing a solid foundation upon which many CymaScope-related research projects can be built in the future. Their novel approach included evaluation of the time that a cymatic pattern takes to reach full expression after the sound is injected into the water. The novel procedure they developed was semi-automated, involving calculating the mean grey value of pixels in each video frame, that were then shown in graphic form.

The term "Faraday Wave patterns" refers to Michael Faraday, mentioned earlier. Faraday began a series of beautiful experiments on acoustic figures in 1831, with the goal of gaining insights into the nature of vibration. Faraday, who experimented with vibrating sand and water on a glass plate, wrote, "On putting a candle exactly below this plate and holding a screen of tracing paper an inch above it, the picture given was beautiful. Each heap [of sand] gave a star…of light at its focus which twinkled…this was exceedingly beautiful and easily rendered visible to a large audience."[8] This statement is remarkably similar to Hans Jenny's thoughts, who wrote over 100 years later: "…periodicity represents an aspect of the world, and at first its mysteriousness always inspires a feeling of the greatest astonishment… In attempting to observe the phenomena of vibration, one repeatedly feels a spontaneous urge to make the processes visible and to provide ocular evidence of their nature. For it is obvious that, by virtue of their abundance, clarity, and conscious nature of the information communicated by the eye, our mode of observation must be visual…"[9]

---

[7] *Determinants of Faraday Wave-Patterns in Water Samples Oscillated Vertically at a Range of Frequencies from 50-200 Hz.* Merlin Sheldrake & Rupert Sheldrake. https://waterjournal.org/archives/sheldrake-summary/ DOI: 10.14294/WATER.2017.6

[8] Faraday, *The Life*, ibid page 235. ISBN: 0-00-716376-2

[9] *Cymatics, Volume 1*, p. 32.

Dr. Ralph Abraham, professor of mathematics at the University of California Santa Cruz, who is mentioned in the Sheldrake paper (and also has a Commentary in this section on page 322, Ed.), considered vertically oscillated water as an analog computer for dynamic catastrophes, just as the CymaScope instrument can be considered a form of analog computer that can be applied to explore many different forms of vibrational phenomena.

In 2018, I collaborated with Professor Sungchul Ji of Rutgers University on a project involving the cymatic differentiation of cancer cells and healthy cells. All cells generate sound as a natural aspect of their metabolic processes, poetically named the "song of the cell" by Dr. James Gimzewski, who first discovered this aspect of cell biology and coined the term "sonocytology" for this new field of science. The concept that Professor Ji and I explored was designed to test whether it was possible, to use cymatics to differentiate between the song of a cancer cell and that of a healthy cell. Our study was published in the Water Journal and titled, "Imaging Cancer and Healthy Cell Sounds in Water by CymaScope, followed by Quantitative Analysis by Planck-Shannon Classifier."[10]

One of the potentials of this technology is the development of a new tool for surgeons. When a surgeon removes a tumor, its margins are not always obviously differentiated from healthy tissue, yet it is vital that the entire tumor be removed, and equally vital that the surrounding, healthy tissue remains unharmed. The creation of a real-time system for surgeons is based on visual data provided by a CymaScope instrument, in which the sounds of healthy and cancerous cells are imprinted onto medical-grade water. Rather like a fingerprint on glass, the imprint leaves a visual signature of the cell sounds. A typical cymascopic image of a healthy cell sound is symmetrical and beautiful, while that of a cancer cell is usually skewed and ugly by comparison. A Raman laser probe would be scanned over the tissues and the resulting cymatic imagery displayed to the surgeon via specially adapted eyewear and would assist the surgeon's decision where to cut. The system could also lead to an AI-based method of early cancer detection and possibly to other methods of cancer cell eradication, in the future.

Music Medicine

In a second collaboration with Professor Ji, in 2020, we investigated the regenerative power of specific sound frequencies and music on old red blood cells. We worked with whole human blood, obtained from a blood bank, and our hypothesis concerned the idea that certain frequencies in music would cause an increase in the longevity of red blood cells. The experimental design involved decanting a test tube of blood into two smaller vials, one placed in an incubator containing a speaker into which music would be played, while the other was our control sample, placed in a second incubator in the lab's Faraday cage. The blood in the music incubator received music for 20 minutes, while the blood in the Faraday cage incubator received no music, having a sound level similar to an anechoic chamber; in other words, very quiet.

The results were astonishing. Every genre of music we tested showed significantly more viable red blood cells in the music incubator, compared with the Faraday cage incubator. The question was why? What was the biological mechanism that underpinned this result? This is where cymatics led the way in providing the answer. We realised that the speaker in the music incubator had a relatively poor bass response, due to it being so small, thus we decided to test the blood in the CymaScope instrument, to see if old red blood cells would respond even better, since the

---

[10] *Imaging Cancer and Healthy Cell Sounds in Water by Cymascope, Followed by Quantitative Analysis by Planck-Shannon Classifier.* John Stuart Reid, et al. https://waterjournal.org/current-volume/reid-summary/
DOI: 10.14294/WATER.2019.6

CymaScope's frequency response permitted the blood to receive frequencies in music as low as 40Hz, whereas the small speaker could reproduce frequencies to around 80Hz at best. The magical moment came when we injected the blood in the CymaScope with a pure 44Hz sound. A beautiful hexagonal cymatic pattern appeared in the blood, and when we removed the sound, the faint image of the hexagon remained imprinted on the blood. This is remarkable because, with cymatic patterns once the sound is removed the image normally vanishes. The fact that we could still see the image meant that the blood had been oxygenated in the antinodal areas of the cymatic pattern, causing the hemoglobin molecules in those regions to absorb oxygen, which in turn caused them to become bright scarlet instead of their quiescent dark red color. This was an exciting result because it pointed us to the mechanism by which old red blood cells can be regenerated by certain frequencies in the music, particularly the low registers that mimic the sounds created by human heartbeats. We concluded that the human heart has two primary roles, first to circulate blood, and second to cause oxygen binding to hemoglobin molecules, due to the low-frequency heartbeat sound pulses. The blood then transports the oxygen to all parts of the body, powering many healing mechanisms. Our cymatics discovery points the way to the power of music to stimulate the body's healing response, and to new clinical therapies of the future, a subject about which Professor Ji and I will publish in the future.

Hans Jenny loved music, and he experimented by playing the music of JS Bach and Mozart into water, revealing intricate patterns. We can only imagine his feelings, had he known that those same cymatic patterns illustrated one of the most powerful clues of healing with music, which I later demonstrated with the voice of Danish vocalist, Anders Holte.

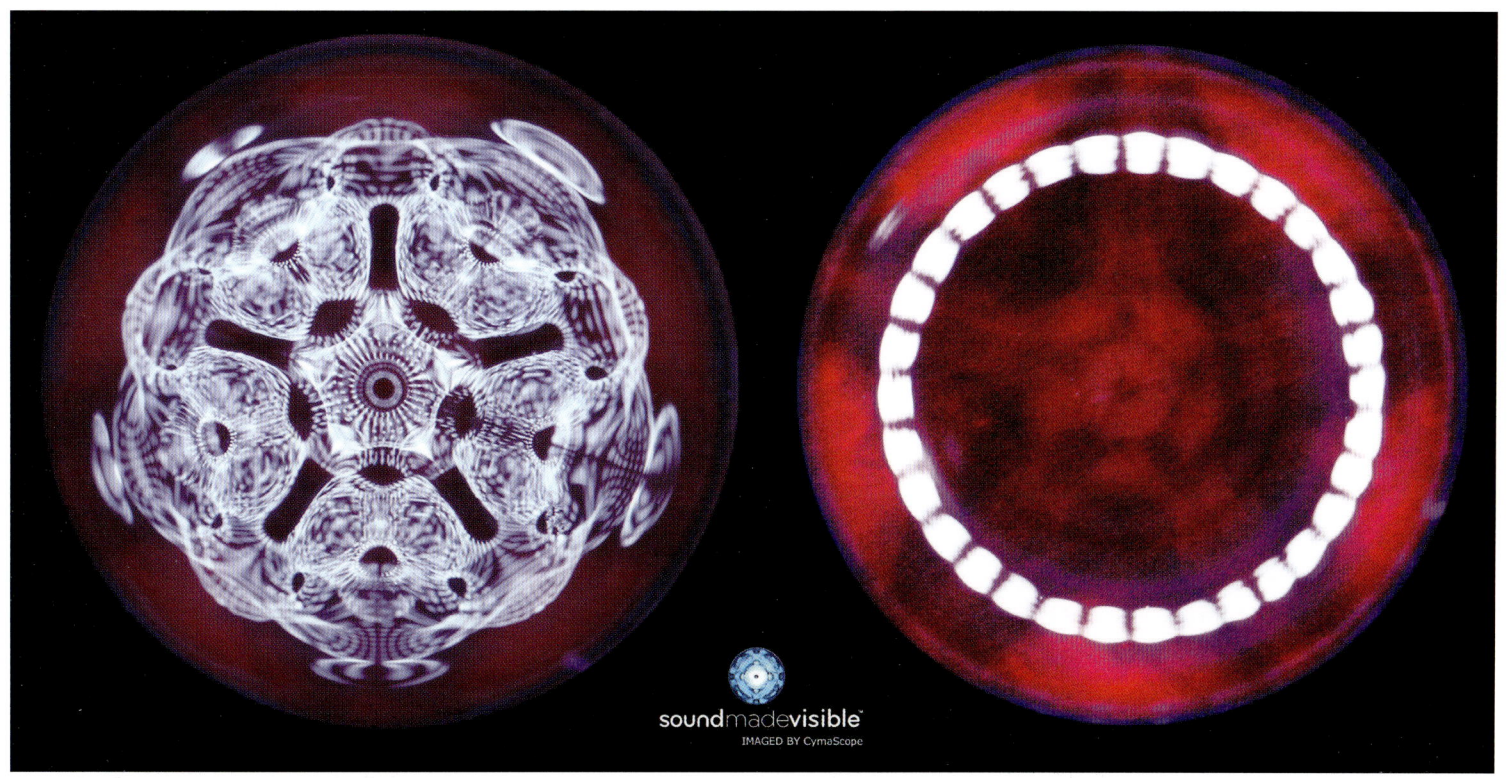

The cymatic imprint of Anders Holte's voice on human blood on the left, and the imprint after removing the input sound on the right.

Materials Technology

Many other studies involving cymatics have been published in the half-century since Hans Jenny's death, but one that is particularly noteworthy is titled, "Cymatics of selenium and telluriam films deposited in a vacuum on vibrating surfaces", by T. Hristova-Vasileva et al, published in the journal: Surface & Coatings Technology.[11] Not only is this an entirely new application of cymatic phenomena, but the title of the study actually contains Jenny's word "cymatics," which will serve to help spread the word about this wonderful aspect of Nature to scientists, worldwide. The essence of their study concerns the fact that films deposited by thermal deposition can be positively affected if waves of certain frequencies propagate through the substrate onto which the film is being deposited. The authors concluded that there is a qualitative relationship between the applied sound vibrations and the morphology of the thin films applied, a technique that they believe will lead to smoother films.

Cardiology via cymatic principles

Cardiologist and professor of medicine, Sean Wum, and acoustics bioengineer, Utkan Demerci, used cymatic principles to pack heart cells very densely and maintain an ability to control and tune their organization. A change in frequency guides the cells to a new position, allowing Wum and Demirci to create any pattern. Cymatic positioning of the heart cells in a tight configuration closely resembles natural cardiac tissue, which they believe may lead to an option for heart patches in patients who have weak cardiac walls or have damage from a heart attack. They also consider that cymatic engineering of heart cells could help foster more realistic cardiac disease modeling.[12]

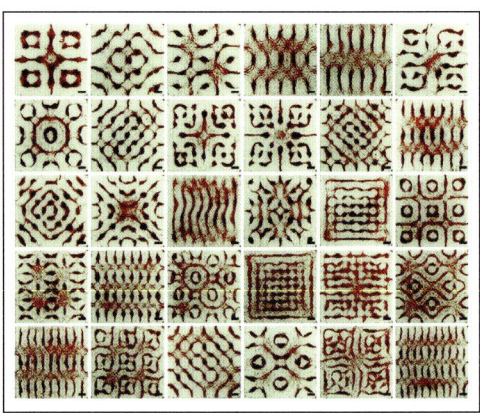
Cymatics positioning of heart cells to create new cardiac tissue

Cymatics in Times Square

When Times Square Arts organization decided to honor New York's essential workers, especially healthcare workers, the result was a series of groundbreaking cymatics videos broadcast in Times Square. The initiative was a collaboration between CymaScope.com and an artist collective known as Gathering of Noetic Generation (G.O.N.G.). In a seamless blend of art and science, a heart-felt line of poetry was made visible in several languages by the CymaScope instrument. A visually dramatic and beautiful message reflected how each one of us, individually and collectively as a society, is deeply indebted for the sacrifices made by those on the frontlines, as well as recognition of those who endured painful loss during the pandemic. One of the world's busiest pedestrian areas, Times Square, epitomizes the city that never sleeps, and the cymatic poetry displayed a beacon of love every night during that unique moment in history. The cymatic techniques used to

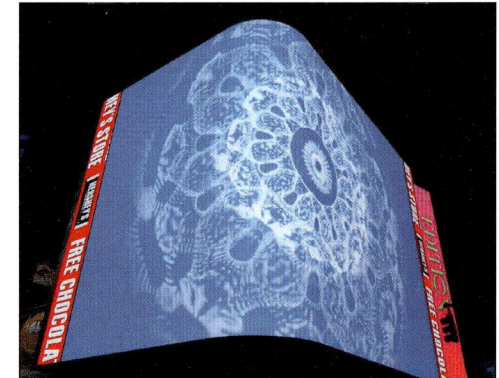
Cymatic imagery on the giant Times Square screen

---

[11] *Cymatics of selenium* and tellurium films deposited in a vacuum on vibrating surfaces. T. Hristova-Vasileva et al. Surface & Coatings Technology. DOI: 10.1016/j.surfcoat.2016.09.042

[12] https://stanmed.stanford.edu/innovations-helping-harness-sound-acoustics-healing/

showcase poetry paves the way for the study of languages via their cymatic forms, and Professor Lila Pine of Ryerson University in Toronto, is already beginning to use a CymaScope instrument to make visible First Nation languages, to help raise awareness of this important aspect of North American heritage.

From Manners to Meridians

This review would not be complete without mentioning the story of Dr. Peter Guy Manners (b 1915, d 2009) the English osteopath who was inspired to incorporate Jenny's research into his therapeutic device. Manners named his device The Cymatic Applicator and treated thousands of patients at his clinic in Bretforton, England, with great success. I interviewed him in 2001, just after he had sold the rights to his Cymatic Therapy technology to Dr. Mandara Cromwell's company (see next commentary), because he realized that his important work needed to be continued after he was gone. Further development led Dr. Cromwell to create the Acoustic Meridian Intelligence range of therapeutic sound devices in what she came to call "Cymatherapy."

Dr. Peter Guy Manners

She commissioned CymaScope.com to make visible some of the many sound codes emitted by her company's devices and the result was some of the most extraordinary CymaGlyphs we have ever witnessed. I have no doubt that Jenny would have been proud to know that is work has inspired so many avenues of research and has even led to devices to support healing for humanity.

John Stuart Reid is an English acoustics pioneer, scientist, and inventor, on a mission to educate and inspire the world in the field of cymatics, the study of visible sound, and to promote the field of sound therapy, supporting one's body to heal naturally. His CymaScope invention has changed our perception of sound forever, and his cymatic research continues to elevate this important new field in the scientific arena.

Much more information is available at http://www.CymaScope.com

Shedding New Light on Sound
Validation of early glimpses into the hidden world of vibration
by Mandara Cromwell

In the mid-1970s, some twenty-five years before I discovered Hans Jenny's work, I lived in India, studying Ayurveda and immersing myself in powerful Vedic healing mantras. I marveled at the ornate geometric patterns that were carved into the walls, pillars, and ceilings of the ancient Hindu temples, and for years thereafter, I wondered how they came to be there and what they represented. Encountering Jenny's *Cymatics* in 2001 shed new light on this enigma. Never before had I held in my hands a book that emitted such a beautiful vibration, evoking both scientific and spiritual modes of perception.

Original 2001 compilation

Contemplating Jenny's *Klang Figuren* (sonorous figures), reminded me of the yantras and other symbols I had seen in those temples. It also sparked kinesthetic sensations akin to those I had experienced as a child while looking at flowers in my rural farming community in northeast Kansas, or while listening to Gregorian chants, or to the massive pipe organ in our local church, where the halos of saints and angels adorned a different type of temple walls. Viewing these images helped me to see, in a vivid, sensorial way, what I could only intuit or imagine until then.

Hans Jenny's writings encouraged me to look at the "bigger picture" as one composite process that was constantly energizing and animating the material world. As a child, I had occasionally been graced to sense this universal dance. Now I could see cymatic principles in action everywhere I looked, just as Jenny had predicted one could. Jenny's research provided a scientific context, confirming my childhood experiences of living in a vibratory world and helping me to better understand it. I felt a strong resonance with his approach as a natural scientist, and I admired his ability to link the microcosmic and macrocosmic levels of existence.

Exploring cymatics in a therapeutic context:
Building on Hans Jenny's experiments and Peter Guy Manners' Compilations

Around the same time that I was devouring and digesting the newly republished 2001 edition of *Cymatics*, I was introduced to Peter Guy Manners, a clinical osteopath from Britain who had developed a device based on radiesthesia. It also incorporated principles of resonance, which cymatics so beautifully illustrates. For this reason, Manners named his clinical practice "Cymatic Therapy." Jenny's eloquent descriptions and awe-inspiring images were so prominent in my thoughts that this seemed like an obvious next step in my exploration of cymatics — further researching and applying cymatic principles in a therapeutic context. Our timing seemed divinely orchestrated; as I visited Manners' clinic in Worcestershire, England, mutual trust and respect quickly developed between us. At this time, Dr. Manners was in the twilight of his long and multifaceted career. When he shared with me that he had been having difficulty getting his device produced according to his specifications, I offered to help with manufacturing — an offer he enthusiastically accepted. Excited to participate in bringing this groundbreaking technology to the world, I coordinated the development and production of a new device, which I named the "Cyma 1000," initiating formal research protocols to document its efficacy. In order to differentiate cymatics— the science of making sound visible, from the therapeutic application of specific audible frequencies, I coined the name "Cymatherapy™."

In late 2001, Peter Guy Manners did me the great honor of bequeathing his life's research to me. As I read through the mass of research papers he had written or collected over six decades, what stood out to me the most was a list of several hundred sound frequencies documented to have been particularly effective in harmonizing physical and emotional states. These "commutations," as Manners called them, were specific combinations of audible sound frequencies that often resulted in relief from pain and even the regeneration of damaged tissues. I was eager to begin looking for ways to make these restorative tones visible and to prove their efficacy… and I was filled with questions. Might these sonorous figures resemble the vibratory fields I saw as a child? Would I discover similarities with the "sound carvings" in the Indian temples? Could these "harmonic patterns" have a healing effect on the observer? I was determined to find out!

This turned out to be a much more involved process than I had initially imagined. Though at first, I was disappointed to discover that Manners' device did not have the capacity to visually portray the sounds it emitted, I soon became totally fascinated as I explored their therapeutic effects on the human body. I was to spend the next four years researching and validating these healing and regenerative frequencies— while completing my doctorate at the Medicina Alternativa Institute, under the tutelage of Dr. Manners. The mission of the Institute, now known as The Open International University, was to bring awareness to complementary therapies, with the goal of "health for all in the world." Members came from a wide variety of scientific disciplines, with an emphasis on clinical research and practice, and included the renowned musician, Ravi Shankar, who held a diploma from the Institute. Upon completing my program there, I was awarded a dual Doctorate of Philosophy and a Doctorate of Cymatic Medicine.

Upon my graduation in November 2004, Dr. Manners shared with me his decision to retire. As it turns out, I was his last student and the torch was fully passed.

Another fortuitous event occurred early in 2005 when Jeff Volk introduced me to John Stuart Reid. A British acoustic engineer, Reid had spearheaded the development of the Cymascope, a 21st-century re-design of Jenny's original electroacoustic tonoscope, into a precision-crafted instrument that can make sound visible in ways that Jenny could have scarcely imagined. My updated version of Manners' therapeutic device, the Cyma 1000, had just been launched in January, and Reid was as captivated by my device as I was by his. That April, I invited John to Atlanta and during this visit, he invited me to his laboratory in the Lakes District of England. Just a few months later, in September of 2005, my yearning to see the regenerative sounds from the Cyma 1000, finally came to fruition.

I will never forget the great sense of anticipation that filled the lab that morning, as I, along with several students, huddled around the Cymascope, while John projected the commutations I had selected. These were the first live cymatic visualizations that I had ever seen, and I watched with wonder as the fine powder danced over the membrane, articulating the patterns of these therapeutic frequencies.

I continued experimenting with hundreds of audible sound frequencies, and over subsequent years of observation and collecting data on a wide range of patients, it became evident that the beneficial effects of the transdermal application of these sounds extended beyond the physical level, to encompass the emotional and psychological realms as well.

My clinical observations of how people respond to Cymatherapy, reflect my understanding of how people heal according to the Ayurvedic perspective; that healing occurs through the physical and subtle layers of our being— known as the 5 *koshas*. I wondered (and still speculate) whether the 5 frequencies in each "Cyma code" could effectively "harmonize" the 5 *koshas* through resonance. My clinical observations seemed to suggest this possibility as I noticed that over the 30-minute sessions that I created by sequencing together short samples of these 5-frequency codes, the "veils" of congestion seemed to peel away, starting with the most subtle until the release became palpable at the physical level.

Most of the patients I work with, even those who are not particularly spiritually inclined, report effects ranging from relief of physical symptoms to experiencing a better quality of sleep, often even recalling their dreams. Many make new, life-affirming choices that they previously hadn't thought possible, such as no longer being as easily triggered by troublesome relationships. All the while, I notice subtle shifts in the patient's vibrational fields, which are akin to what I see in Jenny's experiments, and his descriptions speak to me in a language that goes far beyond rational cognition. Though I may not understand his German, I sense the profound universal principles that underlie the phenomena, and which may be perceived and understood through both intuitive awareness and scientific understanding.

Since 2005, I have visited John Reid's lab many times and he has made videos of many of the healing and regenerative frequencies (commutations) generated by my devices. Just like when I watched Jenny's old 16 mm films, I never tire of seeing how the sounds that I'm hearing can bring matter to life!

We have collaborated on many experiments, examining the frequencies emitted by my devices and other sources. John's modern-day images are quite distinct from the ones Jenny made over 50 years ago, yet each offers its unique perspective. To me, Jenny's work is characterized by deep spiritual undertones that highlight the unifying principles, while John's sophisticated technology, enables us to see a wide range of differentiation, due to its ability to independently control many variables. This allows for the creation of intricate images that can be closely examined, precisely measured, and meticulously documented. Motivated by my interest in the therapeutic implications, as well as his own first-hand experiences, John expanded his research to include medical equipment through which we can observe the effects of cymatic sounds and images on live blood cells, such as improving their oxygenation, cell motility, and viability. Some of these results were verified microscopically by Beverly Rubik, Ph.D., a leading scientist and scholar internationally renowned for her pioneering work in frontier science and medicine. She discovered that Live Blood Analysis under a high-powered microscope potentially reveals inflammation, oxidative stress, dehydration, parasites and uric acid crystals.

In a pre-LBA, the white blood cells appeared lethargic. After applying the AMI 850 Stress Relief treatment, the LBA shows the white blood cells active again, doing their job of cleaning up the cellular environment. With continued technological advancement, there appears to be an ever-expanding frontier for exploration. From macrocosm to microcosm and back again, the thrilling journey continues *ad infinitum!*

A new type of experimental set-up for therapeutic application

Dr. Jenny's meticulous experimentation and his articulate synthesis of what he observed, profoundly influenced my approach to research

To obtain the most precise data in scientific experiments, it is necessary to determine the essential variables, and then modulate them in order to see what changes this brings about. The first thing I needed to do was to create a device that could emit the audible frequencies known to have therapeutic effects, (The *AMI 750*, in 2009). Then, I could begin to validate their effectiveness.

Jenny used a frequency generator and an amplifier in order to generate his sound figures on a steel plate. I envisioned using a pair of gel pads upon which patients placed their feet while the health-promoting frequencies transmitted their vibratory patterns directly into the body's "messaging system," the meridian pathways located on the soles of the feet as depicted in Chinese medicine.

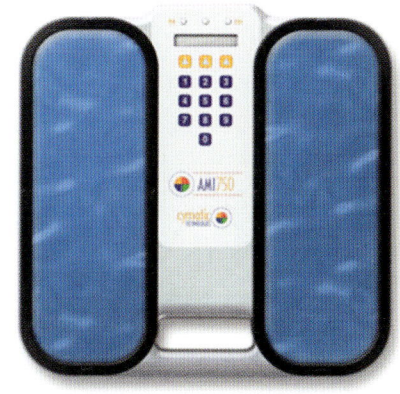

The original Acoustic Meridian Intelligence device

It is well known that sound waves are better conducted through water than through air. Since our bodies are over 70% water, I felt certain that this type of transdermal application through a liquid medium would prove to be a very effective method, and my efforts proved successful.

Since I first read *Cymatics* in 2001, I have viewed diagnostic ultrasound and thermal images of hundreds of subjects suffering from pain and inflammation. With each scan, I hear Jenny's words echo in my mind:

> "Let us look at histology, the science which deals with the structure of tissue. The very origin of the word tissue (Latin, *texere*; to weave) is a significant comment on the prevailing conditions: cells are arrayed in rows, one pattern following another, wherever we look. The intercellular structures take the form of frameworks, networks, grids, families of elements continually repeated and following each other in regular sequence."

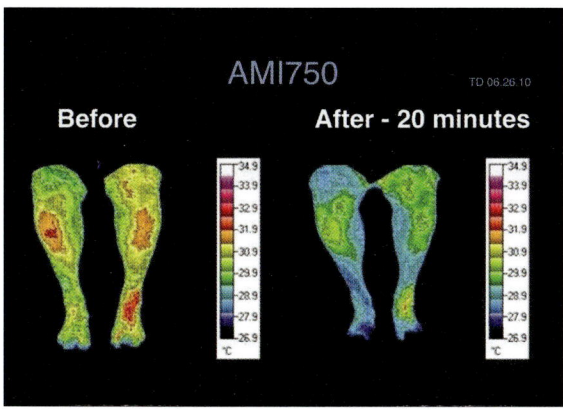

In the baseline images, the injured tissues present themselves in chaotic weaves, while in the images taken after Cymatherapy treatment, they are realigned and restored to their effective patterns of "networks" and "grids." Through countless case studies over the past two decades, this powerful, non-invasive diagnostic approach has proven to be highly effective for many high-performance athletes, as well as animals trained to compete. Amongst them are several champion thoroughbreds, as well as Super Bowl, NBA, and Wimbledon notables. Muscle and tendon injuries, chronic pain and inflammation, and compromised respiratory systems are only a few of the troublesome conditions that have been recognized and mitigated.

This process of re-establishing the formidable "weave," through resonance, is how Jenny's experiments showed that sound can be an organizing force of nature— calling back into coherence a state that had become disturbed.

Toward Cymatic Medicine: Exploring the Healing Effects of Combining Frequency and Imagery

When I first opened *Cymatics*, I spent long hours gazing at Jenny's images. Likewise, I never tired of watching the film footage of his experiments. These mesmerizing visuals had a very relaxing effect on me and I began to wonder: what if the frequency codes within my device could be made visible and become part of the therapeutic regimen? Would they be pleasing to the eye? Would they spark a sense of recognition in the patient viewing them? Could seeing the sound while listening to it create a powerful synergistic effect?

In late 2017, I began looking at new images of sounds produced by my devices to balance the body's systems. I spent hours each week viewing these images that John had scoped with his new 4K video camera. As I researched the hundreds of resonant health commutations passed on to me by Dr. Manners, Jenny's book sat on my desk for regular reference, complementing my previous studies in Ayurveda and Chinese medicine to create a comprehensive plan for wellness and vitality.

I employed a diagnostic device in my research called the *Vedapulse*, which measures heart rate variability and other stress indices. Dissolving the still images I had received from the Cymascope lab, I created a 7-minute visual montage with images of particular sounds known to support the overall energy of the body. I also began testing with the Vedapulse before and after this viewing experience, and it was so successful in reducing stress levels in just 7 minutes that *The Soundflower Experience* has since become an integral part of our product line that we offer through the Cyma Technologies website.

As these frequencies were projected into small samples of water resonating on a vibrating diaphragm within the Cymascope, intricate geometric patterns arose within the water droplets, flowing up from below and blossoming across the surface of the water, like a rose unfurling its delicate petals beneath the summer sun. I was immediately transported back to my childhood days, communing with the flowers, while simultaneously I realized that no one, anywhere on earth, had ever seen this particular set of frequencies made visible before.

I was overwhelmed by the potential of combining cymatic imagery with sound therapy. The possibilities seemed limitless, the scientific and spiritual implications far-reaching. Although I could not fully comprehend everything on a cognitive or conceptual level, the kinesthetic experience I had watching these awe-inspiring sound figures, while simultaneously listening to them, resonated deeply within me.

The SoundFlower Experience

That day in John Reid's lab marked a new chapter of my work: making visible those sound vibrations known to benefit the human body.

I commissioned John to "scope" the frequency that is designed to strengthen our overall vitality. I was astounded by the beauty of the images that emerged in the water samples, revealing pentagonal masterpieces, five-pointed stars, and many other mystical shapes of intricate beauty.

Each "soundflower" gave way to yet another, as the vibrations traveled through the water, culminating as standing waves upon the surface. And not one of these visual patterns was digitally created. Each figure that coalesced into a harmonically oscillating structure was an actual sound made visible. In observing the "cymaglyphs" of these therapeutic frequencies, I could see each sound, in my mind's eye, gently leading the cells back to their natural resonance, like a siren's call back to health. I selected approximately 100 of the glyphs that the Cymascope created out of the tones emitted by my therapeutic device, the AMI 750. *The Soundflower Experience* is the name of the series of videos that I produced based on these "Hot 100" favorite glyphs, and they have been shown to create deep states of relaxation and a profound sense of well-being in subjects in a wide variety of settings. It was this mystical marriage of Jenny's cymatic experiments and Dr. Manners' research that inspired the Soundflower experiment.

An Invitation into Deep Contemplation

My experience of the book you hold in your hands is that, when deeply contemplated, it is capable of transmitting profound insights and wisdom. The science of cymatics opens a window of awareness into the exquisite expression of beauty, which is the infinite unfolding of nature herself. I encourage you to sit with this book, attend to it deliberately, and explore how you might connect with the messages held within. Ponder the relevance of any paragraph that draws your attention and feel how it might relate to your life right now. There's a good chance that you will glean deeper insights into the mysterious workings of creation.

Mandara Cromwell, DCM, has been a researcher on the effects of therapeutic sound since 2001. She has produced international cymatics conferences since 2006, several cymatics videos, and she is the CEO of Cyma Technologies, Inc. She is also the inventor of the AMI (Acoustic Meridian Intelligence) devices, the creator of "The Soundflower Experience," and the author of *Soundflower: The Journey To Marry Science and Spirit.*

Further information may be found at http://www.CymaTechnologies.com

Meeting Hans Jenny
by Ralph Abraham

I met Hans Jenny in February of 1972, just four months before his death. This meeting had a profound effect on my career in mathematics. This is my story.

The lead-up
In the academic year 1971-72 I was a visiting professor at the University of Amsterdam, teaching catastrophe theory. At the same time, I had a visiting position at the Institut des Hautes Etudes Scientifiques (IHES) at Bures-sur-Yvette outside Paris. I used to commute weekly on the train, which I loved. At this time, IHES was newly formed, and had only two permanent professors, David Ruelle and René Thom, both of whom were superb. Thom was one of the great mathematicians of the 20th century and had received the Fields Medal at the International Congress of Mathematicians in 1956 for his work in differential topology. I had met him in 1960 in Berkeley, where we began working together on the foundations of catastrophe theory.

Early in 1972, René and I were both stymied in our work and were browsing the borderlines of science looking for clues. On arriving at IHES one day, I asked René what he was working on. He pulled a book from his desk and began showing me photo after photo of what appeared to be familiar forms from nature: spiral galaxies, cell mitosis, sand dunes, and so on. These forms, he explained, had been photographed in vibrating water. The book was *Kymatik*, or *Cymatics*, by Hans Jenny (1904-1972), who was a medical doctor from Dornach, a suburb of Basel, Switzerland. I was thunderstruck to see images I had seen in my meditations in the pages of a book, especially in support of the vibration metaphor of the Pythagoreans. I immediately called Jenny in Dornach and, to my delight, he agreed to meet me.

The Anthroposophic ambiance
The Anthroposophic movement was founded by Rudolf Steiner (1861-1925), the esoteric Christian student of the *Secret Doctrine*, the basic text of Theosophy, written by Madame Blavatsky around 1900. Among Steiner's contributions to alternative culture were eurythmy, a form of harmonious movement (1911); Waldorf education (1919); Anthroposophic medicine (1921); spiritual science (1923); and biodynamic agriculture (1924).

The meeting
I took the train to Basel, and was met at the station by Jenny's son-in-law, Christiaan Stuten, who drove me to Dornach. Along the way I learned that Dornach was the world headquarters of the Anthroposophic movement. Jenny formulated many of his insights into cymatics from studying Steiner's works while he was living and practicing medicine in Dornach. Jenny greeted me in his home, showing me part of his lab as well as an animated film of some of his experiments in progress. I collected his papers and books and went home to Amsterdam, inspired.

The aftermath
In Amsterdam, while teaching catastrophe theory, I had time to carefully study the photos and the English text of Jenny's first edition, Volume I of *Cymatics*. In the very first chapter, Problems of Cymatics, he gives examples of periodic systems in nature including: circulation, respiration, the heart beat, the nervous system, and the muscular system. It was clear from the start that Jenny's medical training helped shape his views of cymatics.

Jenny followed in the footsteps of these cymatic pioneers, among others:
- 1490, Leonardo da Vinci observed patterns formed in the dust on a tabletop upon which he drummed with his hands.
- 1787, Ernst Chladni, known as the Father of Acoustics, used patterns of sand that arose on vibrating glass plates (later known Chladni patterns) to design a better sounding glass harmonica.
- 1831, Michael Faraday observed standing wave patterns on the surface of beer on the tops of barrels being transported by a horse-drawn wagon.

I left Amsterdam for India in the summer of 1974. I continued my learning and contemplations about vibrations, while determining to begin a formal program of research on sound vibrations upon my return to UC Santa Cruz in the fall. I offered a seminar on vibration that winter, and began construction of a laboratory for vibrations in fluids, based on Jenny's lab, the following summer.

The Jenny Macroscope at UCSC

With the aid of students from my UCSC seminar, I reproduced Hans Jenny's kymatik device in my lab. Our device was larger and less precise than Jenny's. We used a four-inch dish for the water/glycerol solution, four-inch telescope mirrors loaned by Lick Observatory, and a color Schlieren filter developed by Gary Settles. An analog electronic tone synthesizer was built especially for the device, and a Jenny Memoire industrial xenon arc lamp provided the illumination. When finished, I aligned the optics by eye, turned everything on, and glanced at the screen. I was astonished to see a perfect Jenny-style Chladni pattern, in full color. The experience overwhelmed me, and I retired to the corridor outside the lab to recover my composure.

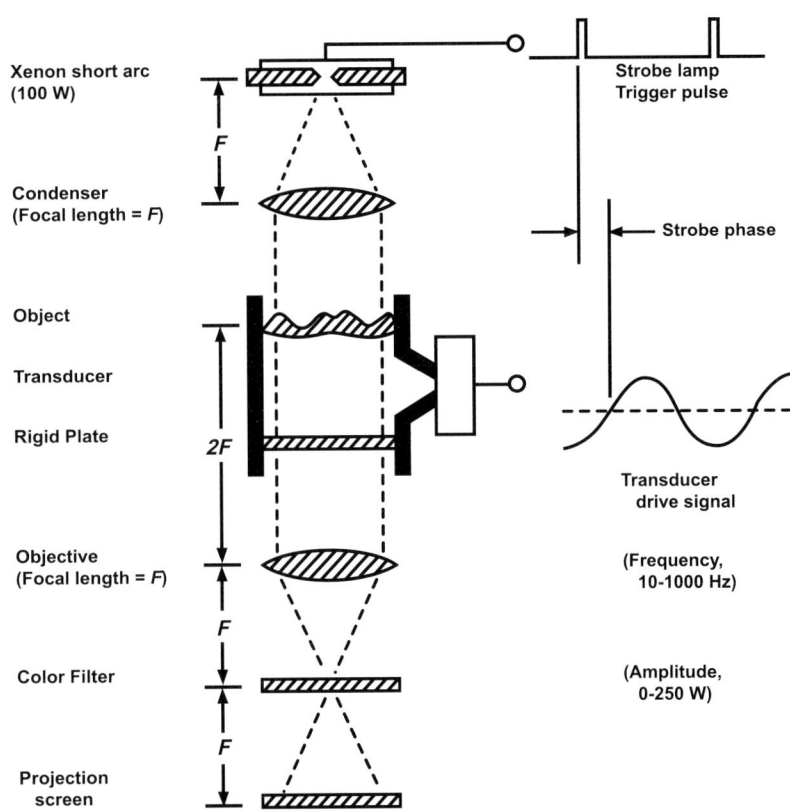

Schematic view of the four-inch macroscope of the University of California, Santa Cruz (Diameter, 4 in.; F, 48 in.)

An official opening was planned for the lab, renamed the Jenny Four-Inch Macroscope, in July, 1974. On impulse, I asked my Indian "music guru," S. D. Batish, to sing at the opening. We attached a microphone to the amplifier that vibrates fluid in the dish, in place of the tone generator. This event provided my first experience of visual music based on Chladni patterns, the essential forms of vibration in three dimensional media. It connected, all at once, my experience with Indian music, Thom's catastrophe theory, and the light shows of the Sixties. Math, music, and mysticism all became one!

Subsequently, until 1979, we made systematic use of the instrument (with the tone generator, not my master's voice) to study the bifurcations of chaotic motions of vibrating waves. The experimental arrangement for the study of Faraday crispations is shown in the preceding schematic. An electrical current delivered to a transducer (loudspeaker) creates a sound wave in the air beneath a plastic plate, inducing and maintaining a pattern of vibration. This in turn activates a related pattern of vibration in the thin layer of liquid resting on the plate, and thus a pattern of Grex vortices arises in the layer of air above the plate. Air, plastic, liquid, air: four layers of coupled pattern formation processes.

Hans Jenny, born August 16, 1904 in Basel – died June 2, 1972 in Dornach.

Ralph Abraham
Professor of Mathematics
University of California Santa Cruz
Mathematician specializing in chaos theory, applications to social problems, and consciousness.

Author
- *Schism: The Madness of Crowds, Toxicity of Social Media, Social Polarization, and Political Violence; A Cybernetic Approach,* 2023
- *Vibrations and Forms: Findings from Psychedelic Adventures,* 2021

Co-author with Terence McKenna and Rupert Sheldrake
- *The Evolutionary Mind: Conversations on Science, Imagination & Spirit,* 2005
- *Chaos, Creativity, and Cosmic Consciousness,* 2001

Learn more at http://www.ralph-abraham.org

Cymatic Images as Vibrational Icons for Contemplation
by David Perez-Martinez, M.D.

As a modern cymatic phenomenologist, it is an honor to contribute this commentary on the visionary work of Hans Jenny, demonstrating the effects of sonic frequencies on the structure, design, and organization of matter. Jenny's work resonates with both my thoughts about matter, energy, frequency, and vibration, and my visual perceptions as I walk through the world of natural and manufactured creations. I specifically set out to portray in nature and human design what Hans Jenny and subsequent investigators have demonstrated in the laboratory; that there is a commonality of repeating patterns, processes, and structures observed in nature and human design that is not due to happenstance but rather to an *order* in the universe that is both a reflection and a manifestation of vibratory frequencies interacting with consciousness. The ability to perceive order is an act of consciousness. It is the nature of human consciousness to create meaning, values, and narratives that attempt to make sense of sensory perceptions. As phenomenologists we allow nature to be our teacher. Everything in nature, from a phonon or photon to a galaxy, carries intrinsic information that reveals its unique vibrational signature in the context of universal vibrational dynamics.

I aspire to capture images of everyday phenomena so as to allow one to contemplate them as vibrational icons that reveal a greater, hidden reality than is apparent in the mere physical forms observed. They contain information about the universe, ourselves, and the vibratory nature of everything that exists, and by attending profoundly to what they portray, they can instruct and delight us as tools for wellness and artistic expression. The universe is a giant library full of information and the history of humanity is one of constantly educating ourselves to be able to read it. We need to look and listen in order to hear and see nature tell its story—our story, revealing the foundation of a new cosmology for these times as the conscious, vibrational beings that we are, in this universe of constant movement and change. Being aware of the role that vibration and sound play in the design and organization of matter, and the role of consciousness in interpreting experiential phenomena into the perceptual and conceptual realities that we create, makes the observance of nature a spiritual exercise of self-exploration, self-realization, and survival. More than ever we are facing potential planetary-wide cataclysmic changes and it is incumbent upon every person alive to develop a new sense of appreciation and respect for the planet, its diversity, and the way that *all life is vibrationally connected and interdependent*.

Dr. David Perez Martinez, MD is an Integrative Psychiatrist, Therapeutic Sound Healing Practitioner and a trained Anthropologist and Ethnomusicologist, currently engaged in private practice, research, and writing about the therapeutic use of sound and cymatic phenomena in nature and human consciousness.

TUNEUP RX
252 7th Ave, #6D, New York, NY 10001
tuneuprx@gmail.com • http://tuneuprx.com
(212) 594-6405 • Fax: (845) 658-4210
davidperezmartinezmd@gmail.com

My Love Affair with Cymatics
by Jodina Meehan
Dreamer, Artist, Explorer, and Catalyst in the Field of Cymatics

Just over twenty years ago, as I was finishing school and getting ready to step forward into my future, I found myself wrestling with two challenges.

The first was that I was deeply dissatisfied with a series of oil portraits I was finishing in preparation for my college graduation exhibition. No matter how good I was at rendering the look of the eye or the tilt of the head, there was always something missing. I was trying, and failing, to capture what I was really interested in painting: the invisible enigma at the heart of their being. Though many admired and loved my paintings, I found them wanting.

My second challenge was that I had just been told it was time to "choose a path" and dedicate myself to it. I was someone with numerous passions: painting, physics, fiction writing, sculpture, music, biology, cognitive psychology, shamanism, and poetry; and I was warned that I would fail if I did not pour all of my creative energy into just one of these pursuits, essentially for the rest of my life.

And so, at a time which should have felt exciting and promising, I was in fact feeling deeply unsettled.

This is when I had a dream, a very strange and disturbing dream, even by my own standards. I was standing in the center of a dark, cavernous room, filled with paintings that I was to be making in the future. They were all black, with amorphous white shapes emerging out of their center. And there were hundreds of them, all standing on little wooden easels, stretching back to infinity.

I woke up feeling baffled, and somewhat angry. Why was this the future of my art? It did not resemble anything I was doing at that time. I had no idea what these paintings were supposed to represent and, to be honest, I was not sure I liked them.

And yet, I had learned to trust the messages from my dreams, so, for the next five years, I attempted to recreate the images I had seen in the darkness. I painted them in charcoal, chalk, oil, and enamel. I painted them large and small, but no matter how hard I tried, I knew I wasn't getting at the essence of what I had seen in my dream.

Then, one day, I came across Hans Jenny's book, *Cymatics*, in my local library. I opened it up to a random page, and…voila! There were the paintings from my dream staring back at me. I looked closer, read the descriptions under the images, and realized these were scientific images of what sound looks like. A-ha! I thought. I am not supposed to be painting with brushes at all. I am supposed to be painting with sound waves. But how the heck do I do that?

I brought the book home and read the whole thing. The images, I learned, were made by a doctor, working after hours in a make-shift lab, using metal, water, light, salts, and sound. They were pure poetry, geometry, physics, and music, all wrapped up in one. I was enamored—and still am—of all things cymatic, including the word itself, which I find beautiful.

Over the past 17 years, I have explored cymatics through numerous lenses, including the artistic, educational, and scientific. Cymatics became my life, and my life's work. It has allowed me to enter into and reveal that "invisible enigma at the heart of being," that universal essence at the core of everything, that mystery that had always eluded me back in school. And it has also allowed me to pursue all my passions without having to choose just one!

During the course of my explorations I have built 5 cymatic instruments, founded the Journal of Cymatics and the School of Cymatics (online), completed cymatic commissions for Cirque Du Soleil and National Geographic's show Brain Games, and am in the process of writing a YA Fantasy Trilogy about a world where "sound is magic," called Cymatica.

I have also served as an advisor to elementary, high school and college students studying cymatics for everything from an entry into their school science fair, to completion of their graduate plan in this exciting new field. I have seen, firsthand, the way people of all ages light up when they get to engage with cymatic processes, even on very simple levels.

So what I would like to do now is to offer a simple way that you, too, could step into this magical world of waveforms, and try it out for yourself - perhaps even inviting the children in your life to jump in and do a project with you.

The earliest device I built (after much trial and error) was something called a "tonascope" which is a way to see what the human voice looks like. I finally landed on a design that you can make at home with materials from your local hardware or craft store, that shouldn't cost more than about $20. So here you go!

How To Make A DIY "Cymatic Tonascope" To See Your Voice

Materials:
- A PVC pipe "elbow," with a 45 degree angle (looks like a check-mark)
- A length of PVC pipe, preferably about 2" in diameter, cut to 3' length
- A large plastic funnel (opening 8-10" across)
- 1-2 large black (or other dark colored) balloons
- Some large rubber bands
- Table salt or white sand
- Super glue or hot glue

Tools:
- Felt marker
- Scissors
- Hack saw

Step 1 - Set your PVC elbow piece on the floor, with the bigger end as a base and the angled part angled up. Insert the end of the straight PVC pipe firmly down into the angled part of the elbow (think of it as building a saxophone). The length of pipe should be going up at an angle now and the wider, open end of the PVC elbow should be facing straight up.

Step 2 - Place the plastic funnel into the wider, upright end of the PVC elbow, so that it is facing up, and looks level. Draw a line around the funnel where it meets the pipe. Lift the funnel out and use the hack saw to cut the smaller tip off the funnel, ¼" below the line you drew. (Note: make sure you make this cut on the smaller side of the funnel, not the larger side, and not on the line. You need enough left to fit back into the pipe again).

Step 3 - Using either hot glue or super glue, put a bead of glue around the top edge of the large PVC pipe opening (not the long pipe, the elbow) and then place the smaller end of the funnel back into this opening, and hold until well set. Again, try to make the top of the funnel as level to the floor as you can. This will be your holding surface.

Step 4 - Cut the "neck" off a balloon at about 3 inches from the blowing end, just enough so that it can stretch to go over the funnel. Now stretch the balloon tightly over the top side of the funnel, so that there are no creases, and use the rubber bands to go around the edge and hold it down. (This is much easier if you have an extra set of hands to help). Use several bands to make sure it stays.

TIP: After the bands are on, you can go around the balloon edge and pull it gently down further to get a smoother fit. It should look like a drum head when you are done. That's it! Now you are ready to play your instrument and see your voice.

How To Play Your Tonascope:
1. Set it upright on the ground or a table so the balloon-covered end is facing up and level like a drum. You will need to hold it like this while playing it.

2. Sprinkle some salt (or white sand) over the surface of the balloon, just enough to lightly cover it over, as if you were dusting a cake with sugar.

3. Now put your mouth up to the other end of the PVC pipe (as if it's a saxophone) and sing and hum. You will see the salt respond to your voice and start to make cymatics patterns like circles, ovals, and snake-like figures.

HINT: If you make an air-seal between your lips and the pipe, and make nice, loud "aaaaa" or "oooooo" or "eeeee" vowel sounds into the pipe, you will get the best shapes. You do not need to be a singer, you just need to make a sound!

I hope you can try this cymatic design yourself and share it with others. Dr. Jenny's *Cymatics* opened a door for me, a way to look at the world that revealed the invisible but implicit magic all around us, the secret language spoken by the Universe — a language that we can now see and read, and speak back with... and try to understand.

Ever since I first read *Cymatics*, I have been inspired to do the same. This is my invitation to you, to wander through the pages of this book, try making sound waves visible yourself, and explore the magical world of sound and vibration that Jenny introduced to the world with his groundbreaking work.

Jodina Meehan - Brattleboro, Vermont - January 23rd, 2022

Jodina Meehan is a dreamer whose interests span the spectrum of the arts and sciences. Her fascination with cymatics has compelled her to explore the magic hidden within the invisible world of sound, and to assist others, especially children, to discover this for themselves, through a variety of modalities, including the following websites and social media, where she hosts, and upon which she posts, regularly:

- The Journal Of Cymatics: http://cymatica.com
- The School Of Cymatics: https://cymatics.ning.com
- Facebook: https://www.facebook.com/jodi.meehan.5
- Email: jodi.meehan@gmail.com

Hard Evidence of Genuine Magic
by Jim Metzner

During a Soundscape workshop, students point out that when they're on a blindfolded "listening-walk," their senses of hearing and smell become more acute. A similar phenomenon happens for the guide as well. It appears that when one's attention is expanded it heightens all of the senses, catalyzing a synesthetic experience. Suddenly we're not only hearing, but we're sensing sound with our entire organism.

Through *Cymatics*, Hans Jenny helps us to rediscover this experience objectively. He reminds us that we are surrounded by vibrations of exquisite intricacy and beauty which we usually cannot see. And because we don't see them, we don't pay them the same kind of attention. The more we become in touch with a global sensation of our bodies, our associative thought machine quiets, and the more possible it becomes to attune to finer vibrations, within and about us. Yet even in my typical unconscious state, I cannot help being touched by Hans Jenny's photographs. Cymatics is hard evidence of genuine magic.

Jim Metzner is a Fulbright Specialist in Media & Communication, Executive Producer of the *Pulse of the Planet* podcast, and the American Soundscape Project. His recording archive has recently been acquired by the Library of Congress in Washington, DC, and his upcoming book will explore the mystery of listening.

pulse@igc.org
http://JimMetznerProductions.com
http://pulseplanet.com
Adventures of a Lifelong Listener

My Journey of Curiosity
by Jacob Lee Adlington

My journey of curiosity began after contemplating a photo by Alexander Lauterwasser. An oval Chladni figure, resembling the same structured patterns as a tortoise shell. The physics of Cymatics compelled me to explore the many questions it posed. What were the enchanting secrets that awaited discovery? In the spirit of Eden Phillpotts - What amount of "magical things were patiently waiting for our senses to grow sharper?"

Countless hours of transformation before my eyes, the unseen becoming seen. From chaos to order, unmanifest to manifest, if only for a brief moment, ultimately to return to chaos. It taught me that there is a unique and repeatable song in everything, if we choose to listen.

The intersection of art and science satisfied both parts of my brain. Though through my journey of cymatics I have come to a differing perspective. I feel I have always been an artist at heart, and where once I was seduced by the importance of frequency alone, I come to view cymatics as more art than science — a poetic alchemy.

The art of cymatics invites me to contemplate my own resonance. Frequency is only ever half the picture. The "medium" (mass of water, metal plate or even YOU) is the secret, the frequency is only the key. Change the medium in any way, and you change its resonant frequency. Imagine yourself as the physical medium, constantly interacting with other "frequencies." The constant movement and interaction with ourselves, finding resonance within and without. Everyone is unique, tuned to resonate in differing ways, each positioned to deeply experience and understand varying aspects of reality. Through the interconnection and sharing with each other, we may begin the journey of putting the pieces together.

It is important for me to explore works that are both intrinsically beautiful and true to nature, which deepens my understanding and appreciation of the natural world.

Though I find the visual beauty of cymatics to be intrinsically meaningful in and of itself, I also believe that there are hidden truths to be discovered through contemplating its metaphors.

In the same way we use musical and physics metaphors in daily life, Cymatics offers a new lens to expand our lexicon, inviting contemplation of harmony, resonance and transformation.

Cymatics is that which hints at the beauty of the horizon, yet beckons beyond.

Jacob Adlington is a visual sound artist and photographer from New Zealand, based in Brisbane, Australia. His deep belief in the power of music and sound to shape our realities is what inspires his passion to bring sound vibrations into visual form. His mission is to share the beauty of sound as art, as a healing pathway to the sacred within. This dynamic dance of the physics of vibration reminds us how magical and interconnected everything truly is.

For more information and to see Jacob's art, visit https://JourneyOfCuriosity.net/

Cymatic Artistry
by Rachael Linton

I was introduced to this fascinating field in 2008, while studying for my Master's in Design in Digital Media Artistry, at Massey University, in Wellington, New Zealand. I had been looking at various types of visual and auditory phenomena proven to have measurable therapeutic effects on the observer, such as geometric shapes; patterns in motion; colors; light; and even what sound might look like, if we could only see it. As fate would have it, a colleague showed me some of Hans Jenny's work in the field of cymatics, a term Jenny had coined in the 1960s. When I first saw those images of Jenny's experiments spreading sand on a steel plate, they reminded me of the sand mandalas used in Tibetan meditation practice, and I realized that this was just the type of imagery I was looking for in my digital therapy. When I came across the vibrational patterns in his book, it was like striking gold.

I have always been an artist and a musician, never a scientist. But Jenny's body of work allowed me to intuitively understand the physics of how sound moves, by making this process visible. His remarkable experiments offered me insights into the cyclical, harmonic oscillations and lattice structures hidden within the sound. Viewing these processes also allowed me to see analogous relationships to patterns found in many different types of natural energetic systems. He also established a linguistic translation of what we see in cymatics, using parlance accessible to both scientific and lay audiences alike. Since multimedia was my main modality, I easily flowed with Jenny's suggestion that film (videography, in the digital age) was an ideal modality through which to study and capture the movement of sound. With my eye glued to a macro lens, I peered into a small sample of water and became transfixed by these cymatic flowforms, as if I were a deep-sea diver discovering things never before been seen by human eyes!

Jenny's books, photographs, films, and observations have had a profound impact on my creative work and research, inspiring me to delve deeper into cymatics and its archetypal geometric patterns. I would meditate over Jenny's pictures and films all day, contemplating how those patterns organized themselves so perfectly and what that might mean in the context of my work as an artist and in visual communication design. Throughout my career, I have focused on creating works of art that explore how frequency, resonance, and oscillation interact with the human bioelectric field. I've experimented with interactive art installations using the most cutting-edge digital technology available to me, with the objective of creating a holistic therapy designed for a 21st-century audience. Jenny's experiments using inert substances provided a visual representation of the way that molecules within the cells of our body could be coherently re-structured by particular sound vibrations. What also caught my attention was how they detailed similarities between a vast array of biological structures and dynamic processes found throughout the natural world.

When two frequencies interact, they combine to create a third unique frequency, and it's my belief that there is great untapped potential in this sonic alchemy. Jeff Volk, the publisher of this current edition, once dubbed me an acoustic alchemist because I use cymatics, (the visual form of sound) to create art that can transmute the perception of the viewer. In theatrical lighting, for example, we mix frequencies of red and blue light to make magenta. Combining one sound frequency with another produces an oscillation, which generates harmonic overtones. Jenny's body of work suggested to me a vast range of implications as to how visual and energetic frequencies can interact with our bioelectromagnetic fields. I feel inspired that cymatic artistry can catalyze advances in biology and consciousness research, yielding innovative discoveries and technologies.

One of my most exciting pieces of experimental art was an immersive installation called the *Aura Illuminator*. In a gallery, my guests would sit at a desk with a speaker placed under a pool of water – a classic cymatics setup. The user presses a slider on a touchscreen to choose a sound within the 1-300 Hz frequency range, as well as a color on the RGB color wheel. Both are controlled via an app, called Sound Vision, which I designed and had custom programmed in 2012. The colored light shines down onto the water illuminating cymatic patterns rippling on the surface. A video camera, placed above the desk, projects these images onto a reflective screen in real-time, where the user can then interact with the projection by bathing their body in the colored cymatic light patterns. In this art experiment, I was hoping that the colored light patterns would interact with the subject's bioelectromagnetic field to illuminate and make visible their own auric field.

The *Aura Illuminator* piece evolved into *Crystal Cymatics*, an immersive cymatic journey, which I exhibited at the Mind Body Spirit Festival at London's Olympia, in 2015. With a team of collaborators, I projected the visual patterns produced by resonating crystal "singing bowls," onto a 360-degree planetarium "dome cinema." The audience was immersed in a 3-D performance of sound and form in movement— a cymatic light healing bath. The crystal bowl musician played six quartz crystal bowls, creating harmonic symphonies, while backstage, our digital technician and I used contact microphones, professional video engineering software, and specialized projection equipment, to transduce the vibrations of the bowls into a pool of water. We focused the camera on the surface of the vibrating water, and then projected the cymatic patterns in real-time, onto the *Soul Dome*, a 360-degree cinema 'screen.'

Feedback from the participants suggested that cymatics may have therapeutic potential. What began as an art installation has led me to consider that cymatic art may have the capability of releasing healing-specific neurochemicals within the brain. Since *Crystal Cymatics*, I have continued to exhibit kinetic art installations across the UK, making amendments to improve them along the way. More recently, I have shifted away from digital, to analog sound, working with *The Golden Water Gong* (one of my sonic artworks), crystal bowls, and the human voice, as I've recognized that these combinations of frequencies seem to have a more organic way of interacting with the human body.

Mixed media with cymatic vibrations on birch wood panels, Rachael Linton

As I was experimenting with ways of solidifying cymatics onto a painting to make art that can be hung on a wall, a colleague brought to my attention the work of Margaret Watts Hughes. She was a Welsh vocalist who in the late 1800s, discovered by chance that she could visualize the sound patterns of her voice. She then developed an acoustic instrument which she called the eidophone. Like me, she had the desire to fuse cymatic patterns directly into works of art, which she did by applying sand, lycopodium, powdered pigments, and liquids onto glass plates. To me, these were as ground-breaking and visually striking as Jenny's work, done 75 years later.

Like Watts Hughes, I also paint abstract, intuitive works and have used my voice and other instruments to capture the visual form of sound. Likewise, we both value improvisation and intuitive painting, which allows an aspect of the unconscious and numinous mind, to enter our work. I have also drawn on my training in gilding techniques and *verre eglomise* to devise a method of solidifying fluid materials onto both glass and wood panel paintings. I use fine art pigments, genuine gold powders, and precious metal leaf, which all conduct electricity and bounce light into the room and into your body. My paintings are intended to transport the viewer to another realm of consciousness, offering a visionary or transcendent experience. They are especially suited to spaces where one wishes to bring a clear, energized frequency.

During my academic career, I was encouraged to follow a scientific line of inquiry in my research. Jenny's work became vitally important to me because he added an anthroposophic perspective to his phenomenological observations. He used a strict scientific methodology to quantify his cymatic experiments with empirical evidence. I look forward to one day testing my own theories— using the latest biofeedback, and sensing technology, and to further explore the effects of projecting cymatic sounds and images onto the human body and its surrounding biofield.

Rachael Linton, MDes, enjoys an inspired and prolific career as an artist. In each of her works of art, she aspires to find new ways to explore the intricacies of cymatics. A painter, musician, and multimedia artist, Linton has exhibited her work throughout her home country of New Zealand, as well as internationally. As a gilder, radiance and illumination are quite familiar to her— and she strives to convey them through her art. She's also developed kinetic art installations that project illuminated imagery of dynamic, cymatic sound-figures, to create immersive and interactive experiences for her audiences. She uses digital video and analog sound to entertain, educate and transform her viewers, weaving harmony and chaos to create ripples in the fabric of space-time.

Find out more about Rachael Linton's art and cymatics work at:

- http://www.tiliagilding.com
- http://www.soundvisionstudio.co.uk
- Join the worldwide Facebook group @Cymatics

Cymatics: A Felt Sense Bigger than Ourselves
by Joshua Leeds

There is something so primeval about cymatic imagery. It enables us to see the awe-inspiring shapes of vibration and to watch how frequencies interact with one another. Through cymatics, we're able to actually look at the effects of sound, to study and compare them visually. Note that I'm interchanging the terms vibration, frequency, and sound in a manner where each is an approximation of the other. More precisely, each reflects a specific aspect of what cymatics reveals to us.

At the atomic level, every aspect of our world is in constant motion, and wherever there is motion, there is vibration and frequency. Across the spectrum of sound—from the movement of atomic particles to the sensory phenomenon we call music—there is this corollary of vibration:

- All atomic matter vibrates
- Frequency is the speed at which matter vibrates
- The frequency of vibration creates sound—heard or unheard
- Sounds can be molded into music

This describes the omnipresent, yet invisible world of sound, which cymatics singularly brings to light.

To see vibration-made-visible evokes a higher-self perspective. This is not surprising given that Hans Jenny was a deeply spiritual man, greatly influenced by Rudolf Steiner. To perceive the different shapes and designs of life as vibration is reminiscent of the universality of everything. It's no wonder that this study of wave phenomena has attracted and inspired so many inquiring minds across the centuries; all the way from Pythagoras in 569 BC to Da Vinci in the 15th century, Galileo (16th), Hooke (17th), Chladni (18th), Faraday and many others (19th), and Hans Jenny (20th). Such an impressive progression, this multi-cultural fascination with the phenomena of waves.

In our lifetime, many researchers have been attracted by the magnetism of cymatics. Informed by Dr. Jenny and his predecessors, publisher, Jeff Volk; acoustic scientist, John Stuart Reid; Cyma researcher, Mandara Cromwell; artist/philosopher, Alexander Lauterwasser; artist/researcher, Gabriel Kelemen; researcher, Jeremy Pfeiffer; and so many others live to tell the tale. And before them, Peter Guy Manners, Olav Skille, Gary Robert Buchanan, and countless sonic explorers dedicated their lives to the study of vibration. Those drawn to corresponding fields, spanning vibroacoustics to bioenergetics, have welcomed this "seeing of sound." Sometimes, words are simply insufficient to communicate that which we can see. That's why we look at cymatic images and simply effuse, "awe (Ah)!" And I say hurrah! Not a moment too soon.

We live in a world no longer on the precipice of climate change, but rather plunging headlong into it. Covid, a once in-a-hundred-year event, showed 8 billion of us, in real-time, what happens when governments and cultures ignore our collective habitation of this planet. I believe that our psychological and physiological relationship to the world, and to each other, changed fundamentally after Wuhan. It has become as evident as the elegant choreography of Jenny's experiments—animating powders, pastes, and liquids—that *we* too are inextricably collective in our human beingness.

Consequently, the questions I now ask lean toward the relevance of ideas, art, and actions in a post-Covid 21st century. The field of cymatics is no exception to my inquiries. After 600 years of progressive exploration, what is the relevance of wave phenomena now? How do we convert the big *aha* moment into something useful, to surmount challenges of such magnitude? It seems that every century brings each generation its own trying times. Odds are pretty good that we'll survive these coming decades and be all the better for it, but it's likely to be a roller coaster of inspiration and heartbreak. Consider, for example, the paradox that we have robotic landers on Mars, yet a dire shortage of life-giving oxygen in Mumbai. Might a new "cymatic awareness" be a game-changer in the 21st century?

In 1972, the year Dr. Jenny died, he concluded his second volume of *Cymatics: A Study of Wave Phenomena*, with similar questions: *"What will come of it,"* he asks? *"Will the mysteries be solved? Shall we come through alive? How will this adventure end?"*

To which he answers: *"…We are, in reality, this mystery; in it we become; it is not that man simply is, he is becoming all the time…"*

So, what are the men and women of planet Earth becoming? How do we listen to each other; to the trees; the animals; ourselves… let alone our cells?

My primary sound mentor, Dr. Alfred Tomatis (1920-2001), was a neighbor of Hans Jenny, separated only by the 525 kilometers between Paris and Dornach, Switzerland. Perhaps they even met? But as far as I'm aware, it's not recorded in either of their writings. Concurrently, they carried out parallel experimentation, both recognizing that the ear and the voice (our organs of hearing and our equally sophisticated capacity to make and shape sound), were profoundly interconnected in some way. But what is more important is the acknowledgment that these two curious men—brilliant medical minds and intrepid researchers—were independently delving into the powers of sound; Jenny, through a new mode of *seeing*, bringing forward the visual understanding of frequency and tone, and Tomatis, through *listening*, by developing therapeutic methodologies employing specific audible frequencies to essentially rewire parts of the neocortex of the brain. Both fervently devoted their lives to empowering humankind to become the best that we can be.

Each of us has the potential to discover a new clue—to contribute something vital, laying the foundation upon which later generations may stand, to reach ever further into the future. Isn't it magnificent the way this is set up? Equipped with the beautifully adaptive mind and our creative imaginations, we balance on the shoulders of those who have come before us, shaping our human understanding to a finer approximation of the truth. This is incredibly exciting, and it has been going on forever! So I ask, upon whose shoulders will the next generations stand?

If awe, inspiration, and a holistic collective memory counts for anything, I'll bet that in this pivotal millennia, cymatics will make a significant difference. Why? Because it is profoundly inspiring and infinitely applicable. John Stuart Reid, creator of the *Cymascope* instrument, might be one example *{as might many of the others who have contributed to this section!, Ed.}*. He envisions micro/macro possibilities: On the one hand, embedding tiny cymatic imprints on cellular membranes to massage the integral proteins of cells—triggering the body's immune response in ways similar to mRNA, but using vibration instead of chemistry! And from the macro perspective, applying Reid's cymatics research into the mystery of dolphin language, he demonstrates that dolphins see with sound, by imprinting cymatic representations of objects onto the tectorial membrane of their inner ear.

To me, this is really fascinating—giving us a glimpse of what just may "be-coming" through us, in the years to come.

Joshua Leeds
Asheville, NC, USA
January, 2022

Joshua Leeds is a music producer and sound researcher specializing in psychoacoustics. His books include *The Power of Sound* (2000/2010) and *Soundwork on a Hot Rock* (2024).

Discover more at http://www.JoshuaLeeds.com

Cymatics and Resonance
by Alexander Lauterwasser

I first encountered a tortoise in the forest as a teenager. Over the years I became more and more fascinated by this wondrous animal, until one day I was confronted with the question of why animals appear the way they do, and how they come to look this way. The common Darwinian answer is that ultimately everything comes about according to some random principle of chance, and that only the most adaptable and thus the 'fittest of all,' would prevail in the struggle for survival. This may appear as a reasonable explanation of life on earth, but this superficial model in no way does justice to the actual phenomenon of existence.

After many years of preoccupation with questions about the formative processes in nature, I happened upon a Swiss magazine with Hans Jenny's "Chladni sound figures" on the cover. I was fascinated by the calligraphy of this "formal language" created by pure vibrations, and I was immediately convinced that these phenomena could be an important key to my deeper understanding of morphogenesis. So I immediately began to collect all of Jenny's writings that I could locate, and to carry out my own experiments in this arena.

In this way, it became increasingly clear to me how important the processes of cymatics were for a deeper understanding of the organizing and shaping processes at the interface between pure vibration and life processes on the one hand, and the material substances that ultimately comprise all-natural designs. Since the sand on the plate may move in the absence of resonances but remains blurred and unstructured, upon careful observation, one comes to understand that ultimately all clearly structured Chladni sound figures are the result of resonances between the effects of the vibratory frequencies and the inherent vibrations of the respective plate. (See composite figures 1-6 on p. 33)

Even before Hans Jenny started working on cymatics, he had already dealt in his very first publication, Der Typus (1954), with fundamental questions about the formative structure of animals— but above all with the cipher of the formative types of different animal species. Just as Jenny was led to the vibrational processes by studying the morphogenesis of natural forms, so was I awakened and stimulated by his work, and through Rupert Sheldrake's articulation of the concept of morphogenetic fields. But Jenny did not stop at animating grains of sand on vibrating metal plates. He continued— proceeding far beyond Chladni— to stimulate various other media through vibration and to observe and describe the processes that occurred. (*The Law of Repetition* 1962; *Cymatics* 1967)

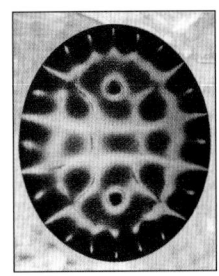 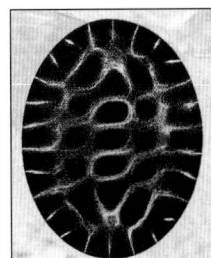

1021 Hertz   2041 Hertz

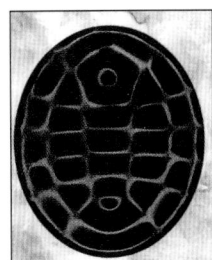 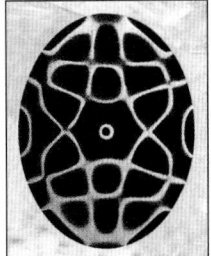

1088 Hertz   1085 Hertz

The tortoise photo and the 4 cymatic analogs were created by Alexander Lauterwasser and were taken from his book, "*Water Sound Images*," first published in 2002, with English language edition published by MACROmedia Publishing in 2006 (See page 351).

The fundamental correspondence between the types of oscillation patterns and natural forms, especially with the forms of living things, becomes more and more apparent. First: the periodic rhythmic alternation of the polar processes of contraction and expansion, of compression and thinning, of systole and diastole (See figures 12 and 13 on p. 36); and second, that design processes do not result from an additive accumulation, or stringing together, of discrete particles— a form is not the sum of its parts. Rather the form is the result of movements, vibrations, and pulsations that holistically structure and organize the material substances. (See figure 24 on p. 42)

Novalis spoke of water as the "sensitive chaos", Theodor Schwenk, of the special "impressiveness" of water, and the biophoton physicist, Fritz A. Popp, of water as the "universal resonator," the medium, or material connection, with the most universal willingness and ability to respond. Water researchers now even characterize water as a quasi-living "bio-molecule."

For me, especially through cymatic work with water, the fundamental importance of the phenomenon of resonance emerged more and more clearly as the central element of many gestalt formative processes. Because the prerequisite for successful resonance is always a general excitability— a vibratory ability of the respective medium, so that it can touch, move, grasp, and be shaken, i.e. to free itself from the mere inertia of gravity, to make itself addressable, to vibrate itself and if necessary, to have the information encoded in the vibratory structures brought into shape, i.e. into a form that then makes a certain way of life possible in the first place.

After my first publication, *Wasser Klang Bilder (2002)**, and my subsequent book, *The Secret of the Tortoise (2009)*, I then tried in *Vibrations Resonance Life (2015)*, to portray resonance as the universal phenomenon that "holds the world together in its innermost being" (Goethe). A few of the underlying fundamental principles are outlined below.

1. Resonance as the basic principle of gestalt formation: "the world as coagulated sound" (Novalis).

2. Resonance is the life principle of all organic forms of existence, which can only succeed due to resonance-based processes of different organs, especially the sensory organs, despite their physically delimited corporeality to relate to the environment. Due to the highly complicated photosynthetic process in chlorophyll, which takes place mainly through resonance, the various plants are able to materialize different aspects of sunlight and can then serve as specific food and remedies due to the different active substances that arise therein. The different animal species internalize different aspects of their respective environment via emotional resonance, into which they are ingeniously organized and thus become mirrors that have taken form, as confidants of the world. As Hans Jenny stated in his exquisite art book, *Tierlandschaften* (Animal Landscapes*): "The animals are hieroglyphs, which come from the world language; they are also seals that lead to the universe and its reflections." The well-known water researcher, Victor Schauberger, is said to have gained all of his knowledge about water from observing how trout navigated the invisible currents.

3. Far beyond the commonly accepted doctrine of chance, adaptation, and natural selection, as the sole evolutionary mechanisms, the unstoppable, unbroken effort of life to increase the complexity of resonance processes and possibilities within one's own organism, could be seen as the impulse and drive underlying *all* evolutionary developments.

---

*The English language edition of *Water Sound Images* was published by MACROmedia Publishing in 2006. A few copies of *Tierlandschaften,* although out of print as of this publication, are available for purchase. See full-color promo pages at the end of this edition for further details about these two books.

4. Using the example of the transformation of a caterpillar into a butterfly, it can be seen that the decisive moment in the process of metamorphosis lies in a passage through a completely fluid stage. In this resting phase within the pupa, the old caterpillar shape and all tissue are completely dissolved, liquefied: without a solid shape—yet still alive, in a state called histolysis. It is precisely this chaotic formlessness, which is characterized by a very special sensitivity and thus a susceptibility to new design impulses, from which the new butterfly shape then emerges. It therefore looks as if, at that very moment of liquefaction, there is a highly increased ability to resonate, a receptivity that opens up the respective substance to the possibility of embodying new form-giving information.

So it is not surprising that in the latest research in the field of quantum biology, *Life on the Edge*; Al-Khalili, McFadden (2014), it is becoming increasingly clear how living things, through the development and differentiation of their own molecular structures, are able to undergo the most amazing energy conversion processes (e.g., photosynthesis), and to accomplish sensory performances (magnetic field compass of migratory birds) by oscillating at the quantum physical level, by way of highly sensitive resonance processes. It is not surprising that analogous synchronous vibratory processes are increasingly appearing to be the basic modalities of our thinking, and perhaps even of consciousness itself.

Hans Jenny had already performed extensive cymatic experiments cataloging the movements of individual sine wave frequencies, as well as more complex and dynamic musical processes. (See figures 70-73 on pp. 77 & 78) Following up on this work, I photographed my own water-sound-images derived from a wide variety of instruments and musical compositions—not to visualize music itself, but rather to demonstrate the resonance-related responses of water to the music. Over the years, I have studied these images in my cymatics laboratory, published them in several books and films, and also projected them onto large screens for live audiences in galleries, theaters and concert halls throughout Europe.

All of my explorations in this fascinating field, was done in a spirit of thankfulness for Hans Jenny and with deep appreciation of the unique impulse he brought into the world. It is my pleasure to contribute further thoughts and impressions on these matters for this exciting new edition.

Alexander Lauterwasser - November 2021

Alexander Lauterwasser's interest in water began early, having grown up on the shores of Lake Constance, where many waters meet at the confluence of Germany, Switzerland and Austria. He, like Hans Jenny, could be described as a natural scientist with an abiding interest in studying how vibrational processes generate prototypical forms of nature. As a philosopher and a third-generation photographer, he has authored detailed studies on animal morphology, including an encyclopedic volume on the genesis and development of the "visual language" of the markings of tortoise shells, and among other things, their similarity to the patterns of the Tao te Ching.

In addition to his wide-ranging research in the life sciences, he works full-time as a psychologist with young people in a drug therapy center in Salem, Germany.

http://www.wasserklangbilder.de • alexander-lauterwasser@t-online.de • 049 170 977 4592

The Universal Vibrational Field as the "World-creating Power"
by John Beaulieu, N.D., Ph.D. and David Perez-Martinez, M.D.

We believe that Hans Jenny would agree when we say that the most extraordinary and transcendental aspect of his work was the fact that it derived from, and was driven by, *his direct observations of phenomena;* how he demonstrated experimentally the role of "periodic phenomena" (movement/momentum, vibration, position, and frequency) in the design and organization of "Nature, animate or inanimate," and the behavior of natural, biological, and psychosocial systems. As Therapeutic Sound Healing Practitioners, we can also summarize that his experimental research created evidence that supports the two basic principles of Therapeutic Sound Healing Practices, as well as the language used by most practitioners in the field of Vibrational Medicine today. These basic principles being:

1. All existence is vibratory in nature.
2. It is the underlying vibratory field that gives matter structure and form.

Of all the accomplishments of Hans Jenny, and they are legion, the one that stands out the most is how he walked in the world; the way he observed everything phenomenologically from the point of view of vibration, recurrent elements and patterns— and the systems and art he created by "getting inside" phenomena with empirical and systematic research. His greatest desire was to influence others to become cymatic phenomenologists and to create their own lines of research and artistic expression. He wished that we could all become keen observers of the ubiquity of vibratory phenomena and let nature "speak" for itself.

The practice of vibrational medicine and the therapies that have evolved— especially the areas of sound-based therapeutics, art, and research, have expanded exponentially throughout the world in the past few years. Ideas developed by Einstein and quantum theorists about the interchangeable nature of matter and energy depending on movement and the particle-wave duality, have come to fruition in Jenny's research, and increasingly so in the consciousness of individuals all over the world in their medical, wellness, and spiritual practices. This is because, in Dr. Jenny's work, the Universal Vibrational Field was not an abstract concept dependent upon a knowledge of quantum physics and more experiments. He intended cymatics to be a gateway for everyone to immerse themselves conceptually and creatively in a vibrational universe. He stated,

> *"If this method can fertilize the relationship between those who create and observe, between artists and scientists, and thus between everyone and the world in which they live, and inspire them to undertake their own cymatic research and creation, it will have fulfilled its purpose."* (p. 136)

Dr. Jenny refers to our intuitive creative self as "the imaginative," and the vibrational principles we learn through the phenomenological observations of sonic design as "the conceptual." He believed that the integration of the imaginative and conceptual was the key to a better life for everyone. Dr. Jenny's sound images are appealing to both artists and scientists because they are based on rigorous experimental designs that can be accessed through intuition. His sonic designs draw the viewer's consciousness into a creative vibrational reality, while his science informs the reader of objective principles of that reality. Jenny states,

> *In Cymatic phenomenology it is the whole phenomenon upon which we concentrate as we follow Nature unswervingly with eyes, ears, and brain.* (p. 297)

For Dr. Jenny the experience of the whole is most important. The whole is what today we refer to as the Universal Vibrational Field, the "causa prima" that creates, forms, and sustains everything. The result of conceptual and imaginative integration is a continual process of experiencing the vibrational nature of all phenomena in our life to include all objects, thoughts, emotions, and sensations. Jenny's vision was to give universal access to what he called a "world-creating power" arising from a holistic experience of a vibrational universe. He states:

> *In our research we move towards a creative world, towards a world-creating power. That in itself— in virtue of its creativity— provides for the investigator, for the artist, and for every truly living man, an element in which he can breathe, live, and fashion and work.* (p. 299)

The importance of the integration of intuition and creative thinking with objective science is echoed today in many scientific and artistic circles. In *Conceptual Revolutions In Science*, A.E. Dorffman states:

> *If these bridges are not built, then science will wander aimlessly and will not understand how to benefit humanity until we know and understand how we create suffering on this planet, and until the outer world is informed by the intelligence of the inner world.*[1]

Dr. Leonard Shalin, a surgeon, suggests that artists, poets, composers, and musicians often foreshadow the theories and discoveries of scientists.[2] Albert Einstein acknowledged that the intuitive mind was the sacred gift, and the rational mind should be its faithful servant.[3]

Cymatics is ultimately about our inner ability to reduce all phenomena to vibration through a continual balance of the conceptual (science) and the imaginative (creative intuitive). Everything in life requires a specific balance between conceptual and imaginative. When the process is working, there is a continuous expansion of consciousness through increased cognitive capacity. As Jenny states in this volume:

> *We are, in reality, this mystery; in it we become; it is not that man simply is, he is becoming all the time with an ever fuller and clearer consciousness.* (p. 299)

---

[1] Dorfman, A. E. (2015). Conceptual revolutions in science. Boulder, CO: Relentlessly Creative Books.[Google Scholar]--- p. 152
[2] Shlain, Leonard. (1991). Art and Physics "Parallel Visions In Space, Time, and Light, New York, NY. William Morrow.
[3] Calaprice, Alice. (Ed.). (2000). The Expanded Quotable Einstein. Princeton, N. J.: Princeton University Press.

The mystery Jenny is referring to is consciousness. Today there is a growing body of research and philosophical writing seeking to understand consciousness as a fundamental manifestation of a Universal Vibrational Field. Cymatics can make a substantial contribution to our phenomenological understanding of consciousness, through the relationship of nodes to wave dynamics. Nodes are locations on a vibrating surface where sound waves have no amplitude, and they are fundamental to the creation of a cymatics pattern. Again, Jenny says,

> *…. there is nothing to be seen where the nodal lines are.* (p. 40)

For example, a pattern is created in the sand spread on a vibrating plate, as it is attracted to, or comes into resonance with, the invisible nodal lines. When the frequency is changed, the invisible nodal lines change their location on the plate, creating a different pattern. If one tries to change the pattern without changing the frequency the original pattern will always return. If the sound ceases, the plate will stop vibrating, the nodal lines will disappear, and the sand will not be able to maintain the pattern. Today we see a similar relationship in the micro world of quantum vibrations, where a quantum node is conceived of as a place where all points have no density. Quantum particles, much like cymatic patterns, do not exist in nodes. However, an understanding of nodes is necessary to define harmonics, electron potentials, and where a quantum particle might be found.

Jenny refers to invisible nodes as "the creating," and cymatics patterns as "the created." It is the relationship of the creating and the created which produces consciousness. We can only relate to waves and nodes as vibrations through consciousness. We can talk about and imagine waves and nodes separately, however, if they are theoretically separated, the Universal Vibrational Field and consciousness would cease to exist. When our imaginative intuitive self is in balance with our objective conceptual self, the result is a resonant coherence with what we are experiencing, that allows us to experience "pure consciousness." Dr. Jenny references Goethe, the famous German philosopher, when he describes the inner experience of "pure consciousness."

> *Through our process of conceptual and imaginative thinking, we perform on phenomena, what Goethe called "the highest operations of the spirit."* (p. 297)

Phenomenologically subjecting everything in our life to an understanding of this vibrational relationship, is fundamental to Dr. Jenny's cymatics and healing. The embodiment of consciousness in a "sound state of mind," is what is most important: "Mens sana in corpore sano," (a sound mind in a healthy body). Our ability to come into coherent resonance with the vibrational flow of life reunifies us with the Universal Vibrational Field. Through this experience, we become whole and complete. This is the basis of holistic medicine and the practice of sound healing. As Dr. Jenny states:

> *What is the status of the parts, the details, the single pieces, the fragments? In the vibrational field, it can be shown that every part is, in the true sense, implicated in the whole.* (p. 141)

John Beaulieu, N.D., Ph.D. and **Dr. David Perez Martinez, MD**
Their bios appear on pp. 20 and 326 respectively.

Cymatics: Making Waves Around the World

How cymatics is turning up in our contemporary culture; in the arts and sciences; the mystical and magical

Gabriel Kelemen

Gabriel Kelemen, Ph.D., is an artist and researcher, lecturer, and head of the Art History and Theory department of the Faculty of Arts and Design at West University of Timişoara, Romania. For almost 40 years he has studied the phenomenon of how stationary waves arise in liquids as they are stimulated by audible sound frequencies. He is truly a pioneering figure in the field of Cymatics— experimenting and illustrating not only the diverse geometric patterns that occur on the surface of the liquids, but also clearly illustrating the complex dynamic forces that generate these intricate and orderly patterns both experimentally, and throughout the vastness of nature.

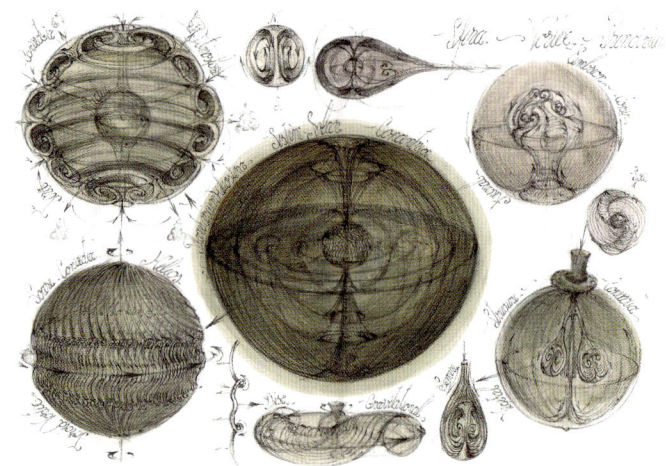

Dr. Kelemen's experiments are performed with the rigor and discipline of a physicist, while his artistic sensitivity brings a playfulness and an inventiveness that often yield surprising results— for instance, his startling 3-dimensional cymatic "sculptures" that mimic complex natural structures, such as flowers, biological organisms, and even the human face itself!

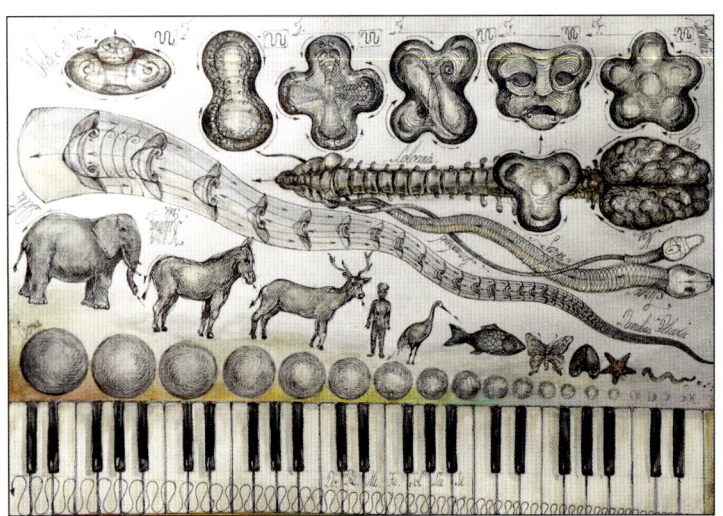

Focusing particularly on the interaction between the structure of the sphere and the spiral vortex, and toroidal motions that are fundamental in structuring matter into form, his in-depth studies have led to the articulation of an original theory— *The Universality of the Vortex-Sphere Archetype*, which was first published in 2015 as an elaborately illustrated oversize edition by ArtPress Publishing House, in Romania.

Dr. Kelemen's work has garnered much respect in scientific and artistic circles within his native Romania, and his drawings, photos, videos, and sculptures have been widely exhibited throughout Europe. His invitation by Jeff Volk, to lecture at the Tenth International Cymatics Conference in Atlanta, produced by Mandara Cromwell, and also at The Open Center and the Rudolf Steiner Center in New York City, culminated in his first US tour in October of 2015, for which Volk produced a short promotional video, *To Illustrate the Universe* (see http://www.CymaticSource.com/video)

To see examples of his illustrations and cymatics images, please visit:

 http://www.cosminnasui.com/2013/12/gabriel-kelemen/

 https://500px.com/kelemengabi

 http://fineartamerica.com/profiles/gabriel-kelemen.html

Water into Chaos: Phases of chaos and reintegration in an oscillating water sample

Filmed in August, 1999 at The Heureka Exhibition in Zurich and at the Technorama Science Museum in Winterthur, Switzerland, this 4-minute program reconstructs one of Hans Jenny's experiments using his original equipment to oscillate a small sample of water contained upon a lens, with audible sound frequencies. Light shines through this lens from beneath and is projected onto a screen, illuminating the complex geometries that arise as standing waves within the water sample. Also features 16mm footage from Reinhard Eichelbeck that was filmed in super high-speed and then slowed down to reveal astounding patterns that mirror crystalline structures as if the water had been transformed into a frozen state. Complex harmonic patterns arise and are explained in detail in the narrative.

 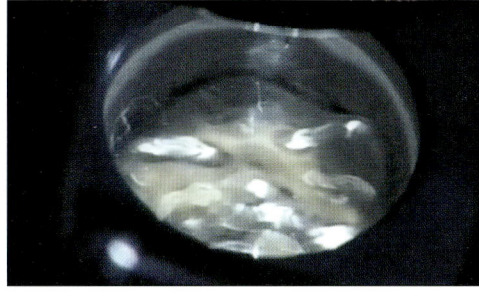 

 Watch the video at
https://vimeo.com/730183683/59bb8d2ba9

The First Museum Installation of the Cymascope

The first CymaScope to be installed in a museum was in late 2006 at explora, a museum for young creators and explorers in Albuquerque, NM. This hands-on interactive exhibit allowed participants to observe and create visual analogs of recorded sounds of local fauna, from rattlesnakes to roadrunners, cattle to coyotes. The exhibit at explora also incorporated a microphone, allowing visitors to experiment further by generating their own sounds while viewing the resulting patterns. This became one of the museum's most popular exhibits and has remained part of their permanent collection to this day!

1701 Mountain Rd NW, Albuquerque, NM 87104
(505) 600-6072 | Email: gmoran@explora.us

  Visit explora at
http://www.explora.us

 Visit x studio at
http://www.explora.us/xstudio

Christiaan Stuten and the Tonoscope

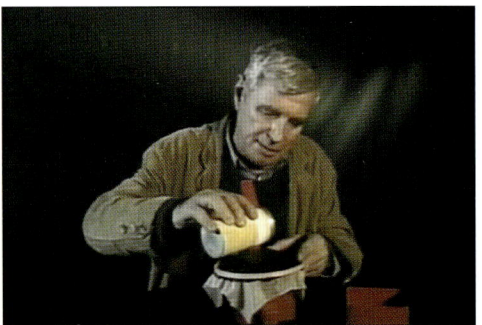

A demonstration of Hans Jenny's original acoustic tonoscope, showing how the human voice can create harmonic patterns in sand spread over a taut rubber membrane. Christiaan Stuten was Jenny's assistant for 14 years, and photographed most of the images in the book *Cymatics: A Study of Wave Phenomena and Vibration*.

 Watch the video at
https://vimeo.com/843711920/380f21f468

Cymatics in Lord of the Rings

The main title sequence for Amazon Studios/Prime Video's hit series "The Lord of the Rings: The Rings of Power" took deep inspiration from cymatics. The work was conceived by Mark Bashore and Katrina Crawford of agency Plains of Yonder. Their creative director, Anthony Vitagliano, had this to say about the work.

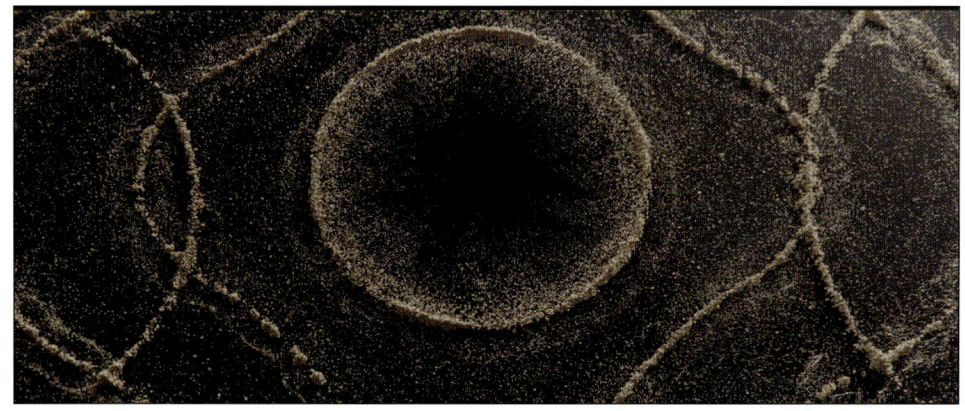

Taking inspiration from J.R.R Tolkien's Ainur, immortal angelic beings that sing such beautiful music that the world is created from their very sound, we conceived of a main title sequence "built from the world of sound."

Cymatics is a natural phenomenon that makes sound visible to the eye. Vibrations of fine particles on a flat surface display striking symmetrical patterns that reflect audio frequencies. Cymatics is understood by physicists and mathematicians, but to us mere mortals, they are nothing short of magic.

The sequence conjures an ancient and invisible power struggling to be seen. Symbols form, flow, push, and disappear as quickly as they arose. The unknowable realms of sound create fleeting visions of conflict and harmony that move in lockstep with Howard Shores' opening title score.

 To learn more about this project, visit
https://plainsofyonder.com/work/lord-of-the-rings

Cymatics: Science vs. Music

Nigel Stanford is a New Zealand composer, who after watching a documentary on synesthesia, got to thinking that it would be cool to make a music video where every time a sound plays, you see a corresponding visual element. Many years later, he saw some videos about cymatics—the science of visualizing audio frequencies—and the idea for the video "Cymatics" was born.

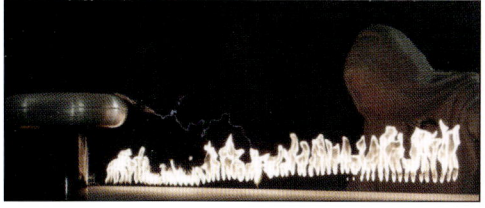 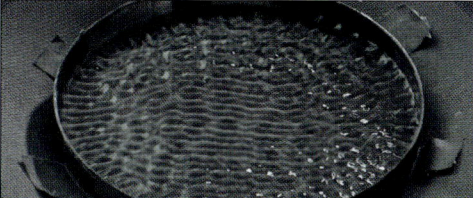 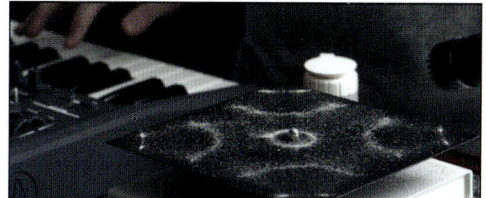

 Watch the video and remarkable journey to create it at https://nigelstanford.com/Cymatics/

Shaped by Water

Water was a powerful inspiration for some of Google's recent hardware designs showcased in April 2023 during Milan Design Week. "Shaped by Water" was an immersive and intimate sensorial collaboration presented by Google Design Studio and co-created by Google's vice president of Hardware Design, Ivy Ross, and the renowned water, light, and sound artist, Lachlan Turczan. One particularly revealing installation featured how popular Google hardware products, such as the Pixel Watch, reflected the unique characteristics of flowing water in their design. Turczan further elaborated on his own experience as an artist; "I started wanting to shape water so that I could create light projections. For me, sound was the best way of doing that. Cymatics is this incredible science of visualizing sound and using sound as a sculptural tool to shape water. It has been a very long learning experience of figuring out how to coach water and bring it to the beautiful state you want."

And Ross clearly echoes his sentiments! "I think art will move toward more of this type of experiential work that wants to wake up or enliven our sensory systems. And so I think this is the beginning of this kind of art using movement, sight, and sound. As a culture, we are craving those kinds of experiences."

 Learn more about the exhibit at https://shapedbywater.withgoogle.com/

Wavespace by Lachlan Turczan
Photograph by Edoardo Delille

You've never seen (((sounds))) like these before!

**Insights into Sound—
as pulsation, pattern and process.**

In these films, the still photos from the *Cymatics* book spring to life, adding a whole new dimension to these phenomena, illustrating the effects of sound over matter!

Watch as complex and intricate forms, often resembling living organisms, arise from inert powders, pastes, and liquids, solely through the influence of audible sound frequencies.

DVD or online streaming options available.
Visit our website for pricing.

MACROmedia Publishing
*Enlightening the Questing Eye
Awakening the Discerning Ear*

www.CymaticSource.com

350

CYMATIC SOUNDSCAPES AND BRINGING MATTER TO LIFE WITH SOUND
56 Minutes • DVD or online streaming

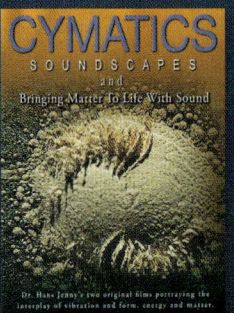

These dynamic processes are vividly brought to life when one sees the experiments in motion. It is for this reason that Dr. Jenny produced a series of 16 mm films documenting his experiments. They are now available for streaming, and two of these 28-minute films comprise a composite DVD, *Cymatic Soundscapes* and *Bringing Matter to Life with Sound*.

Part I of this two-part series highlights Dr. Jenny's pioneering experiments using audible sound frequencies to excite inert substances into life-like flowing forms. The stunning array of images reflects a variety of patterns found throughout nature, and viewing them in motion elicits an intuitive understanding of the invisible forces that underlie natural processes, from the movements of amoebae to the formation of galaxies.

Part II details a variety of patterns created by vibrating lycopodium powder. Audible frequencies give rise to circulatory processes which closely resemble sunspots and solar flares. Other frequencies generate flowing patterns analogous to the formation and rotation of spiral galaxies. Unusual three-dimensional forms arise, seeming to defy gravity and laws of motion, but all adhere strictly to the dynamics of the generating tone. The closing sequence features liquid kaolin paste as it solidifies under the effect of vibration, creating intricate geomorphic patterns.

OF SOUND MIND AND BODY:
MUSIC AND VIBRATIONAL HEALING
72 Minutes • DVD or online streaming
Winner of the Hartley Film Award through the Institute of Noetic Sciences

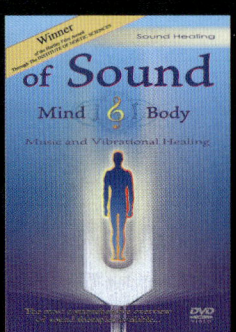

World-class composers, performers, health practitioners and leaders in mind-body medicine, convey a wealth of information ranging from music therapy to chanting and overtone singing. This evocative, stimulating and visually beautiful production explores a wide range of therapeutic applications of sound, from medieval plainsong to cosmic music!

A segment on Cymatics featuring commentary by **Deepak Chopra, Jill Purce, Rupert Sheldrake** and **Kay Gardner**, is illustrated by many cymatic images, including some not seen in any of Dr. Jenny's original films. These vivid representations of harmonic resonance depict how audible sounds can structure matter into coherent states, making visible the principle of entrainment, which is fundamental to all sound therapies.